TD 407 .H68 2007

Houben, Georg.

Water well rehabilitation
and reconstruction

New England Institute
of Technology
Library

Water Well Rehabilitation and Reconstruction

ABOUT THE AUTHORS

GEORG HOUBEN, PH.D., is a senior hydrogeologist at the Federal Institute for Geosciences and Natural Resources in Hanover, Germany and a lecturer in civil engineering at the University of Hanover. Previously, he worked as a consultant and researcher. Dr. Houben is the author of numerous publications in the field of hydrogeology, including many on well ageing and rehabilitation.

CHRISTOPH TRESKATIS, PH.D., is a senior consultant for hydrogeology and well construction, and a lecturer at the Technical Universities of Aachen and Darmstadt in Germany. He is the author of a wide range of publications on hydrogeological topics, including papers on well ageing and rehabilitation.

Water Well Rehabilitation and Reconstruction

Georg Houben

Christoph Treskatis

New York Chicago San Francisco Lisbon London Madrid
Mexico City Milan New Delhi San Juan Seoul
Singapore Sydney Toronto

The McGraw·Hill Companies

CIP Data is on file with the Library of Congress.

Copyright © 2007 by The McGraw-Hill Companies, Inc. All rights reserved. Printed in the United States of America. Except as permitted under the United States Copyright Act of 1976, no part of this publication may be reproduced or distributed in any form or by any means, or stored in a data base or retrieval system, without the prior written permission of the publisher.

1 2 3 4 5 6 7 8 9 0   DOC/DOC   0 1 3 2 1 0 9 8 7 6

ISBN-13: 978-0-07-148651-4
ISBN-10: 0-07-148651-8

The sponsoring editor for this book was Larry S. Hager, the editing supervisor was David E. Fogarty, and the production supervisor was Richard C. Ruzycka. It was set in Century Schoolbook by International Typesetting and Composition. The art director for the cover was Anthony Landi.

Printed and bound by RR Donnelley.

 This book was printed on recycled, acid-free paper containing a minimum of 50% recycled, de-inked fiber.

McGraw-Hill books are available at special quantity discounts to use as premiums and sales promotions, or for use in corporate training programs. For more information, please write to the Director of Special Sales, Professional Publishing, McGraw-Hill, Two Penn Plaza, New York, NY 10121-2298. Or contact your local bookstore.

Information contained in this work has been obtained by The McGraw-Hill Companies, Inc. ("McGraw-Hill") from sources believed to be reliable. However, neither McGraw-Hill nor its authors guarantee the accuracy or completeness of any information published herein, and neither McGraw-Hill nor its authors shall be responsible for any errors, omissions, or damages arising out of use of this information. This work is published with the understanding that McGraw-Hill and its authors are supplying information but are not attempting to render engineering or other professional services. If such services are required, the assistance of an appropriate professional should be sought.

Contents

Preface xi
Acknowledgments xiii

Chapter 1. Introduction — 1

1.1 History of Well Ageing — 1
1.2 History of Well Rehabilitation and Reconstruction — 2
1.3 Types and Relative Importance of Well Ageing — 3
1.4 Definitions of Terms — 9
1.5 Legal Background — 10

Chapter 2. Elements of Well Hydraulics and Well Operation — 17

2.1 Well Types — 17
    2.1.1 Introduction — 17
    2.1.2 Dug wells — 18
    2.1.3 Vertically drilled wells — 19
    2.1.4 Horizontal and radial collector wells — 19
    2.1.5 Directionally drilled wells — 22
2.2 Well Hydraulics — 23
    2.2.1 Radial symmetric flow to wells — 23
    2.2.2 Aquifer yield and intake capacity of wells — 27
    2.2.3 Flow to wells under natural conditions — 30
2.3 Dimensioning of Wells — 37
2.4 Well Operation and Maintenance — 40
2.5 Well Monitoring — 42
2.6 Influence of Ageing on Well Hydraulics and Operation — 51

Chapter 3. Chemical Ageing Processes — 55

3.1 Corrosion — 55
    3.1.1 Terms and definitions — 55
    3.1.2 Electrochemical corrosion — 55
    3.1.3 Microbially induced corrosion — 57
    3.1.4 Ageing of nonmetallic material — 58

Contents

- 3.2 Ochre Incrustations: Iron and Manganese Oxides ... 59
  - 3.2.1 Geochemistry of iron and manganese ... 59
  - 3.2.2 Oxidation of iron(II) and manganese(II) ... 67
  - 3.2.3 Auto-catalysis of iron(II) and manganese(II) oxidation ... 70
  - 3.2.4 Mass balancing of ochre buildup: example calculations ... 73
  - 3.2.5 Mineralogy of ochre ... 75
  - 3.2.6 Microbiology of ochre formation ... 76
  - 3.2.7 Ageing (recrystallization) of iron and manganese oxides ... 82
  - 3.2.8 Interactions of ochre minerals with trace elements ... 86
  - 3.2.9 Structure and texture of ochre incrustations ... 93
- 3.3 Carbonate Incrustations (Scale) ... 94
- 3.4 Aluminum Hydroxide Incrustations ... 98
- 3.5 Metal Sulfide Incrustations ... 99
- 3.6 Bioclogging ... 102
- 3.7 Special Incrustation Types ... 104
- 3.8 Passively Incorporated Components of Incrustations ... 106
- 3.9 Spatial and Temporal Distribution of Incrustations ... 107

Chapter 4. Mechanical Causes of Well Ageing ... 117

- 4.1 Suffossion and Colmation ... 117
  - 4.1.1 Mobilization and transport of particles in the subsurface ... 117
  - 4.1.2 Mechanical filtration processes ... 119
  - 4.1.3 Filtration as a cause of mechanical well clogging ... 125
- 4.2 Abrasion ... 132
- 4.3 Sand Intake ... 134
- 4.4 Damage by Plant Roots ... 139
- 4.5 Ground Movement ... 140
  - 4.5.1 Soil settling and subsidence ... 140
  - 4.5.2 Earthquakes ... 144
- 4.6 Vandalism ... 144
- 4.7 Ageing Processes in Infiltration Wells ... 145
- 4.8 Ageing Processes in Open Borehole Wells ... 150

Chapter 5. Identification of Ageing Processes and Performance Assessment of Wells and Well Rehabilitation ... 153

- 5.1 Camera Inspections ... 153
- 5.2 Step-Discharge Tests (Well Performance Tests) ... 155
- 5.3 Borehole Geophysics ... 163
- 5.4 Measurements of Incrustation Deposition Rates ... 172
- 5.5 Sampling and Investigation of Incrustation Samples ... 172
- 5.6 Particle Counting ... 174
- 5.7 Mass Balancing of Removed Material ... 176
- 5.8 Calculation of Saturation Indices ... 179
- 5.9 When to Rehabilitate? ... 183
- 5.10 Rehabilitation or Reconstruction? ... 185

Chapter 6. Economics of Well Rehabilitation and Reconstruction ... 187

- 6.1 Economic Principles and Evaluation Standards for Wells ... 187
- 6.2 Financial Mathematical Principles ... 189

6.3		Economic Appraisal of Wells Applying the Annuity Method	192
6.4		Economic Appraisal of Wells Applying the Discounting Method	193
6.5		Economic Considerations for Well Rehabilitation and Reconstruction	197
6.6		The Market for Rehabilitation and Reconstruction— Prices and Volumes	199

Chapter 7. Mechanical Rehabilitation — 203

7.1	Processes of Mechanical Rehabilitation	203
7.2	Brushing	206
7.3	Hydraulic Methods	209
	7.3.1 Background: hydraulic suffossion	209
	7.3.2 Surge blocks	209
	7.3.3 Intense pumping	210
	7.3.4 Closed-circuit pumping (chambered systems)	213
	7.3.5 Jetting	216
	7.3.6 Jetting spears	221
7.4	Thermal Methods	222
	7.4.1 Carbon dioxide (CO_2) freezing	222
	7.4.2 Steam injection ("geyser")	226
7.5	Impulse Methods	227
	7.5.1 Background: impulse generation and attenuation	227
	7.5.2 Explosive charges	231
	7.5.3 Explosion of gas mixtures	233
	7.5.4 Release of compressed fluids	234
	7.5.5 Ultrasound	235
	7.5.6 Enforced jet cavitation	240
7.6	Comparison of Methods	241

Chapter 8. Chemical Rehabilitation Techniques — 245

8.1	Legal Constraints	245
8.2	Reaction Mechanisms	245
	8.2.1 Introduction	245
	8.2.2 Dissolution of iron and manganese oxides	247
	8.2.3 Dissolution of carbonates	254
	8.2.4 Dissolution of aluminum hydroxides	255
	8.2.5 Dissolution of metal sulfides	256
	8.2.6 Oxidation/dissolution of biomass	256
	8.2.7 Removal of drilling fluid remnants	258
8.3	Combinations of Chemicals	260
8.4	Additives	260
8.5	Comparison of Rehabilitation Chemicals	261
8.6	Chemical Rehabilitations in the Field	264
	8.6.1 Introduction	264
	8.6.2 Safety precautions during transport of chemicals and on-site	266
	8.6.3 Do we really need chemical rehabilitations?	270
8.7	Disposal of Liquid and Solid Waste	271
	8.7.1 Introduction	271
	8.7.2 Neutralization of inorganic acids	273
	8.7.3 Neutralization and disposal of organic acids	274
	8.7.4 Neutralization of bases	274
	8.7.5 Disposal of reductants	274
	8.7.6 Disposal of sludges and solid waste	274

Chapter 9. Repair, Reconstruction, and Decommissioning of Wells — 277

- 9.1 Definition of Terms — 277
- 9.2 Reconstruction — 284
  - 9.2.1 Partial reconstruction — 284
  - 9.2.2 Complete reconstruction — 289
- 9.3 Decommissioning — 295

Chapter 10. Practical Well Rehabilitation — 303

- 10.1 Preparation — 303
  - 10.1.1 Problem definition — 303
  - 10.1.2 Tender documents — 303
  - 10.1.3 Selection of the executing company — 304
  - 10.1.4 Approval by authorities — 305
- 10.2 Execution — 306
  - 10.2.1 General recommendations for rehabilitation sites — 306
  - 10.2.2 Occupational safety — 306
  - 10.2.3 Progress control — 311
  - 10.2.4 Waste disposal — 311
- 10.3 Sustainability of Well Rehabilitation Measures — 312
- 10.4 Rehabilitation-Induced Problems — 315
  - 10.4.1 Changes to water chemistry induced by rehabilitation — 315
  - 10.4.2 Secondary microbial pollution — 316
  - 10.4.3 Mechanical damage and well collapse — 320
- 10.5 After the Well Rehabilitation — 320
  - 10.5.1 When is the well due for its next rehabilitation? — 320
  - 10.5.2 Insufficient yield gain after rehabilitation—what are the next steps? — 321

Chapter 11. Prevention — 323

- 11.1 Microbicidal Techniques — 323
  - 11.1.1 Introduction — 323
  - 11.1.2 Chemical disinfection — 323
  - 11.1.3 Thermal disinfection ("Pasteurization") — 325
  - 11.1.4 Ionizing irradiation — 326
- 11.2 Chemical and Physical Methods — 328
  - 11.2.1 Addition of reaction inhibitors — 328
  - 11.2.2 Acidification — 329
  - 11.2.3 Electrochemical methods — 329
  - 11.2.4 Magnetic methods — 331
- 11.3 Prevention through Planning, Construction, and Operation — 331
  - 11.3.1 Drilling techniques, materials, and operation — 331
  - 11.3.2 Suction flow control devices (SFCDs) — 334
  - 11.3.3 Inert gases — 337
  - 11.3.4 Subsurface iron removal — 338
  - 11.3.5 Gravel pack flushing devices — 339

Chapter 12. The 10 Dos and Don'ts of Well Rehabilitation — 341

References 343

Appendix A. Conversion of Units 357

Appendix B. How to Read a Box-Whisker Plot 365

Appendix C. Example of a Tender Document for Well Rehabilitation 369

Index 379

Preface

In many developed countries, the technical infrastructure has reached a state in which maintenance of existing buildings and facilities is becoming more important than new constructions. The same applies to water supply and one of its most prominent features, water wells. An increasing number of publications, patents, and conferences over the last years demonstrates the increasing importance of well ageing and rehabilitation in many countries.

Much effort has been invested into finding appropriate techniques for rehabilitation, reconstruction, and ageing prevention. Today, wells are such a commodity that one tends to forget that they are constantly interacting with their natural environment in many ways. These interactions can affect their performance and may decrease the well's yield or even endanger its structural integrity. Even though the interactions proceed well hidden in the subsurface, we are now able to identify and assess many of the processes involved.

This book is based on a German precursor published in 2003. During translation it was updated and expanded. It summarizes the current state of the technical, scientific, and legal background of well ageing and the remedies to overcome this problem. Included are the results of several years of the authors' work on this topic in research and practice. The reader will find many German and some Dutch references cited in this book. Although many readers will not be able to understand these, we have included them and tried to extract the main findings for the benefit of a larger community.

Wells almost have to be considered individual beings. Their technical properties and their ageing processes are influenced by a multitude of parameters related to the applied drilling type, technical equipment, and most of all by the hydraulic and hydrochemical properties of the aquifers they tap. Because of this complexity there is no perfect remedy for ageing wells. Methods that prove useful for one well may be useless or even dangerous for the next one. In this book, all methods known to the authors are presented unprejudiced, with all their possibilities, limits, and potential drawbacks. Knowing the technical and scientific background helps to select useful methods.

New methods are being developed all the time, older ones are being constantly improved, and research is going on. The authors are grateful for any corrections and hints at new developments and references which can be included in later editions.

We have used the SI system for units throughout the book. Wherever possible, values in U.S. customary units are also given in parentheses. Readers who are more familiar with nonmetric units are also referred to an extensive collection of conversion tables in Appendix A. The currency unit used throughout the book is the euro, but again, values in U.S. dollars are given in parentheses wherever possible (1 € = $1.26291).

Legal Disclaimer

The authors do not warrant or assume any legal liability or responsibility for the accuracy, completeness, or usefulness of any information, apparatus, product, or process described in this book. We further do not accept responsibility for any loss or decrease of yield, damage, or malfunction resulting from the application or use of any of the machinery, concepts, and methods described in this book. The reader is advised that concepts, apparatus, techniques, products, and chemicals mentioned in the text of this book may be protected by national and/or international patents and trademarks.

Acknowledgments

The authors are indebted to many persons for their help and encouragement during the preparation of this book. The authors' interest in the topic of well ageing and rehabilitation was originally raised by the late Dieter Leda of the Städtische Werke Krefeld (SWK) AG. Since his death, his employer has continued his exemplary support of our applied research.

The manuscript has profited greatly from assistance by our Dutch colleagues, Bert-Rik de Zwart (Technical University Delft), Kees van Beek, Jan-Willem Kooiman, and Marc Balemans, (KIWA, Nieuwegein). Section 4.1 contains many findings from Bert-Rik de Zwart's Ph.D. thesis, and Sec. 5.6 was essentially written by Kees van Beek.

The following persons, institutions, and companies assisted through helpful discussions, supply of samples, and by making available data, figures, and photographs (in alphabetical order):

Karsten Baumann, BLM GmbH, Storkow, Germany

Bieske & Partner Beratende Ingenieure GmbH, Lohmar, Germany

Brunnen- und Pumpenservice GmbH, Pulheim/Köln, Germany

Dieter Eichhorn, Deutsches Grundwasserforschungszentrum (DGFZ), Dresden, Germany

Hermann Etschel & Xaver Stiegler, E & M Bohrgesellschaft mbH, Hof (Saale), Germany

Gamma-Service Recycling GmbH, Leipzig, Germany

Toine Jongmans, Wageningen University, Wageningen, The Netherlands

Andreas Kappler, ZAG, University of Tübingen, Germany

Stephan Kaufhold & Axel Schippers, BGR, Hanover, Germany

Udo Kleeberger, Wasserwirtschaftsamt Nürnberg, Nuremberg, Germany

Robert McLaughlan, University of Technology, Sydney, Australia

Harald Munding, Aquaplus, Kronach, Germany

Lutz-Peter Nolte & Sven Tewes, NBB, Hamburg, Germany

Ate Oosterhof, Vitens Friesland, Leeuwarden, The Netherlands

Robert Plängsken GmbH, Neukirchen-Vluyn, Germany

Walter L. Plüger, Technical University (RWTH) Aachen, Germany

Päivi Puronpää-Schäfer & Brigitte Kretzer, Cleanwells, Rottweil, Germany

Clem Rowe, Inflatable Packers International Pty. Ltd, Perth, Australia

Stuart A. Smith, Upper Sandusky, OH, USA

Sonic Umwelttechnik GmbH, Bad Mergentheim, Germany.

Sven Steussloff & Andreas Wicklein, pigadi GmbH, Berlin, Germany

Harrie Timmer, Hydron Zuid-Holland, Gouda, The Netherlands

Jürgen Wagner, Grundwasser-und Geo-Forschung, Neunkirchen, Germany

Ulrich Weihe, Oldenburgisch-Ostfriesischer Wasserverband (OOWV), Brake, Germany

This book contains some results from M.Sc. theses by Sabine Merten, Christiane Schröder and Zeycan Tutar-El Beqqali of the Technical University of Aachen (RWTH). We would like to thank them for the fruitful cooperation.

Many of the drawings were expertly handled by Ulrich Gersdorf (BGR). The authors would like to thank the Federal Institute for Geosciences and Natural Resources (BGR), Hannover, Germany, for permission to use these figures. The symbol **BGR** designates all figures copyrighted to the BGR.

Further graphical assistance by Angelika Peter and Christiane Nienhaus (both Bieske & Partner GmbH) is gratefully acknowledged. Finally, we would like to thank Susanne Stadler and Lynne Lackenbach for their editorial review of the manuscript.

Water Well Rehabilitation and Reconstruction

Chapter 1

Introduction

1.1 History of Well Ageing

In regions where river or spring water is absent or available only during certain periods, humans were quick to learn that the underground often contains water. The construction of water wells began with the deepening of waterholes but quickly developed further. During Neolithic times (~7,000 years before the present) elaborate wells were already common (LVR 1998). The ageing of wells has been of concern to people ever since they began constructing them, as well ageing potentially endangers access to their no. 1 vital necessity. Evidence of cleansing, maintenance, and even reconstruction of wells has been found for the earliest Neolithic wells (LVR 1998). Over 2,300 years ago, Aristotle described deposits of sulfur bacteria in wells ($\rightarrow$ 3.1.3, 3.5, 3.6). Massive deposits of travertine, a calcium carbonate deposit, often hampered the operation of Roman pipelines. These deposits sometimes reached sufficient thicknesses to be used as decorative stone.

In the small German town of Duderstadt, a certain Mr. Barckefeldt complained in 1683 that "wells and water channels are constantly being contaminated by rubbish, sweepings and other refuse" (Veh, in Porath & Rapsch 1998). To overcome this problem, already in the year 1518 the city had introduced well cleaning measures at regular intervals of two to three years. These cleanings were always taken as an occasion for a "well cleaning fest." This nice tradition was abolished in 1724, because the amounts of beer (5 barrels) served—free of charge—during this festivity caused "on the one hand much exuberance, fighting and desecration of the holy days, on the other hand also the ruin of citizens and neighbours."

The occurrence of iron oxides (ochre) in water installations is not a modern problem. In 1545 a hospital warden in Nuremberg (Germany) complained about a "strange red slime which is rather unhealthy to drink" (Wetzel 1969). Iron bacteria were identified as the cause of the plugging of pipelines by Brown in 1904 (Ellis 1919).

Both the reasonable well rehabilitation interval of two to three years and the "well cleaning fest" were lost in the course of history. What is left are the many well owners around the world who complain about their imminent financial ruin due to the cost of maintenance and rehabilitation of their wells.

1.2 History of Well Rehabilitation and Reconstruction

The advent of the steam engine and motor-driven drilling rigs in the nineteenth century allowed access to aquifers in ever-increasing depths compared to the previously dug wells. Deeper wells always mean a higher chance of encountering reduced water containing iron, manganese, and sulfide. The resulting corrosion and formation of ochre inflicted severe impediments on the operation of wells. Intelligent people immediately took up the task of inventing methods to remove ageing products and to prevent their genesis. The first German patent, by Heinrich Böttcher (1905, patent no. 181 578), dealt with the "cleaning of tube wells by means of hot steam." While some patented ideas, such as the early concept of using divers to clean well interiors, have never gained much popularity, others have led to the development of elaborate machinery for rehabilitation and reconstruction.

A total of 102 patents related to "well ageing and rehabilitation" were granted between 1900 and 2005 by the Federal German Patent Office (patent class E 03 B 3/15). This number also includes patents granted in East Germany (GDR) between 1949 and 1990 and also some European and International patents which came into effect in Germany. The trickle of patent applications prior to 1980 has since developed into a constant stream. More than half of all patents date from the 1980s and later (Fig. 1-1). Of all patent applications in Germany relating to water well rehabilitation, 57 deal with mechanical, 34 with chemical rehabilitation methods, and 11 with prevention techniques.

Literature on well ageing and rehabilitation, in both professional journals and monographs, has multiplied since the beginning of the 1990s. The authors come from many countries and report on laboratory investigations and practical case studies from all continents. The references in this book give a good overview of the wealth of available information. In 1990, an entire conference in Cranfield, UK, was dedicated to the topic (Howsam 1990a). Some authors from Europe, North

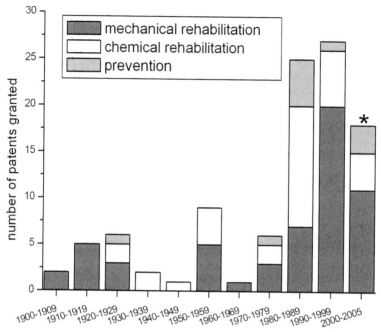

Figure 1.1 Development of the number of patents granted in Germany related to "Well Rehabilitation" (Group E03B 3/15) (partially after Data by Paul 2000). *Note that the last column represents only five years rather than 10 years.

America, and Australia have dedicated whole books to the topic (Jäkel 1958; Krems 1972; Olsthoorn 1982; McLaughlan et al. 1993; Howsam et al. 1995; McLaughlan 1996; Mansuy 1998; Alford & Cullimore 1998; Cullimore 1999; McLaughlan 2002; Houben & Treskatis 2003, 2004; Wicklein & Steußloff 2006). The increase in books published since the late 1990s is obvious.

1.3 Types and Relative Importance of Well Ageing

Well ageing, that is, the deterioration of the material and of the performance of a well, is caused by physical, chemical, and biological processes acting on the well and its immediate surroundings (Fig. 1-2). The following processes—or combinations thereof—are the main causes:

- Material deterioration (corrosion) of casing, screen, pump, pipes, and annular filling
- Buildup of mineral incrustations
- Biomass accumulation (biofouling)

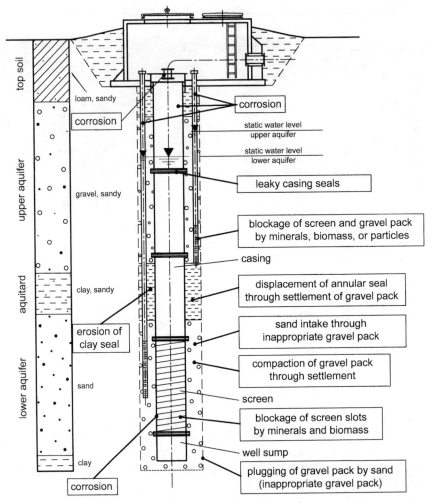

Figure 1.2 Components of a well that may be affected by ageing processes.

- Particle mobilization (sand intake) and particle accumulation (mechanical clogging)
- Ground movement (soil settlement, earthquakes)

The causes will be discussed in detail in Chaps. 3 and 4.

In 1998 the DVGW (German Technical and Scientific Association for Gas and Water) conducted a survey among its members, both utility companies and drilling/rehabilitation contractors, on the topic of well ageing and rehabilitation (Niehues 1999). A total of 507 utility companies returned the questionnaires, a number roughly equal to half of all DVGW members.

These companies operate about 12,000 wells. Of these, about 3,080 wells had been rehabilitated during the preceding five years (ca. 620 per year). This number is equal to about 5% of all wells being rehabilitated each year. A simultaneous survey of rehabilitation companies showed that they rehabilitate about 2,400 wells per year. This substantially higher number reflects the limited response rate of the questionnaires by the utility companies and the large number of wells used for industrial, agricultural (irrigation), and private (gardening, pools) purposes. Between one-quarter and one-third of all rehabilitations involve the use of chemicals ($\rightarrow$ 8). The economic impact of well ageing is discussed in Chap. 6.

A total of 347 utility companies answered the questions on ageing causes (Niehues 1999). Sand intake and corrosion were responsible for a mere 9% and 4%, respectively. The overwhelming majority, 87%, of aging causes was related to incrustations (Fig. 1-3). A total of 281 utility companies were able to provide more detailed information on the type of incrustation involved (Fig. 1-4). Iron and manganese oxide incrustations, summarized under the term "ochre" here and throughout the book, make up the vast majority.

During the last few years, the authors of this book have collected a large number of well-incrustation samples for detailed geochemical and mineralogical analysis (Houben 2003a). Samples were supplied by both utility companies and rehabilitation contractors. Most were collected from material recovered during mechanical well rehabilitations, but some came from the redrilling of abandoned wells. Samples originated

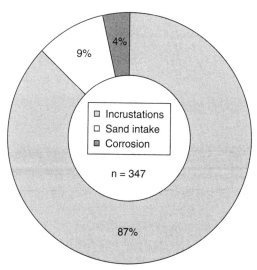

Figure 1.3 Relative importance of causes for well ageing in Germany according to a DVGW survey of utility companies (after data from Niehues 1999).

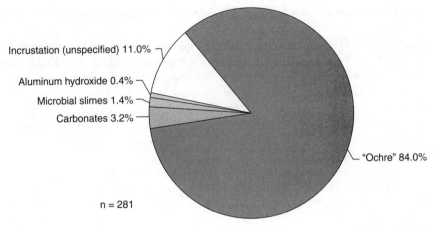

Figure 1.4 Distribution of incrustation types in wells in Germany according to a DVGW survey of utility companies (after data from Niehues 1999).

from 62 locations throughout Germany and covered most of the different geological regions in this country. These locations included the glacial plains of Northern Germany, the alluvial terraces of the Lower Rhine and the Elbe Rivers (and other rivers), karstic limestone aquifers, and fractured sandstone and crystalline-rock aquifers of Bavaria and Saxony. We therefore consider this data set to be fairly representative. As can be easily seen from Fig. 1-5, our investigations agree well with the results of the DVGW survey. Ochre is by far the most abundant incrustation type. Our analyses provide more detailed insight into the types of incrustation. Iron oxide incrustations are far more abundant than those of predominantly manganese oxide composition. Mixed iron and manganese oxide incrustations are rare. Carbonates are responsible

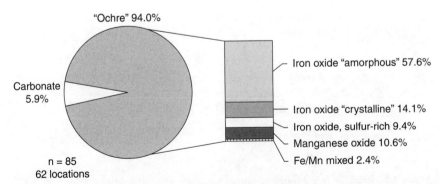

Figure 1.5 Distribution of well incrustation types in Germany according to chemical analysis of samples collected throughout the country.

for about 6% of all incrustations. Not surprisingly, most of these samples originate from karstic limestone aquifers. Knowing that karst aquifers provide about 7% of all groundwater abstracted in Germany, we see immediately that our sample set is indeed quite representative for the conditions of German aquifers.

The small set of well incrustation samples from Australia investigated by McLaughlan et al. (1993) also shows a predominance (11 of 15 samples) of iron oxide incrustations (Fig. 1-6). Carbonate and aluminium hydroxide samples (two each) account for the remainder. The "higher" relative abundance of the latter could be a result of the widespread occurrence of aluminium-rich soils (e.g., laterite) in Australia.

In his book on well rehabilitation, Smith (1995) presents the results of a "private" survey of contractors and utility companies in the United States. The contractors stated that two-thirds of ageing problems are related to incrustations of some kind (Fig. 1-7). Similar to Germany, iron oxides make up the majority of the incrustations, but sulfides are more abundant in the United States. The relative amount of sand intake is similar to the German situation, while corrosion seems to be more abundant in the United States. The view of the utility companies, however, seems to be somewhat different (Fig. 1-8). In their eyes, sand pumping and corrosion seem to be more important, at 31% and 25%, respectively. Incrustations are held responsible for less than half of the observed causes of ageing.

The Dutch water research institute KIWA performed a survey on well ageing and its causes in the Netherlands in 2001 (KIWA 2002). The five major utility companies involved represent about 1,000 wells, about half of which suffer from clogging. The most common cause is clogging of the borehole wall (68%), which is usually attributed to mechanical

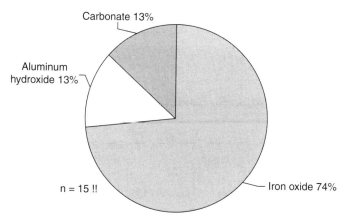

Figure 1.6 Distribution of incrustation types in wells in Australia (after data by McLaughlan et al. 1993). Note the limited number of samples on which this graph is based.

8 Chapter One

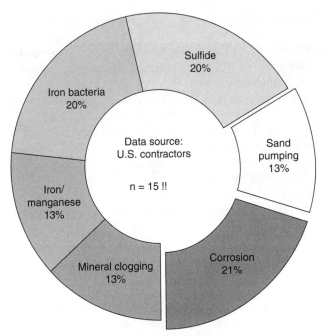

Figure 1.7 Distribution of causes of well ageing in the United States according to well rehabilitation contractors (after data from Smith 1995). Note the limited number of contractors on which this graph is based.

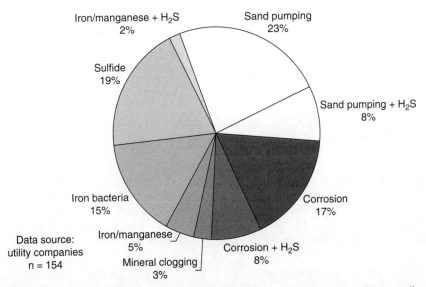

Figure 1.8 Distribution of causes of well ageing in the United States according to utility companies (after data from Smith 1995).

clogging by particles. The remaining causes are distributed among screen (17%) and combined screen/borehole wall clogging (14%) and are often related to buildup of mineral incrustations. The difference from the situation in Germany, one of the neighboring countries, is striking but yet unexplained. Whether different geology, drilling, or well completion techniques are the cause for the higher abundance of mechanical clogging in the Netherlands remains unclear.

1.4 Definitions of Terms

The following terms will be used throughout this book and therefore require proper definition. They are based mainly on DVGW recommendation W 130.

Well ageing. Well ageing is defined as the sum of all processes affecting the structural integrity and the yield of a well during its lifetime. It includes processes ranging from material deterioration, such as corrosion, to mechanical and chemical plugging of pathways for water inflow. The effects of ageing lead to a performance decrease in practically all cases.

Cleaning. Cleaning of a well is defined as the removal of mineral and organic deposits from the inner well casing and inner screen and the removal of deposits from the well sump. Brushing and subsequent clearing of the sump is a classical example. Cleaning does not act on the outside of the well casing (annulus).

Rehabilitation. Rehabilitation of wells comprises all measures aimed at the removal of mineral and organic deposits and plugging particles from the well interior, the annulus including gravel pack, and—if possible—from the adjacent geological formation. The final target is the restoration of the hydraulic function of the well. All rehabilitation schemes comprise three main steps (Fig. 1-9; DVGW W 130):

- *Step I: separation* (of incrustations or particles from casing, screen, and pore spaces)
- *Step II: removal* (of loosened incrustation or particles from the well)
- *Step III: monitoring* (of the rehabilitation process and its success)

Reconstruction. The reconstruction of a well can aim at the (re)establishment of (a) its functional efficiency and (b) its productivity, or (c) both. Unlike rehabilitation, reconstruction requires intervention into the constructional state of the well. This may include an exchange of the annular filling and/or the casing, resealing of annular seals or casing connections, and may go as far as complete redrilling. Causes for reconstructions include inappropriate dimensioning of the well, deficiencies in the construction, or

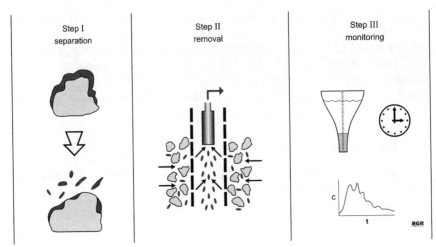

Figure 1.9 Main process steps in well rehabilitation.

material defects (DVGW W 135). Reconstruction is often necessary to prevent the seepage of polluted near-surface waters into deeper layers or to overcome hydraulic "short circuits" between different aquifers.

Decommissioning. Decommissioning comprises all measures necessary to safely and permanently put a well out of operation. In many cases the above-ground and the subsurface installations have to be dismantled and the well has to be backfilled. Leaky annular seals have to be reinstated prior to final decommissioning.

1.5 Legal Background

Legislative bodies, such as parliaments, may pass laws that regulate water use (statutory laws). In many countries, such water laws regulate the use of all types of water bodies, including groundwater. Although these laws initially dealt mainly with water quantity, they now include issues of pollution and water protection. In the European Union, framework directives may be issued, which have to be converted into national laws by the member states. The full text of such laws is often available on the Internet (Table 1-1).

In addition, legislative bodies may empower administrative bodies, such as a ministry or an environmental protection agency, to issue rules and regulations that have power of law (administrative law). These may include:

- Permissible concentrations of chemicals in water
- Lists of approved chemicals and processes allowed for water treatment
- Regulations governing water rights (licenses), including the requirements a well owner must meet to obtain or maintain a license

TABLE 1.1 Selected Water and Environmental Laws Available on the Internet

Country	Type of law	URL
Australia	All environmental laws	www.aph.gov.au/library/intguide/law/envlaw.htm
Canada	Water act	http://laws.justice.gc.ca/en/C-11
	Environmental protection act	http://laws.justice.gc.ca/en/C-15.31
European Union	Water framework directive	http://europa.eu.int/comm/environment/water/water-framework/index_en.html
France	Water law	www.senat.fr/dossierleg/pjl04-240.html
Germany	Water law	http://bundesrecht.juris.de/bundesrecht/whg
	Soil protection	http://bundesrecht.juris.de/bundesrecht/bbodschv/index.html
Netherlands	Water law	http://wetten.overheid.nl/cgi-bin/deeplink/law1/title=WET%20OP%20DE%20WATERHUISHOUDING
UK	Water act	www.defra.gov.uk/environment/water/legislation/default.htm
USA	All environmental laws	www.epa.gov/epahome/laws.htm

In federal republics, such as Australia, India, the United States, and Germany, the individual states may issue their own statutory or administrative water laws and rules. In this case, the federal laws are usually a framework which leaves room for the states to pass legislation and regulations adapted to their individual needs and concerns. Variations from state to state may be substantial, so that we cannot present all state laws and regulations here. They are often posted on the Internet. Local administrations may be empowered by law to issue even more specific rules and regulations (bylaws). Based on the constitution, courts may rule over water-related law suits (common law). The decisions are commonly based on precedent and can be overturned by higher courts.

Administrative laws and local regulations are usually the legal level that well owners and contractors have to deal with. These may include:

- Drilling permits
- Permits to operate wells and abstract or inject a quantity of water
- Regulations for the reconstruction of wells
- Permits required to abandon wells (and obligations about how to seal them)
- Permits for the injection of chemicals into the subsurface, e.g., during chemical rehabilitation
- Permits required to use explosives, e.g., during well rehabilitation

National and international standards such as ISO, EN, ASTM, DIN, AFNOR, etc., rarely address the topics of well rehabilitation and reconstruction. Some standards do exist for drilling techniques, well casings, and corrosion (Table 1-2). Most of the full text of the listed standards is available on the Internet at www.astm.org and www.din.de, but charges may apply. Some of the DIN standards are available in English.

Professional and scientific organizations (Table 1-3) often provide recommendations on the "best available technology" or "best available practice" in certain water-related fields. They can often be downloaded from the Internet, although charges may apply. They are mainly worked out

TABLE 1.2 Selected ASTM and DIN Standards on Water Well-Related Issues

	Standard	
	ASTM	DIN
Drilling techniques	D6286-98 D5092-04e1 D5781-95(2000) D5782-95(2000) D5783-95(2000) D5784-95(2000) D5787-95(2000) D5872-95(2000) D5875-95(2000) D5876-95(2000) D6169-98(2005) D6724-04 D6725-01	4021 4924
Well casing, steel	A589-96(2001)	4922
Well casing, plastic	F480-02	4925
Corrosion, chemical	G15-05 (and further standards cited therein)	50929 50930
Wear and erosion	G40-05	—
Specific well yield Well efficiency	D5472-93(2005) D6034-96(2004)	—
Measurement of well discharge	D5737-95(2000)	—
Geophysical borehole logging	D6031-96(2004)	—
Maintenance and rehabilitation of monitoring wells	D5978-96(2005)	—
Development and monitoring of wells	D5521-94	—
Decommissioning of wells	D5299-99(2005)	—
Location of abandoned wells	D6285-99(2005)	—

TABLE 1.3 Selected National and International Professional and Scientific Water Associations

Country	Acronym and name of organization	URL
Australia	Australian Water Association (AWA)	www.awa.asn.au
Austria	ÖVGW (Austrian Technical and Scientific Association for Gas and Water)	www.oevgw.at
Belgium	BELGAQUA (Belgian Federation of Water Services)	www.belgaqua.be
Canada	CWWA/ACEPU (Canadian Water and Wastewater Association)	www.cwwa.ca
Czech Republic	SOVAK (Association of Czech Waterworks)	
Finland	IIWA (Finnish Water and Waste Water Works Association)	www.vvy.fi
France	ASTEE (Association Scientifique et Technique pour l'Eau et l'Environnement) (formerly AGHTM)	www.astee.org
Germany	DVGW (German Technical and Scientific Association for Gas and Water)	www.dvgw.de
	BGW (Bundesverband der deutschen Gas- und Wasserwirtschaft)	www.bgw.de
India	IWWA (Indian Water Works Association)	www.emcentre.com/iwwa
Italy	Federgasacqua	www.federutility.it
Japan	JWWA (Japan Water Works Association)	www.jwwa.or.jp
Luxemburg	ALUSEAU (Association Luxemburgeoise des Services d'Eau)	www.aluseau.lu
Malaysia	MWA (Malaysian Water Association)	www3.jaring.my/mwa
Netherlands	VEWIN (Vereniging van Waterbedrijven in Nederland)	www.waterleiding.nl
	NVA (Nederlandse Vereniging voor Waterbeheer)	www.nva.net
New Zealand	NZWWA (New Zealand Water and Wastes Association)	www.nzwwa.org.nz
Norway	Norsk Vannforening	www.vannforeningen.no
Poland	Izba Gospodarcza "Wodociagi Polskie"	www.igwp.org.pl
Portugal	APESB (Associação Portuguesa de Engenharia Sanitária e Ambiental)	www.apesb.pt
Slovak Republic	ACE SR (Association of Wastewater Treatment Experts of the Slovak Republic)	www.vuvh.sk
South Africa	WISA (Water Institute of Southern Africa)	www.wisa.co.za
Spain	ADECAGUA (Asociación para la defensa de la calidad de las aguas)	www.adecagua.org

(*Continued*)

TABLE 1.3 Selected National and International Professional and Scientific Water Associations (*Continued*)

Country	Acronym and name of organization	URL
Sweden	VAV (Swedish Water and Wastewater Works Association)	www.vav.se
Switzerland	SVGW (Swiss Technical and Scientific Association for Gas and Water)	www.svgw.ch
UK	Water UK	www.water.org.uk
	CIRIA (Construction Industry Research and Information Association)	www.ciria.org.uk
Ukraine	UWA (Ukrainian Water Association)	www.cleanwater.org.ua
USA	AWWA (American Water Works Association)	www.awwa.org
	AWRA (American Water Resources Association)	www.awra.org
International		
European Union	EWA (European Water Association)	www.ewaonline.de
	EUREAU (European Union of National Associations of Water Suppliers and Waste Water Services)	www.eureau.org
Africa	UAWS (Union of African Water Suppliers)	www.irc.nl
	IAEH (International Association for Environmental Hydrology)	www.hydroweb.com
	IWRA (International Water Resources Association)	www.iwra.siu.edu
	IWA (International Water Association)	www.iwahq.org.uk
	IAHS (International Association of Hydrological Sciences)	www.cig.ensmp.fr/~iahs/index.htm

by technical committees comprised of experienced members. Prior to final publication, the papers are usually prepublished for review by the interested community. Although these documents have no official legal status, they are still considered to be quasi-standards in many fields, especially when no administrative regulation is available. In Germany, the DVGW regulations have attained such a status for gas and water. The U.S. Army Corps of Engineers has published some Engineer Manuals (EM), Engineer Pamphlets (EP), and Technical Instructions (TI) on construction, operation, and maintenance of wells (e.g., EM 1110-3-161, EM 1110-1-4000, EM 1110-2-1914, EP1110-1-27, TI814-01). They can be downloaded free of charge from the Internet (www.usace.army.mil). The U.S. Environmental Protection Agency (US EPA) also provides some recommendations on the same topics (e.g., EPA-570/9-75-001), also available free of charge at the US EPA homepage (www.epa.gov). The British Construction Industry Research and Information Association (CIRIA) has published a multitude of commercially available reports that provide "best practice" guidance on many construction-related topics. Some of them address groundwater, and one (R 137, Howsam et al. 1995) even deals with monitoring, maintenance, rehabilitation, design, and construction of wells (www.ciria.org.uk/acatalog/R137.html).

It may be a wise idea to follow recommendations by professional organizations as closely as possible, even though they cannot be enforced. In the event of a lawsuit, one has to be aware that the judges are legal experts but not experts in the field of water wells. Often they will base their judgment on formalistic criteria—i.e., if and how regulations and best practices were followed.

Some professional organizations also offer the accreditation of utility companies and contractors, based on recognized standards of best available practices. Accreditation is achieved (or declined) through an independent assessment (audit) of service quality and management efficiency. Audits have to be repeated at regular intervals. Accreditation provides "a credible check signifying to customers and all stakeholders that the utility is doing the 'right things' in its quest to provide high quality service. By accrediting their operations, utilities will be able to assure that they are conforming to 'standard practices' without the need for additional regulation" (AWWA). It should be noted that accreditation is a lengthy, continuous, and also costly process.

Chapter 2

Elements of Well Hydraulics and Well Operation

2.1 Well Types

2.1.1 Introduction

Groundwater is less prone to pollution than surface waters, and in general its formation materials have a certain protective capacity against undesirable chemical constituents. Groundwater is also usually a substantially larger fresh water resource than surface waters. Of the world's fresh water, about 33% is groundwater while barely 0.33% is found in rivers and lakes. The remaining 69% is stored as polar ice or in permafrost (UNESCO-WWAP 2003). In many countries groundwater, bank filtrate, and artificial groundwater recharge are therefore the most important sources for public and sometimes industrial water supplies. Installing wells is a common method of accessing this underground water resource. It is estimated that in Germany alone, about 25,000 wells are used for public water supply. In addition, wells are commonly used for the dewatering of construction sites and mining operations. It should be noted that some wells are used for other purposes than obtaining groundwater, such as for injecting excess (waste) water and exploiting geothermal resources.

For all types of well uses, outages or defects can seriously affect water supply schemes and construction or mining operations. The protection, regular inspection, and maintenance of wells is hence an important part of maintaining a safe operation. Processes leading to outages and defects must be identified, and proper countermeasures need to be applied.

We can distinguish between horizontal and vertical wells. While the former are restricted to rather shallow groundwater occurrences (commonly <50 m (150 ft) depth), the latter allow access to almost any depth. Today,

18   Chapter Two

well depths of several hundred meters are common and depths of more than 1,000 m (3,300 ft) are technically feasible although rarely installed.

This book does not intend to give an overview of all drilling and completion techniques available today. This task would easily require the compilation of several hundred pages. For this, the reader is referred to the classic textbooks by Driscoll (1989), Roscoe Moss (1990), ADITCL (1996), and Bieske et al. (1998).

2.1.2 Dug wells

As the simplest type of vertical well, the dug well was invented early in the history of modern humans and is still in use in many developing countries. One of the oldest known wells (from around 5,090 BC)—found near the small German town of Erkelenz in 1989—was a dug well that tapped a gravel aquifer at a depth of about 13 m (42 ft) (Weiner, cited in LVR 1998). Due to the instability of the formation, the well was stabilized using a rectangular frame (3 m × 3 m [9 ft × 9 ft]) of wooden planks. Nowadays concrete shaft rings or brickwork are used in unstable formations (Fig. 2-1). Since ancient shaft wells had to be excavated by hand, the diameters were—and still are—quite large. Water enters into dug wells usually through the bottom face or through a few perforations in the walls. Most often they only skim the surface of the water table, which is of course the first to be affected by anthropogenic contamination. The

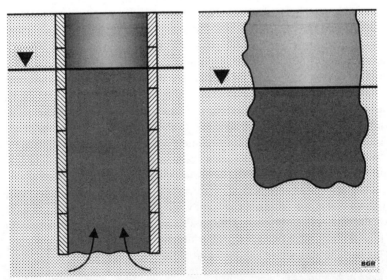

Figure 2.1 Schematic illustration of dug well types. Left = with constructed shaft (here: shaft rings), right = without shaft stabilization.

yield is usually quite low, and decreasing water tables often cause them to fall dry. Another significant disadvantage is related to their large diameters. This makes them prone to pollution from the surface. In Kabul (Afghanistan), about 45% of all shallow drilled wells equipped with hand pumps show pollution by fecal bacteria—which is bad enough—but more than 76% of all (open) dug wells are affected by the same problem (Timmins 2002, cited in Banks & Soldal 2002).

2.1.3 Vertically drilled wells

Although the ancient Chinese already drilled vertical wells to depths of several hundred meters into consolidated rock, the introduction of steam-powered machines gave a serious boost to the drilling of vertical wells. Because of the smaller diameters and the possibility to access deeper and thus well-protected aquifers, the concept of drilled vertical wells still prevails today. Figure 2-2 shows the main components of a modern vertical drilled well. Vertical wells can be adapted to a variety of geological, hydrological, and technical settings, leading to a large variability in subtypes and construction details (Fig. 2-3).

On the other hand, deeper aquifers often encounter reducing hydrochemical conditions and thus elevated concentrations of iron, manganese, and sulfides. Contact with oxygen and mixing in long screens promotes the precipitation of mineral phases which may clog the well ($\rightarrow$ 3).

In consolidated rocks, the well is often simply an open borehole equipped with a pump (Fig. 2-3). Sometimes soil and weathered semi-consolidated upper parts of the aquifer need to be stabilized by a casing—and backfilling of the annulus—while the aquifer remains uncased (Fig. 2-3). Since the yield of unconsolidated gravel and sand aquifers is often much higher than that of consolidated rocks—with the notable exception of karst aquifers—such formations are often the target of groundwater exploration. Because of their instability, the whole borehole must be stabilized. The tubing is divided into perforated parts (screen) which take in the groundwater and nonperforated parts (casing) intended to seal off nonaquiferous formations or zones of undesirable water quality. The remaining space between the tubing and the borehole wall (annulus) is filled with permeable material (filter or gravel pack) around the screen and with impermeable material (bentonite, cement, grout) around the casing (Fig. 2-3).

2.1.4 Horizontal and radial collector wells

The simplest type of horizontal well is the drain, basically a perforated pipe laid into a trench, often embedded in a filter bed covered by impermeable material. It is often used for dewatering purposes, e.g., of swampy areas. Its use for water supply schemes is very rare.

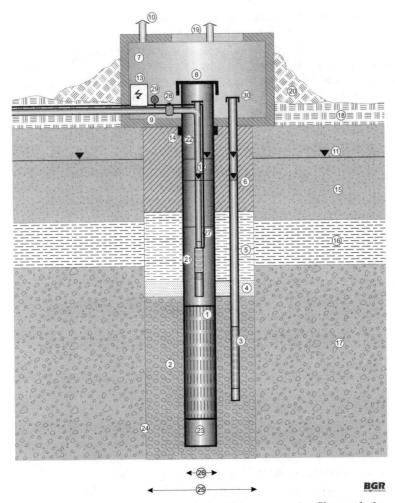

Figure 2.2 Main components of a vertical well: 1 = screen, 2 = filter pack, 3 = piezometer, 4 = sand filter, 5 = annular seal, 6 = annulus filling, 7 = well head, 8 = protective cover, 9 = pipeline, 10 = air vent, 11 = water level of upper aquifer, 12 = water level of lower aquifer, 13 = electrical installations, 14 = foot cementation, 15 = upper aquifer, 16 = aquitard, 17 = lower aquifer (production aquifer), 18 = soil, 19 = access door, 20 = soil backfilling, 21 = riser pipe, 22 = casing, 23 = sump, 24 = borehole wall, 25 = drilling diameter, 26 = well diameter, 27 = electric cable, 28 = backflow preventer valve, 29 = water meter, 30 = protective cap of piezometer.

A more common type of horizontal well is the radial collector well (RCW). From a water-tight, large-diameter shaft (2.5–5.0 m [8–15 ft]), up to eight screen tubes up to 80 m (260 ft) in length are drilled radially into the aquifer (Fig. 2-4). The water may flow by gravity into the shaft and can be pumped from there ("wet shaft"), or each screen

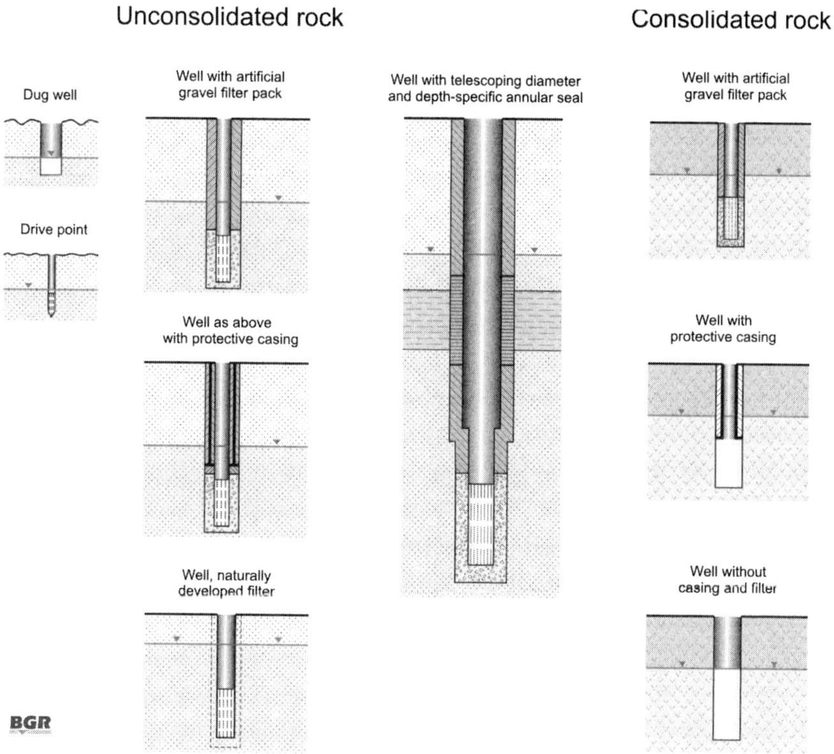

Elements of Well Hydraulics and Well Operation 21

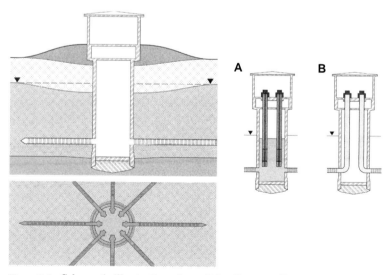

Figure 2.3 Some basic types of vertical wells.

Figure 2.4 Schematic illustration of a radial collector well with eight screens. A = "wet" shaft, B = "dry" shaft.

may be sealed off and pumped individually ("dry shaft"). Because of the elaborate shaft construction, the maximum depth is around 50 m (160 ft). Although they are substantially more expensive than vertical wells, radial collector wells have some notable advantages:

- Very high intake capacity (one RCW can replace dozens of smaller vertical wells)
- Shorter pipeline length (because there are fewer well sites)
- Lower and more evenly spread drawdown
- Less mixing of incompatible waters (→ 3.2)

The listed advantages, especially the last two, may result in lesser degrees of ageing and thus lower maintenance costs (Pluijmackers et al. 2005).

2.1.5 Directionally drilled wells

Directed drilling, originally developed in the oil industry and for trenchless cable and pipeline installation, has found its way to water well construction over the last few years. Horizontal directionally drilled (HDD) wells are constructed using a directional drilling technique. The emplacement of the screen can be adapted to the subsurface morphology of the aquifer while drilling. HDD wells are particular useful to exploit shallow aquifers of low thickness, where the operation of several vertical wells would be hampered by the (overlapping) cones of depression. The directionally drilled tubing section can be up to 350 m (1,150 ft) long and have diameters of up to 200 mm (8 in). HDD wells are often drilled from the surface into a central collector shaft (Fig. 2-5). The

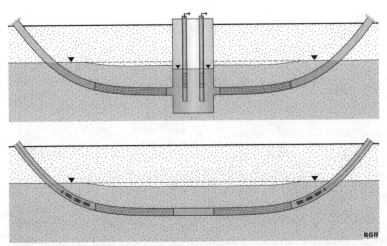

Figure 2.5 Schematic illustration of horizontal directionally drilled wells (HDD) with (upper) and without (lower) central collector shaft.

water flows into the shaft and and can be abstracted from there by pumps of any diameter. Otherwise, pumps have to be installed in the inclined tubing, which significantly limits the possible pump diameters (up to 150 mm, [6 in]).

2.2 Well Hydraulics

2.2.1 Radial symmetric flow to wells

Most theoretical considerations of well hydraulics focus on a single well with radial symmetry of flow. This implies that both the amount and the velocity of flow are equal in all directions. The removal of water through the well leads to an acceleration of flow toward the well and changes in groundwater levels (cone of depression). The latter are usually expressed as drawdown compared to the original water levels, i.e., prepumping levels. Figures 2-6 and 2-7 show the geometry of a depression cone and the flow velocity in a flow field of radial symmetry. According to the continuity equation, the Darcy velocity v (= Q/A) increases with decreasing distance r from the well due to the associated decrease in area A, while the flow rate Q is constant.

The concept of radial flow symmetry around wells forms the basis of the analytical solutions derived for the evaluation of pumping tests (Walton 1988; Domenico & Schwartz 1990; Krusemann & de Ridder 1994; Batu 1998). Most of these calculations rely on the following assumptions:

1. Infinite spatial extent of the aquifer
2. Horizontal groundwater flow

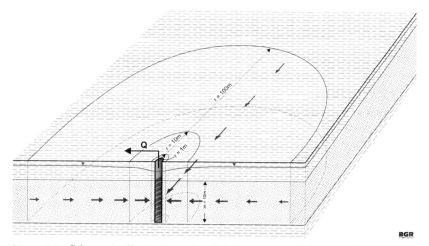

Figure 2.6 Schematic illustration of radial flow to a well in a confined aquifer (r = radius from well axis, m = aquifer thickness).

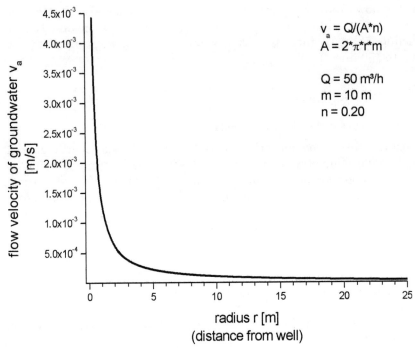

Figure 2.7 Groundwater flow velocity as a function of distance from a pumping well assuming radial flow symmetry (A = area of cylinder, Q = pumping rate, m = aquifer thickness, n = porosity).

3. Steady-state flow conditions
4. Continuous transition from the cone of depression to the natural groundwater surface
5. Homogeneous and isotropic aquifer
6. Incompressible groundwater and aquifer matrix

In nature, assumptions 1, 2, and 6 are never fulfilled, assumption 5 only very rarely, assumption 3 only after a substantial period of pumping, and only assumption 4 is realistic. Nevertheless, the analytic solutions are still quite useful for many basic calculations of well hydraulics.

With v_r as the radial Darcy velocity at a distance r from the well, the continuity equation requires steady-state flow in a confined aquifer.

$$v_r \rightarrow 2\pi r m = \pm Q \rightarrow v_r = \frac{\pm Q}{2\pi m}\frac{1}{r} = -k_f \frac{dh}{dr} \qquad (2.1)$$

where Q = pumping (or infiltration) rate, L^3/T
m = aquifer thickness, L

h = hydraulic head, L
k_f = hydraulic conductivity (permeability), L/T

The water table elevation (hydraulic head) at any distance r around the well can be calculated by the integration of Eq. (2.1).

$$h_2 - h_1 = \frac{Q_E}{2\pi T} \ln\left(\frac{r_2}{r_1}\right) \qquad (2.2)$$

where T = transmissivity = $k_f * m$, L^2/T
r_1, r_2 = radial distances from the well with $r_2 > r_1$, L
h_1, h_2 = water table elevation (hydraulic head) at distances r_1 and r_2, L

Equations (2.1) and (2.2) describe the steady-state flow field that develops after an extended period of pumping time. The time span required to reach this state is a function of the aquifer permeability, thickness, and storage capacity, and of boundary conditions. Nevertheless, the equations are quite useful to answer a lot of questions on well hydraulics which are of concern for well ageing processes.

The increase in flow velocities close to the well screen is often assumed to cause a turbulent flow regime. Turbulent flow is undesirable because of the inherent energy loss, which results in elevated drawdown and thus higher production costs. In addition, turbulence is often attributed to negative chemical effects. Turbulent flow can lead to a degassing of solution-promoting gases (e.g., carbon dioxide → 3.3) and facilitates mixing of incompatible dissolved species from different aquifer levels (→ 3.2). The dimensionless Reynolds number Re [Eq. (2.3)] is usually employed to assess the degree of flow turbulence. Low Re values (<2) indicate laminar flow and a dominance of viscous over inertial forces. Darcy's law is fully valid in this range. At higher Re values, turbulent flow and inertial forces become increasingly important. Darcy's law is no longer valid since in this case the flux density is not a linear function of the hydraulic gradient. There is no sharp boundary between laminar and turbulent flow. While a Re of 100 is considered to indicate fully turbulent flow, values between 2 and 100 describe the gradual transition between laminar and turbulent flow. For the near-field of wells, Re values of 6 to 60 are considered critical (Truelsen 1958; Vukovic & Soro 1992). The former value is certainly a conservative estimate, while the latter is rather optimistic. Roscoe Moss (1990) thus favors a value of 30.

$$\text{Re} = \frac{\text{kinetic energy}}{\text{friction losses}} = \frac{\frac{1}{2}\rho v^2}{(v/2r)\eta} = \frac{2r\rho v}{\eta} \qquad (2.3)$$

where r = radius of capillary, L
  v = flow velocity of fluid, L/T
  ρ = fluid density, M/L^3
  η = dynamic viscosity of fluid, $M/(L \cdot T)$

The development of the Reynolds number as a function of distance from a pumping well is shown in Fig. 2-8. Since the flow velocity is a major factor in its calculation, it of course mirrors the flow velocity distribution depicted in Fig. 2-7. Only very close to the well will the Reynolds number reach values indicating a transition toward turbulent flow. Nevertheless, such simple calculations are worth the effort for the dimensioning of wells prior to their construction.

One application of the Reynolds number to well hydraulics is given by Williams (1985; see also Roscoe Moss 1990). He tried to calculate the critical radius of a well, defined as the distance from the center of the well to the interface between laminar and turbulent flow. At this interface, the hydraulic gradients ($\partial h/\partial r$) for laminar flow [Eq. (2.4a)] and for turbulent flow [Eq. (2.4b)] are equal.

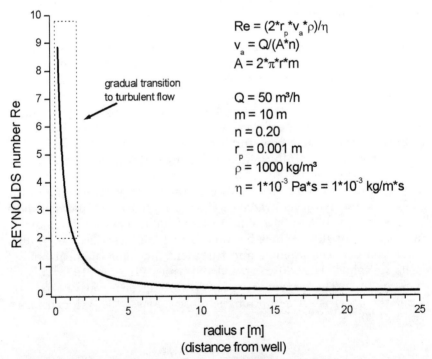

Figure 2.8 Reynolds number Re as a function of distance from a pumping well assuming radial flow symmetry.

$$\frac{\partial h}{\partial r} = \frac{a_1 v^2 \cdot \text{Re}}{gkd} \tag{2.4a}$$

$$\frac{\partial h}{\partial r} = \frac{a_2 \cdot v^2 \cdot \text{Re}}{g \cdot k \cdot d} \tag{2.4b}$$

where h = hydraulic head, L
r = radius, L
Re = Reynolds number
v = kinematic viscosity of fluid, $L^2 \cdot T^{-1}$
g = acceleration of gravity, $L \cdot T^2$
k = intrinsic permeability, L^2
d = mean grain diameter (d_{50}), L
a_1 = constants (a_1 = 1 for laminar flow, $a_2 = \text{Re}_c^{-1}$ for turbulent flow)
Re_c = Reynolds number at critical point (transition from laminar to turbulent flow)

The critical Reynolds number was then determined experimentally by measuring the hydraulic gradient at varying distances from the well for different flow rates (Williams 1985). In a plot of the calculated Reynolds numbers versus the measured gradients, a change in slope from 1 to 2 was observed at Re values around 30. This delineated the transition between turbulent and laminar flow [see the exponents of Re in Eqs. (2.4a) and (2.4b)]. Using the value of $\text{Re}_c = 30$, the critical radius r_c can be calculated from the dimensionally nonconsistent equation (2.5):

$$r_c = 0.9587 \cdot \frac{(Q/L) \cdot d}{n} \tag{2.5}$$

where r_c = critical radius, L (in)
Q = pumping rate, $L^3 \cdot T^{-1}$ (gal/min, gpm)
L = saturated aquifer thickness = length of well screen, L (ft)
d = mean grain diameter (d_{50}), L (in)
n = effective porosity

(In standard metric units (m, s), the factor 0.9587 has to be to replaced by 0.0002.)

2.2.2 Aquifer yield and intake capacity of wells

Based on the concept of radial-symmetric steady-state flow to wells presented in Sec. 2.2.1, Dupuit (1863) and Thiem (1870) derived equations which describe the flow of groundwater to confined and unconfined

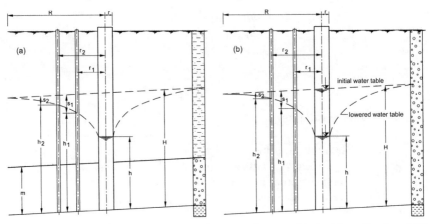

Figure 2.9 Effects of well operation on the groundwater table in (a) confined and (b) unconfined aquifers (m = aquifer thickness, R = radius of influence, r = well radius, $r_{1,2}$ = radial distances from well ($r_2 > r_1$), H = initial aquifer water level (hydraulic head), $h_{1,2}$ = water levels at distances r_1 and r_2, $s_{1,2}$ = drawdown at distances r_1 and r_2). Modified after Bieske et al. (1998).

wells. Despite their underlying simplified assumptions, the equations are particularly useful in assessing the technical and hydrogeological aptitude of a well site and the maximum permissible pumping rate. The general geometric boundary conditions are presented in Fig. 2-9. The most important difference is the variability of aquifer thickness in unconfined aquifers induced by pumping.

The term *aquifer yield* describes the volume of water that can be pumped as a function of the lowering of the water table [Eqs. (2.6) and (2.7)]. As such, it is independent of the well type. Aquifer characteristics alone do not adequately describe the removable amount of water. The technical constraints of the well need be considered as well. To do so, the term *intake capacity* has been introduced [Eqs. (2.8) and (2.9)]. It describes the possible pumping rate under steady-state flow conditions.

Aquifer yield Q_a in unconfined aquifers, after Dupuit (1863):

$$Q_a = \pi k_f \frac{h_0^2 - h_w^2}{\ln R/r_w} \qquad (2.6)$$

Aquifer yield Q_a in confined aquifers, after Dupuit (1863):

$$Q_a = 2 \pi k_f m \frac{h_0 - h_w}{\ln R/r_w} \qquad (2.7)$$

Intake capacity Q_F in unconfined aquifers, after Sichardt (1928):

$$Q_F = 2 r_w \pi h_0 v_{max} \quad (2.8)$$

Intake capacity Q_F in confined aquifers, after Sichardt (1928):

$$Q_F = 2 r_w \pi m v_{max} \quad (2.9)$$

where r_w = effective well radius, L
m = aquifer thickness, L
k_f = hydraulic conductivity, L/T
v_{max} = maximum permissible entrance velocity of water, L/T
h_0 = static water level, L
h_w = pumping water level, L
R = radius of influence, L

Taking into account the simplified boundary conditions of a radial-symmetric flow field and a point discharge source, the aquifer yield increases with increasing drawdown, though nonlinearly for unconfined aquifers. At the same time, the flow velocity increases due to the decreasing flow-through area (Fig. 2-7). The aquifer yield plots as a straight line for confined and as a parabola for unconfined conditions (Fig. 2-10). Applying the empirical Sichardt velocity (Sichardt 1928),

$$v_{max} = \sqrt{k_f}/15 \quad (2.10)$$

(units are m/s)

allows us to calculate a k_f-dependent well intake velocity v_{max} as maximum for a resulting "permissible" slope of the cone of depression. This maximum slope should not be exceeded, to prevent suffossion from the

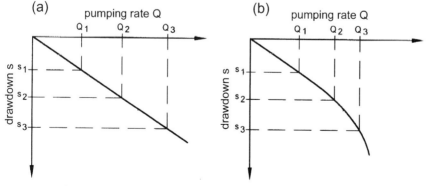

Figure 2.10 Aquifer yield in (a) confined and (b) unconfined aquifers.

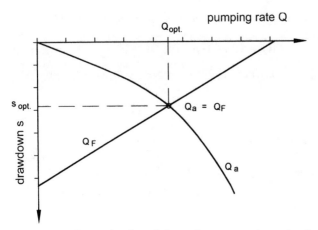

Figure 2.11 Determination of the optimum pumping rate of a well from aquifer yield and well intake capacity (Q_a = aquifer yield, Q_F = well intake capacity, Q = pumping rate, s = drawdown, s_{opt} = optimum drawdown, Q_{opt} = optimum pumping rate).

aquifer and well entrance losses which cause elevated drawdown. More detailed descriptions of critical entrance velocities can be found in Sec. 2.2.1. Combining aquifer yield and intake capacity of the well allows selection of the optimum pumping rate (Fig. 2-11).

Pressure losses at interfaces (screen/gravel pack, borehole wall/gravel pack) cause differences in water levels across the interfaces (Fig. 2-12). They can be quantified by comparing water levels in the aquifer, in the annulus, and in the well interior. For this reason, a complete well requires two observation wells, one installed in the gravel pack and the other installed in the immediate vicinity within the aquifer. The total head loss is the sum of the head losses caused by the

- Aquifer
- Zone influenced by drilling (excavation damaged zone, EDZ)
- Gravel pack
- Screen

2.2.3 Flow to wells under natural conditions

Laboratory-scale experiments and real-life experiences have shown that groundwater flow in the immediate vicinity of wells, especially near the screen, is far from being homogeneous (Petersen et al. 1955). The distribution of flow velocities along the screen is a function of:

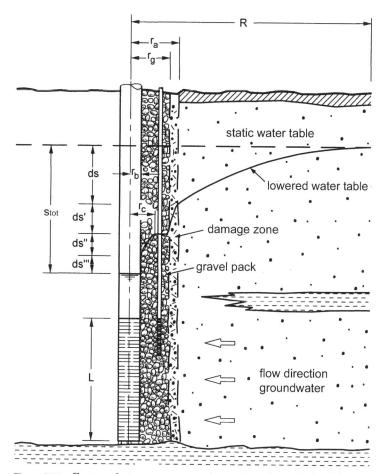

Figure 2.12 Factors that cause pressure loss at interfaces (= head losses) in vertical wells. L_s = length of screen, R = radius of influence, r_b = radius of well tubing, r_c = distance from well axis to axis of annular observation well, r_g = distance from well axis to annulus, r_a = distance from well axis to well damage zone (EDZ), s_{tot} = total head loss, ds = aquifer head loss, ds' = EDZ head loss, ds'' = gravel pack head loss, ds''' = screen head loss, EDZ = excavation damaged zone).

- The position of the submersible pump
- Vertical inhomogeneities of aquifer permeability and thus inflow
- Suffossion of aquifer material
- Colmation and incrustation of screen and gravel pack

Numerical flow models are more useful tools for simulating realistic flow fields in the vicinity of wells than the approaches applied in

32   Chapter Two

Sec. 2.2.1. Models for both horizontal ("plan view") and vertical ("cross-section") flow were performed by Houben (2006). In both cases, steady-state flow was assumed. Instead of assuming radial flow symmetry, for the horizontal case, a well was superimposed onto a natural flow field. For more details the reader is referred to Houben (2006).

For the horizontal flow, a two-dimensional aquifer in plan view was assumed (Fig. 2-13). Water enters the flow field from the right through a constant-head boundary and flows to the left to another constant-head boundary. The "well" is approximated as a quasi-circular arrangement of constant-head cells. They are surrounded by a double layer of

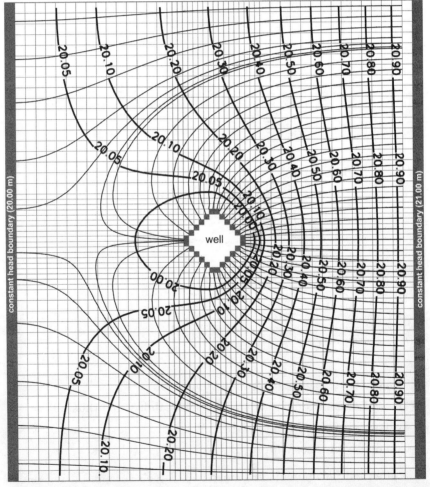

Figure 2.13 Horizontal distribution of hydraulic heads and flow paths (plan view) of a pumping well, calculated using the hydraulic flow model PMWIN. From Houben (2006).

cells of elevated permeability, representing the gravel pack. The resulting flow field is shown in Fig. 2-13. The inflow into the well from the right-hand direction is clearly more pronounced than that from the left. The relative proportions of inflow into the well were calculated for both sections. The section facing to the right—that is, toward the "natural" flow direction (upgradient)—receives much more inflow than the downgradient section. This also means that the majority of all dissolved constituents enters the well through the section facing the natural flow direction. Therefore the predominance of incrustation buildup in this section should be no surprise (→ 3.9). The relative amount of upgradient versus downgradient inflow is a function of the natural gradient of the flow field surrounding the well (Fig. 2-14).

To investigate the vertical distribution of inflow over a screen, a vertical profile of an aquifer in cross-sectional view was modeled (Fig. 2-15). Water enters the flow field from the right through a constant-head boundary and flows to the left in the direction of the partially penetrating well. The well has a total depth of 50 m (164 ft), of which the lower 30 m

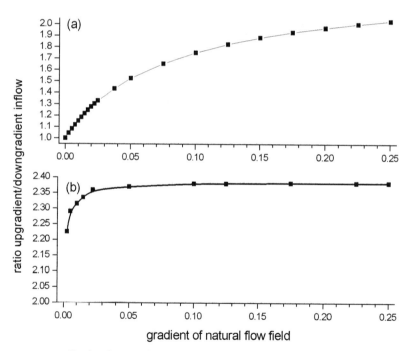

Figure 2.14 Ratio of upgradient and downgradient inflow into a well screen. (a) Hydraulic head in pumped well (19.90 m [65.29 ft]) lower than head in left (lower) constant-head boundary (20.00 m [65.62 ft]), inflow from left boundary into well possible). (b) Hydraulic head in pumped well equal to head of left (lower) boundary (20.00 m [65.62 ft]), inflow from left boundary impossible). Calculated from PMWIN models in Fig. 2-13. From Houben (2006).

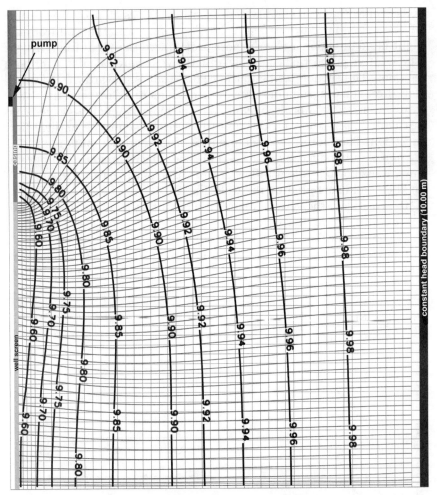

Figure 2.15 Vertical distribution of hydraulic heads and flow paths (cross section) in the vicinity of the screen of a partially penetrating pumping well, calculated using the flow model PMWIN. From Houben (2006).

(98 ft) are screened. The pump is represented by a single constant-head cell. The cells representing the impermeable well casing are surrounded by a double layer of cells of very low permeability (annular seal). The screen below is surrounded by a double layer of elevated permeability (gravel pack). Flow inside the well interior is not considered, because it is most probably turbulent and cannot be assessed with a model of the type used.

The resulting flow field is presented in Fig. 2-15. The vertical flow component close to the top of the screen well is clearly visible. This implies that much water from close to the water table enters the well at the top of the screen. Since this water most likely contains dissolved oxygen due to exchange reactions with atmospheric air or soil air, it plays an important part in ochre formation (→ 3.2). The highest velocities also occur at the top of the screen. Locally elevated flow velocities can enhance well ageing by accelerating several processes:

- Mobilization of sand and silt particles (suffossion) → sand intake or particle clogging (→ 4.1)
- Degassing of dissolved carbon dioxide → carbonate precipitation (→ 3.3)
- Increased mixing, e.g., of oxygen and ferrous iron → ochre formation (→ 3.2)

The vertical distribution of inflow over the screen was calculated. Figure 2-16 shows that while the lower sections of the screen have an almost homogeneous inflow, the upper third shows a distinct increase of inflow nearer to the screen top. The upper third of the screen provides almost 50% of the total water inflow! The uppermost meter of the screen provides more than three times more water than the lowermost meter (Fig. 2-16).

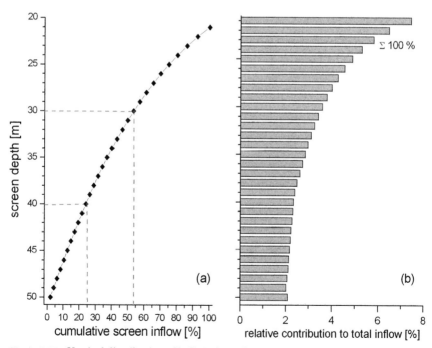

Figure 2.16 Vertical distribution of inflow along the screen depth of a partially penetrating pumping well, calculated from the flow model in Fig. 2-15. From Houben (2006).

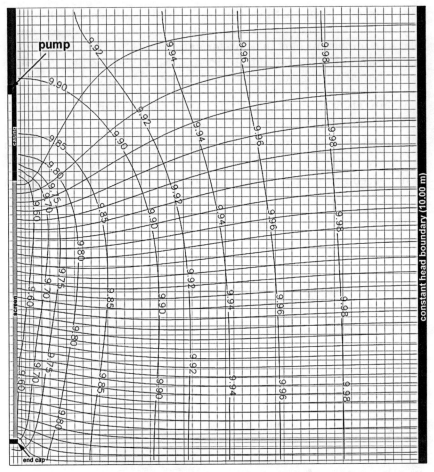

Figure 2.17 Vertical distribution (cross section) of hydraulic heads and flow paths in the vicinity of the screen of a partially penetrating pumping well not screened to the bottom of the aquifer, calculated using the flow model PMWIN.

A model of a partially penetrating well not fully screened to the bottom of the aquifer was also calculated. Three rows of unscreened aquifer cells were added at the bottom. The bottom of the well was assumed to be impermeable (end cap). The flow field is slightly different in that a vertical flow component is visible from the aquifer area below the screen (Fig. 2-17). This flow terminates in the bottom parts of the screen and causes an elevated inflow rate there (Fig. 2-18). This distribution closely resembles results from flow meter investigations of real wells (→ 11.3.2).

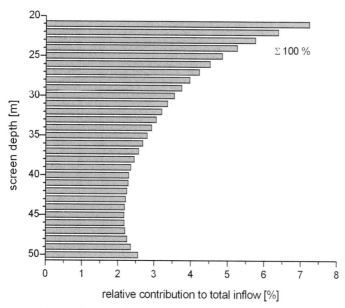

Figure 2.18 Vertical distribution of inflow over the screen of a partially penetrating pumping well not screened to the bottom of the aquifer, calculated from the PMWIN flow model in Fig. 2-17.

2.3 Dimensioning of Wells

Well dimensioning defines the adjustment of the intake capacity of the well to the natural groundwater supply of the aquifer and its hydraulic properties. Again, it is not within the scope of this book to present a comprehensive overview of this topic. The reader is referred to current textbooks on well construction such as those of Driscoll (1989), Roscoe Moss (1990), Vukovic and Soro (1992), ADITCL (1996), and Bieske et al. (1998).

A properly dimensioned well must be able to supply sand-free water (no suffossion) but at the same time minimize hydraulic entrance losses. The latter is a function of the flow velocities in the near-field of the well and in the screen slots, which should thus be minimized as well (Williams 1981; Driscoll 1989). The following parameters have to be considered:

- Length, diameter, and open area of the screen
- Size and open area of screen slots (adjusted to grain size of gravel pack or aquifer)
- Grain size of gravel pack (adjusted to grain size of aquifer)

The screen diameter is often determined by the size of the pump and the riser pipes to be used. Usually the following relation is applied:

$$D_s = D_P + 100 \text{ mm} \quad [\text{mm}] \quad (2.11)$$
$$(100 \text{ mm} = 0.328 \text{ ft})$$

where D_s = screen diameter, mm
D_P = pump diameter (pump and riser pipe), mm

Flow velocities should not exceed 0.5–1.0 m/s (1.6–3.3 ft/s) in the screen (interior) and 1.5–2.0 m/s (5–6.6 ft/s) in the casing close to the pump (Balke et al. 2000).

Groundwater entering a well has to pass two interfaces during its entrance (Fig. 2-19): the borehole wall, which is the contact between aquifer and gravel pack; and the screen. The flow velocity should not exceed a critical value, to prevent turbulence and resulting negative consequences:

- Head losses
- Suffossion of fine particles from gravel pack and aquifer
- Enhanced degassing
- Enhanced mixing of incompatible waters
- Enhanced corrosion

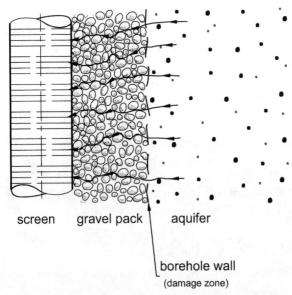

Figure 2.19 Schematic cross section through a well showing the interfaces for water flow.

The value of this critical flow velocity is often assumed to lie between 0.03 and 0.08 m/s (0.1 and 0.25 ft/s), with a value of 0.03 m/s (0.1 ft/s) often used as the standard reference (Smith 1963; Driscoll 1989). Some authors question this figure and claim that even much higher velocities of 0.6–1.2 m/s (2–4 ft/s) still do not cause excessive head losses at the screen (Roscoe Moss 1990; see also www.roscoemoss.com/modern_ tech.html).

The length of the gravel pack can be calculated as

$$H_S = \frac{Q}{\pi \cdot D_B \cdot v_{crit}} \quad (2.12)$$

where Q = pumping rate, L^3/T
D_B = borehole diameter, L
v_{crit} = critical flow velocity at the borehole wall, L/T

The effective length of the screen, L_F, can be calculated according to Balke et al. (2000) using

$$L_F = \frac{Q}{\pi \cdot D_F \cdot v_{crit,a}} \quad (2.13)$$

where Q = pumping rate, L^3/T
D_F = outer diameter of screen, L
$v_{crit,a}$ = critical flow velocity at the outer side of the screen, L/T

Equation (2.14) allows the calculation of the well screen open area, $A_{S,open}$:

$$A_{S,open} = \frac{Q}{v_{crit,F}} \quad (2.14)$$

where $v_{crit,F}$ = critical flow velocity in filter slot, L/T

The open area of the well screen is also a topic of some debate. Some authors suggest that the open area should be designed as high as possible to minimize losses in the screen. This could be interpreted as a recommendation for wire-wound screens. Others claim that the head losses in the screen are negligible when compared to the losses in the formation and in the gravel pack. With entrance velocities of less than 0.75 m/s (2.5 ft/s), the open area of the screen only needs to be higher than 3–5% to obtain sufficiently low screen losses (Williams 2002).

The dimensioning of the grain size (distribution) of the gravel pack is a very important task in determining the well yield and its susceptibility for ageing. The gravel pack must fulfill two requirements:

- Restraining fine-grained formation (and gravel pack) material from entering the well
- Minimizing head losses

Whereas the first requirement necessitates fine-grained material, too finely-grained material will significantly affect the hydraulic performance of the well (→ 2.2). Therefore, it is essential to find an ideal grain size (distribution) to fulfill both tasks. All recommendations currently in use are based on the principle that the mean grain size of the gravel pack should be about four to six times coarser than that of the aquifer material (DVGW W 113; US EPA-570/9-75-001; U.S. Army Corps of Engineers TI 814-01). The underlying theory is discussed in more detail in Sec. 4.1.2.

Another important issue is the thickness of the gravel pack. In theory, a thickness of two to three grain diameters is sufficient to retain fine grains. Nevertheless, such narrow gravel envelopes are practically impossible to construct under realistic circumstances, especially in deep wells. Some well operators actually believe in using very thick gravel packs. This enlarges the interface between the outer gravel pack and the aquifer—minimizing head losses there—and provides a large storage volume for incrustations, which delays their negative effects on the overall well hydraulics. On the other hand, large diameters require more costly drilling and negatively affect desanding, rehabilitation, and reconstruction measures (→ 7.6).

2.4 Well Operation and Maintenance

The terminology of DVGW W 614 and DIN 31051 for technical installations can be applied for wells as follows:

Operation. All activities required for and during regular well operation.

Maintenance. This general term comprises all measures aimed at retaining the nominal condition of a technical installation as well as measures to investigate and assess the current condition. This includes inspection, servicing, and repairs.

Inspection. Measures to investigate and assess the current condition of a well. They may include (visual) inspection, monitoring of operating parameters (e.g., pumping rate, drawdown, water quality), and comparison of measured data to nominal conditions. Inspection and monitoring can be performed continuously or at intervals. Continuous measures comprise all measurements of operating parameters, whereas discontinuous measures include all singular visual, manual, chemical, and physical inspections. Inspection measures comprise:

- Compilation of inspection schemes
- Preparation of inspection measures (technical installations, logistics, and human resources)
- Execution of inspection and data collection
- Documentation of results
- Interpretation and evaluation of results, comparison to nominal conditions
- Derivation of measures to be taken

Servicing. Servicing comprises all measures intended to maintain and conserve the nominal conditions and comprises:

- Setup of a maintenance plan
- Preparation of servicing procedures
- Execution
- Comparison of nominal and actual performance

Repair. Repairs include all measures intended to solve problems and to reestablish the nominal conditions. They also include rehabilitation and reconstruction measures:

- Planning of individual measures (including assessment of alternatives) and requirements
- Decision on practicable and economically viable solutions
- Preparation: time scheduling with internal and external participants, allocation of personnel and materials, applications for permits (if applicable)
- Documentation of results
- Assessment of results, functional testing, and acceptance/rejection
- Authorization for reconnection to water supply network

Three main strategies for maintenance measures exist for wells (Janssens et al. 1996). According to DVGW W 614 we can distinguish:

1. *Preventive, status-oriented, and interval-dependent maintenance measures:* The well is inspected at regular intervals according to a maintenance plan with predefined parameters to be investigated.

2. *Damage-oriented maintenance:* Rehabilitation and reconstruction measures are invoked after a disruption of operation or after a significant decrease in yield.
3. *Breakdown-oriented maintenance:* This concept is often applied for wells which cannot be rehabilitated and are expected eventually to collapse. The well is then reconstructed after a total failure of operation.

The aim of maintenance concepts is to preserve both the functionality and the yield of a well. Many ageing processes have natural causes and will therefore proceed steadily. Such ageing processes cannot be entirely prevented but should be slowed down as much as possible. Timely countermeasures require good knowledge about the processes and their influences on yield and function of the well. Causes of well ageing include:

- Corrosion
- Clogging by particles
- Incrustation
- Technical failure (e.g., of a pump)
- Decreasing aquifer yield (e.g., overexploitation)

Efficient well maintenance requires basic information on the well construction site:

1. Dimensioning characteristics (well log, geological log, static and dynamic water level, pumping rates, hydraulic dimensioning)
2. Hydraulic properties (well test)
3. Water quality data
4. Temporal variations of hydraulic, hydrologic, and water quality parameters

Successful well maintenance is often a combination of preventive and damage-oriented measures. Breakdown-oriented maintenance is strongly discouraged.

Maintenance plans aid in the conservation of sustainable well operation and must include an individual inspection schedule. Tables 2-1 through 2-4 show examples of maintenance plan excerpts for a well that is prone to incrustation buildup.

2.5 Well Monitoring

Well operation and maintenance should be accompanied by inspections and data collection at regular intervals. Since well yield is a

TABLE 2.1 Overview of Inspections During Well Operation

Type of inspection	Target	Aim	Recommended inspection frequency
Visual inspection	Inner protection zone	Groundwater protection	Weekly
	Well head	Functional efficiency Groundwater protection	Weekly
	Well casing and screen	Construction state Identification of ageing	Semiannually to annually (depending on ageing potential)
	Pump, riser pipe	Construction state Identification of ageing	Semiannually to annually (depending on ageing potential)
Function check	Screen	Evaluation of: Ageing Entrance resistance Specific well yield	Semiannually to annually
	Pump	Pumping rate Energy consumption Tightness of riser pipe	Annually
Operational data	Water levels in well and observation well	Development of drawdown (as a function of pumping and hydrogeological boundary conditions)	Daily
	Pumped volume	Legal restraint of water right (not to exceed legal abstraction limit) Also: basic data for calculation of specific well yield development	Daily
	Water chemistry	Legal restraint of water right also: basic data to evaluate mineral incrustation potential	Semiannually
	Operating hours, energy consumption, and pressure head	Utilization ratio of well; functional efficiency	Daily

function of the hydraulic and hydrochemical properties of the aquifer and the technical properties of the well, both factors require consideration. Well monitoring is thus the continuous or discontinuous surveillance of all systemic well data and its hydrogeological environment.

TABLE 2.2 Example Maintenance Schedule for a Vertical Well

Inspection frequency	Inspection targets	Inspection strategy
Weekly	Pumped volume and pump operating hours	Calculation of hourly and daily pumping rates Comparison to water rights Even distribution of pumping in well gallery
	Water levels in well and observation well (in gravel pack) after at least 2 h of pumping	Calculation of entrance resistance (after DVGW W 117)
Semiannually	Water sample analysis	Comparison to previous analyses
Annually	Well test	Comparison to well test performed immediately after construction
Annually	Suspended sand content	Comparison to previous data
Annually	Pumped volume and operating hours of pump	Calculation of pumping rates Comparison to water rights
Annually	Evaluation of energy consumption	Comparison to data from previous year(s)
Annually	Camera inspection	Video, DVD
Annually	Maintenance of equipment	Status and function control (including documentation)
Every 2 to 3 years	Well rehabilitation (if necessary)	Comparison of well yield before and after rehabilitation
Every 3 to 5 years	Borehole geophysical investigation of annular seal and casing connections	Check for tightness of annular seal and casing connections

Data collection can be summarized into three well-monitoring strategies (Detay 1997):

1. Quantitative well monitoring (hydraulics)
2. Qualitative well monitoring (water chemistry)
3. Constructive well monitoring (structural integrity)

The aim of well monitoring is to establish a systematic data collection that comprises all performance-related data as a function of time. The data can be used for both well operation and maintenance. Driscoll (1989) recommends well-specific checklists of basic data which should be renewed at regular intervals:

- Static water level [in "meters below surface" or "above sea level (asl)"]
- Pumping rate (during normal operation)

TABLE 2.3 Example of a Well Data Sheet

Well name: TB1

Water work: East   Technician:....................

Aquifer: Main gravel aquifer (3rd aquifer)   Page 1/.........

Date	Apr 27, 1998	May 4, 1998	May 11, 1998	May 18, 1998	May 25, 1998
Water meter old	96,711	112,133	126,758	143,531	156,790
Water meter new	112,133	126,758	143,531	156,790	170,788
Difference [m^3/wk]	15,422	14,625	16,773	13,259	13,998
Nominal pumping rate [m^3/h]	130	130	130	130	130
Operating hours	122	113	134	106	110
Current pumping rate [m^3/h]	126	129	125	125	127
Static water level (reference) [m below measuring point]	24.82	24.82	24.82	24.82	24.82
Water level in observation well [m below measuring point]	36.86	37.14	36.89	36.88	37.09
Water level in pumped well [m below measuring point]	36.99	37.27	37.03	37.01	37.32
Drawdown in well [m]	12.17	12.45	12.21	12.19	12.50
Specific well yield [m^3/h/m]	10.35	10.36	10.24	10.25	10.16
Energy consumption [A]	43	44	43	43	44
Special investigations	None	None	None	Water sample	Sand content: <0.01 g/m^3 at $Q = 127$ m^3/h

- Dynamic water level (e.g., after 1, 2, 3, or 4 h of operation)
- Current well yield (comparison to well yield at time of construction)
- Sand intake and turbidity
- Depth of well (sedimentation in sump)
- Energy demand
- Circuit switching interval
- Downtimes during a day
- Long-term development of water levels in well and neighboring piezometers
- Influence of neighboring wells on drawdown
- Changes of water chemistry as a function of pumping rate

Significant changes in well yield, energy demand, or energy consumption indicate that the well—or the pump—requires maintenance measures.

TABLE 2.4 Checklist for Problem Identification During Pump Replacement

No.	Question	Answers (to be filled in by personnel)	Actions to be taken
1	Observable phenomena during well-head disassemblage?	Corrosion:...................... Leaks:...........................	■ Remedy of defects ■ Control of tightness
2	Can pump and riser pipe be lifted and pulled without resistance [drag?]?	Tilt:............... Drag along casing:........... Metallic sounds:.............	■ Interupt pulling ■ Check lifting force ■ Camera inspection when cause cannot be identified[or: found]
3	What lift force is expected to pull pump and riser pipe?	Nominal value:............[t] Actual value: ....[t] Difference:....................[t]	■ Negative difference: find cause ■ >120% of nominal value: interruption of pulling and camera inspection
4	Status of riser pipe connections?	Mechanical defects:......... Status of seals:............... Status of screws, bolts, nuts, etc.:.......................	Replacement of ■ Seals, screws, nuts, etc.
5	Mineral incrustations on pump?	Type of deposit:............... Thickness:.......................	■ Check need for rehabilitation (well yield)
6	Gravel/sand around position of pump intake?	Type of gravel (filter gravel aquifer material)............. ,	■ Check cause (gravel intrusion through leaks, fine grains from gravel pack or aquifer); camera inspection necessary
7	Holes in riser pipe?	Number and size of holes:.......................... Location:........ [m below well head]	■ Further investigation of causes: Corrosion/cavitation: camera inspection
8	Sedimentation in well sump?	Nominal depth .... [m below well head] Measured depth .... [m below well head] Difference nominal/current..............	■ Further investigation of causes (camera inspection)
9	Further observations?	Deposits on riser pipes?............................ Sand or rust in riser pipe?.............................	■ Cleaning of riser pipes ■ Detailed investigation of causes recommended

All measures of *quantitative well monitoring* target both aquifer yield and the capacity of the well as a function of operating time. The most important tool is the well yield test ($\rightarrow$ 5.2 for a more detailed description of the evaluation methods).

Natural and pumping-induced fluctuations of the water table can be monitored by measuring the water levels in pumping and observation

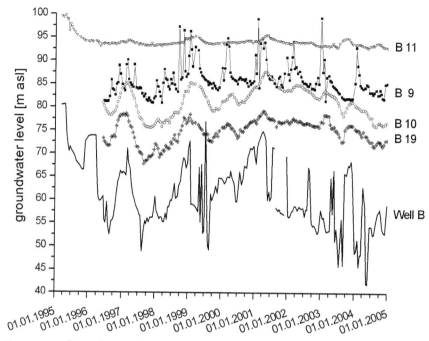

Figure 2.20 Groundwater levels in a well and some surrounding observation wells (carboniferous limestone aquifer, Germany).

wells. Figures 2-20 and 2-21 show the water levels in observation wells located at different distances away from pumping wells. The influence of the pumping well on its surroundings is clearly visible. Figure 2-22 shows the temporal development of groundwater levels in observation wells screened in different aquifers around a well gallery. Although the abstraction from the fourth aquifer remains fairly constant, groundwater levels in the deeper aquifer show a decreasing tendency from about 1988 onwards. In this case this can be attributed to the dewatering scheme of a lignite open-pit mine located several tens of kilometers away.

Qualitative well monitoring includes sampling and analysis of groundwater at regular intervals (Table 2-5). For most drinking-water wells, regular sampling has to be performed to meet legal requirements. The parameters to be investigated are often legally defined. Wells tapping groundwaters with high incrustation or corrosion potential ($\rightarrow$ 3) should be sampled more frequently. Anomalies in the water chemistry can reveal defective designs or constructional flaws. Sudden increases in certain chemical species or microorganisms can be an indicator of the inflow of (surface) waters, e.g., through corrosion cavities or leaky casing connections. Again, only time series can provide the necessary background information to identify causes of such outliers.

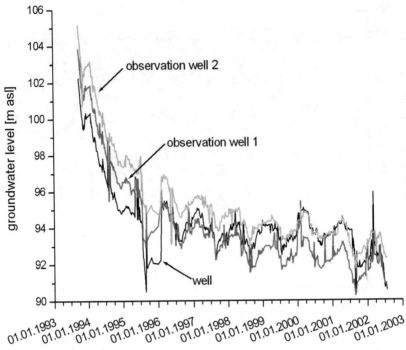

Figure 2.21 Groundwater levels in a well and two observation wells located 10 m (33 ft) (# 1) and 200 m (656 ft) (# 2) away.

Both quantitative and qualitative monitoring data need to be visualized to be of use. Time-variation graphs—nowadays no big problem thanks to personal computers—allow comparisons to previous data and also the identification of trends (Figs. 2-20, 2-21, 2-22).

The assessment of the constructional state of a well is called *constructive well inspection*. All the components depicted in Fig. 2-2 suffer from ageing processes and therefore require regular inspections. Monitoring should be performed every second year and should include checks for:

1. Corrosion of casing and screen (thickness, coating, etc.)
2. Condition of casing connections
3. Condition of screen and screen slots
4. Gravel pack
5. Annular seals

Optical and geophysical methods are particularly suitable for these tasks. While the former can be used directly to evaluate items 1–3, the

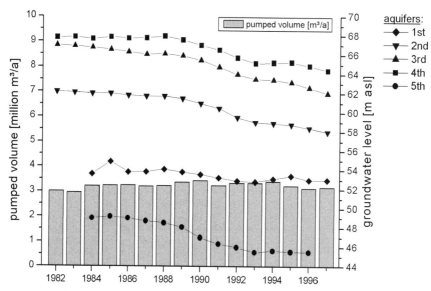

Figure 2.22 Example of well monitoring (pumped volume and groundwater levels) from a multiaquifer system at the German–Dutch border near Geilenkirchen. Water is extracted from the fourth aquifer (aquifer numeration from top to bottom). Modified from Houben et al. (2004).

latter are able to "look behind the screen" and can thus account for items 4 and 5. Further details on camera inspections and geophysical methods can be found in Chap. 5.

Maintenance intervals can often be established individually for specific wells or well groups only. Table 2-6 lists some general recommendations for different types of aquifers and their most common problems. The reader should be aware that local deviations may occur and that time and experience will reveal the right interval for a specific site.

TABLE 2.5 Key Hydrochemical Parameters Indicative of Ageing Processes

Changes in...	...as indicator for...
Dissolved iron (Fe), manganese (Mn)	Changes in redox conditions (e.g., caused by inflow of shallow groundwater through casing leaks or drawdown toward to the screen)
pH Bicarbonate (HCO_3^-)	Changes to the carbonate equilibrium (e.g., caused by drop in pressure and subsequent degassing of carbon dioxide)
Specific electric conductivity (EC)	Changes of dissolved salt content (inflow of shallow, polluted water; also, via defective casing connections)
Temperature	Inflow from other aquifers via leaky seals, etc.

TABLE 2.6 **The Most Common Ageing Phenomena in Wells and Maintenance Intervals**

Aquifer type	Most common problems*	Maintenance frequency**
Alluvial (porous unconsolidated gravel-sand formations)	■ Silt, clay, sand intrusions ■ Mineral incrustations (ochre) ■ Biofouling ■ Casing failure (corrosion)	2–5 years
Sandstone (fractured-porous aquifer)	■ Fissure plugging ■ Casing failure (corrosion) ■ Sand intrusion (suffosion)	5–10 years
Limestone (karst or fractured aquifer)	■ Fissure plugging by clay, silt ■ Carbonate incrustations	6–12 years
Basaltic lavas	■ Fissure and vesicle plugging by clay, silt ■ Mineral incrustations	6–12 years
Interbedded sandstone and shale	■ Low initial yields ■ Fissure plugging by clay, silt ■ Casing failure (corrosion)	4–7 years
Metamorphic rocks	■ Low initial yields ■ Fissure plugging by clay, silt, or mineral incrustations	12–15 years
Consolidated sedimentary rocks	■ Low initial yields ■ Fissure plugging by clay, silt, or mineral incrustations (ochre)	6–8 years
Semiconsolidated sedimentary rocks	■ Silt, clay, sand intrusions ■ Mineral incrustations (ochre) ■ Biofouling ■ Fissure plugging	2–5 years

*Excluding pumps and declining water table.
**Shorter intervals in wells with high incrustation potential is recommended.
Modified after Driscoll (1989).

The DVGW survey in Germany described in Chap. 1 also addressed the motivation and intervals of rehabilitation (Niehues 1999). More than half of all waterworks (52%) rehabilitate "on demand," usually after significant decreases of well yield were noticed or outages occurred. A further third rehabilitates at regular intervals of <5 years (9%) and <10 years (23%). Nine percent of all participating well operators use intervals between 11 and 25 years. This illustrates the large variability in hydrogeological settings and the differing standards of well monitoring. Areas with elevated incrustation potential, of course, require shorter monitoring—and subsequently rehabilitation—intervals. Monitoring and prevention are so far common only in regions where ageing is known to be an issue.

2.6 Influence of Ageing on Well Hydraulics and Operation

The buildup of mineral incrustations or particles in a porous medium naturally affects its hydraulic conductivity. A discussion of the different approaches to quantify this problem can be found from Saripalli et al. (2001).

The classical equations developed independently by Hagen (1839) and Poiseuile in the early nineteenth century describe the amount and velocity of flow through capillaries [Eqs. (2.15) and (2.16). They found that the flow rate Q shows a cubic dependency on the radius of the capillary. It is therefore easy to anticipate that a buildup of incrustations or a deposition of sand in capillary pores will lead to a reduction of permeability and flow (Fig. 2-23).

$$Q = \frac{\rho \cdot g}{\eta} \cdot \frac{r^3}{12} \cdot l \cdot \text{grad\_}h \qquad (2.15)$$

$$K = k \cdot \frac{\rho \cdot g}{\eta} = \frac{r^2}{12} \cdot \frac{\rho \cdot g}{\eta} \qquad (2.16)$$

where Q = flow rate, L^3/T
  r = radius of capillary, L
  v = flow velocity of fluid, L/T
 grad_h = hydraulic gradient
  η = dynamic viscosity of fluid, $M/(L \cdot T)$
  ρ = density of fluid, M/L^3
  g = acceleration of gravity, L/T^2
  k = intrinsic permeability, L^2
  K = permeability, L/T

A similar dependency of permeability and pore space was developed by Snow (1968) for fractured aquifers [Eq. (2.17)]. As the name implies,

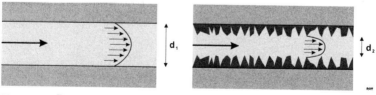

Figure 2.23 Decrease of capillary width and permeability by incrustation buildup ($d = 2\,r$).

this *cubic law* states that permeability shows a cubic dependency on the opening width of the fractures.

$$K = N \cdot \frac{\rho \cdot g}{\eta_w} \cdot \frac{b^3}{12 \cdot f} \qquad (2.17)$$

where K = permeability, L/T
N = number of fractures
B = opening width of fracture, L
η_w = dynamic viscosity of fluid, $M/(L \cdot T)$
ρ = density of fluid, M/L^3
g = acceleration of gravity, L/T^2
f = roughness of fracture surface, L

The Hagen-Poiseuile concept is only applicable to small uniform capillaries and is therefore difficult to use in real porous media. Another way to describe the permeability as a function of porosity is the Kozeny-Carman equation [see Chapuis & Aubertin (2003) and

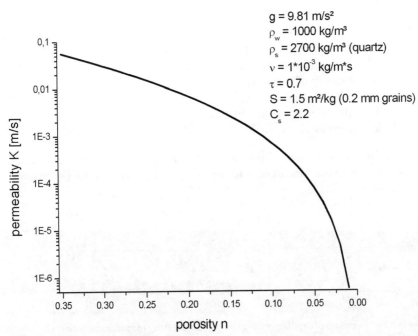

Figure 2.24 Decrease of permeability as a function of decreasing porosity, calculated using the Kozeny-Carman equation (see text).

Hansen (2004) for a detailed description of the equation and its application range].

$$K = \left[\frac{n^3}{(1-n)^2}\right] \cdot \left[\frac{\rho_w \cdot g}{\tau^2 \cdot (\rho_s \cdot S)^2 \cdot C_s \cdot \eta_w}\right] \quad (2.18)$$

where K = permeability coefficient, L/T
n = porosity
g = acceleration of gravity, L/T^2
ρ_s, ρ = density of solid and water, M/L^3
S = specific surface area of granular material, L^2/M
τ = tortuosity ($\tau < 7$)
η_w = dynamic viscosity of water, $M/(L \cdot T)$
C_s = geometry factor (C_s = 2 for tubular pores, C_s = 3 for pores between platelets)

Similarly to Hagen-Poiseuile, the Kozeny-Carman equation postulates a cubic dependency of hydraulic conductivity on porosity but applies better to real porous media. It is useful to determine the development of the permeability of granular porous media such as filter gravel. Figure 2-24 shows the development of permeability over a range of porosities common in (clogging) filter gravel. The permeability curve closely resembles the yield curves of many ageing wells ($\rightarrow$ 10.3). Therefore we presume that the equation is well suited to describe the clogging behavior of wells. On the other hand, some of the required parameters are difficult to determine, for example, S, τ, and C_s. A collection of clogging particles would—at least initially—increase the surface area S (McDowell-Boyer et al. 1986), which would cause a further reduction of permeability. This effect was not included in the calculations presented in Fig. 2-24.

Chapter 3

Chemical Ageing Processes

3.1 Corrosion

3.1.1 Terms and definitions

Corrosion is defined in DIN 50 900 as "the reaction of a metallic component with its environment which causes a measurable change to it and leads to adverse effects to the component or a whole system." Corrosion can affect an entire metal surface or can occur locally (pit corrosion). A detailed description of metal corrosion can be found in several special publications, e.g., Baboian (2004). Corrosion of well material is discussed in detail by Driscoll (1989) and McLaughlan et al. (1993).

Corrosion must not be confused with purely mechanical attacks on material, e.g., that caused abrasion and cavitation, which will be discussed later in this book ($\to$ 4.2).

3.1.2 Electrochemical corrosion

The corrosion of metallic well materials (tubing, screen, pump, and riser pipe) is an electrochemical process that involves the development of a galvanic cell. The cell comprises two areas with different electrical charges, and an electrical current (electron transfer) flowing via a conductor (metal or conductive fluid) that connects them. The strength of the cell depends on the position of the materials in the electromotive series. Material of lower normal potential (less noble) is called the anode and loses electrons to that of higher potential (more noble), which is called the cathode. The anode thereby suffers material loss while the cathode becomes encrusted by reaction products. Figure 3-1 shows this

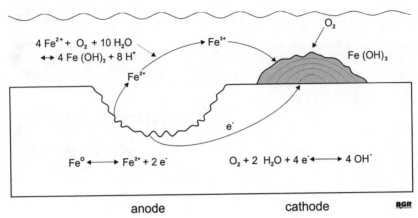

Figure 3.1 Schematic diagram of processes occurring during electrochemical corrosion.

concept for the most common corrosion type ("rust"), in which an iron metal surface (anode) is corroded by dissolved oxygen. Since air contains 21% oxygen, the strongest corrosion occurs at the interface between air and water ("splash zone") due to the constant resupply of oxygen.

The salt content of groundwaters is generally sufficient to act as a conductor, but chloride-rich water, e.g., seawater, strongly enhances corrosion rates due to increased conductivity (e.g., Barnes & Clarke 1969).

Electrical potential gradients can also occur at the interface of different metals ("contact corrosion"). A classical example of this is stainless steel screens connected to carbon steel tubing "to save money." The less noble carbon steel tube acts as an anode and suffers material loss. As long as the material loss is distributed over the entire surface, damage is usually minor. If it is restricted to limited areas or small parts, e.g., casing connections, weld seams, flange bolts, and nuts, those can be destroyed within a few years and may endanger the structural integrity of the whole system. Potential electrochemical gradients can occur locally on one piece of metal, as a result of slight differences in material properties induced by, e.g., welding or cutting of screen slots.

Many metals and alloys are protected against corrosion by a natural thin layer of corrosion products ("passivation"). On steel surfaces this layer consists of a mixture of hydrogen (H_2) and iron oxides. On stainless steel the passivation layer consists of different metal oxides. Technical corrosion protection can be achieved by the application of an additional protective layer (e.g., paint, polymer coating, phosphatizing). If this layer is damaged by mechanical or chemical impacts, a local anode may develop at this site. This can lead to locally enhanced corrosion with substantial material losses ("pit corrosion," Fig. 3-2). Mechanical damage can occur prior to and during construction (loading, handling, and installation of

Figure 3.2 Casing and screen heavily affected by corrosion. Recovered from wells of the lignite open-pit mine Wackersdorf (Bavaria, Germany). *(Photograph: E & M Bohrgesellschaft mbH, Hof / Saale.)*

screen and tubing, backfilling of gravel pack) as well as during well operation (run-in and recovery of pumps, mechanical rehabilitation). The naturally occurring protective layers on steel and stainless steel are prone to chemical attack by acids and chloride-rich fluids. For this reason, the application of acids and especially hydrochloric acid during chemical rehabilitation is problematic.

The mineralogy of rust is quite complicated; common minerals in well corrosion include goethite (α-FeOOH), lepidocrocite (γ-FeOOH), magnetite (Fe_3O_4), and maghemite (γ-FeOOH). The latter two are magnetic.

3.1.3 Microbially induced corrosion

Microbially induced corrosion plays a major role in the destruction of metallic material. The most common cause of anaerobic corrosion is the presence of sulfate-reducing bacteria (SRB) such as *Desulfovibrio* (Ehrlich 2002). A more detailed discussion of these organisms will follow in Sec. 3.5. "Anaerobic" means that oxygen has to be absent (the bacteria involved do not tolerate oxygen).

Steel exposed to acids forms a hydrogen gas (H_2) film [Eq. (3.1)]:

$$4\ Fe + 8\ H^+ \Leftrightarrow 4\ Fe^{2+} + 4\ H_2 \qquad (3.1)$$

This hydrogen film usually protects the iron from further corrosion. Many sulfate-reducing bacteria can utilize this hydrogen as an energy source instead of organic carbon [Eq. (3.2)]:

$$4\ H_2 + SO_4^{2-} \Leftrightarrow H_2S + 2\ H_2O + 2\ OH^- \qquad (3.2)$$

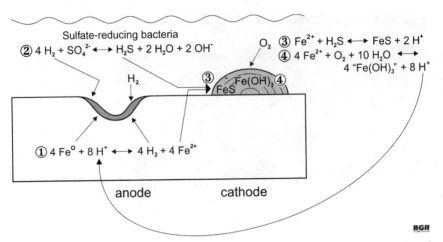

Figure 3.3 Schematic diagram of processes occurring during microbially induced corrosion involving sulfate-reducing bacteria.

Hydrogen sulfide (H_2S) generated during this process reacts with ferrous iron present in groundwater [or via Eq. (3.1)]. This process leads to the generation of more acid [Eq. (3.3)], which can again attack the iron so that a continuous reaction cycle develops as shown in Fig. 3-3.

$$4\ Fe^{2+} + H_2S + 4\ H_2O + 2\ OH^- \Leftrightarrow FeS + 3\ Fe(OH)_3 + 6\ H^+ \quad (3.3)$$

The main solid reaction product is a blackish to reddish mixture of iron oxides and iron sulfides. The latter mainly comprise monosulfides such as mackinawite (FeS) and greigite (Fe_3S_4). Their identification is fairly easy: dripping a few drops of dilute acid onto the sample will result in a strong odor of "rotten eggs" (hydrogen sulfide, H_2S). Since sulfate-reducing bacteria (SRB) cannot grow under oxic environments, they often "hide" in deeper layers of biofilms underneath oxygen-tolerant bacteria. Corrosion rates caused by SRB can be substantial and can reach several millimeters per year.

3.1.4 Ageing of nonmetallic material

In a strict sense, the definition of corrosion presented in Sec. 3.1.1 excludes all nonmetallic materials. Nevertheless, these materials also suffer from ageing processes, that is, interactions with their environment which may negatively affect their performance.

Wood-based materials are prone to degradation by specialized bacteria and fungi. These organisms usually feed on the cellulose contained in the wood and leave a weakened lignin framework. Repeated drying and wetting cycles promote this process significantly, e.g., as triggered by fluctuating water levels.

Stoneware is prone to weakening by partial dissolution, especially when it is exposed to acidic water. As might be expected, this process is promoted when the glaze has been damaged, e.g., by mechanical impacts.

Polyvinylchloride (PVC) is a very common material used in the construction of both production and observation wells. The eponymous polyvinylchloride in its pure form is brittle and thermally quite unstable. PVC therefore contains additives to improve its properties. Commercially available PVC may contain several percents of plasticizers (usually phthalates), stabilizers (e.g., for UV protection), and color pigments, as well as remnants from the polymerizing process. In well construction, "plasticizer-free" PVC is used in most cases. Over the course of years, additives may diffuse out of the PVC, especially at high temperatures. The PVC becomes increasingly brittle and can break easily with impacts or during soil settlement. We all know this phenomenon from observing old plastic toys, which tend to break after some years.

When thinking of ageing materials, smaller parts also need to be considered. Rubber or plastic tubing seals can become brittle and porous due to processes similar to those described for PVC.

Annular sealing materials such as swelling clays (bentonite) or cement suffer from constant contact with acidic groundwaters or repeated contact with high concentrations of acids (or hydrogen peroxide) injected during chemical well rehabilitation. This causes shrinking of the bentonite and dissolution of the cement, which may lead eventually to failure of the seal.

3.2 Ochre Incrustations: Iron and Manganese Oxides

3.2.1 Geochemistry of iron and manganese

Most aquifers show a vertical hydrochemical zonation—even when no impermeable beds are intercalated (Fig. 3-4). This is the result of a series of reactions of the infiltrating soil water within the aquifer matrix. Since the reactions depend on the residence time in the aquifer, hydrochemistry varies over depth. A water sample from a long well screen is therefore always a mixture from different hydrochemical zones and thus cannot be considered representative. The most obvious hydrochemical zonation is due to redox processes. The infiltrating groundwater recharge carries some dissolved oxygen into the shallow groundwater. This uppermost part of the groundwater column is therefore called the oxic zone. Since oxygen is a strong oxidant, it is used by certain microorganisms living in the aquifer to oxidize organic matter to gain energy. After depletion of the limited oxygen input, the water percolates farther downward, where other bacteria attack the next available oxidant, which is nitrate.

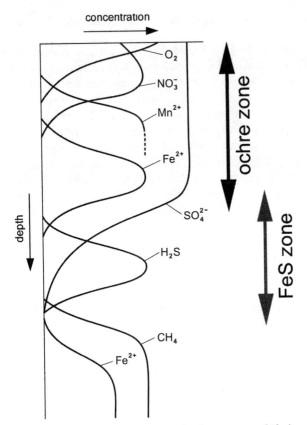

Figure 3.4 Vertical distribution of redox zones and their indicator species in groundwater. Screens overlapping two or more geochemical zones can trigger the formation of ochre or iron sulfide (FeS). *Modified after Appelo & Postma (1996).*

After full denitrification has been achieved, other bacteria deeper in the aquifer dissolve solid manganese and, later, iron oxides. This causes elevated concentrations of Mn^{2+} and Fe^{2+} in the water. These three latter zones are often summarized under the term "post-oxic." The next zone is caused by the activity of sulfate-reducing bacteria (SRB). As the name implies, they utilize sulfate and transform it into (hydrogen) sulfide. This compound is easily recognizable by its very strong odor of rotten eggs. Methanogenic bacteria living at the very bottom then take over what is left by the others. As we might expect from this sequence, the amount of energy obtainable becomes lower with increasing depth. This may remind us of a child in front of a bowl of mixed candy, who will first pick out his or her favorites and then proceed to the less desirable ones.

The redox zones can be sorted according to the potential energy gain for the oxidation of organic matter at pH 7 (Stumm & Morgan 1996):

	kJ/eq (kilojoules per molar equivalent)
Oxygen reduction	−125.3
Denitrification, heterotrophic	−119.0
Manganese(IV) reduction	−85.0
Iron(III) reduction	−29.2
Sulfate reduction	−24.7
Methane fermentation	−23.5

All redox zones are characterized by the presence of chemical indicator species (oxygen, nitrate, etc.). Figure 3-4 shows the idealized hydrochemical zones of a hypothetic aquifer and their indicator species.

It is important to note that ferrous iron can appear both in the postoxic and in the methanogenic zone (Fig. 3-4). These two zones can be separated by a sulfide zone, in which ferrous iron is removed through precipitation as iron sulfides (→ 3.5). As we are interested in ochre incrustations because of their high abundance, it is worth having a closer look at the geochemistry of iron and manganese.

The geochemical behavior of iron and manganese is quite similar. They have similar molar masses (Fe, 55.85; Mn, 54.94 g/mol) and both form solid oxide phases under oxic redox conditions. For brevity we will always talk about "iron oxides" and "manganese oxides" in the following chapters, although real ochre often comprises a multitude of oxides, hydroxides, oxyhydroxides, and other mineral types (→ 3.2.5). Mobilization of the dissolved bivalent aquatic species (Fe^{2+}, Mn^{2+}) occurs only in reducing environments. The geochemical cycles illustrated in Figs. 3.5 and 3.6

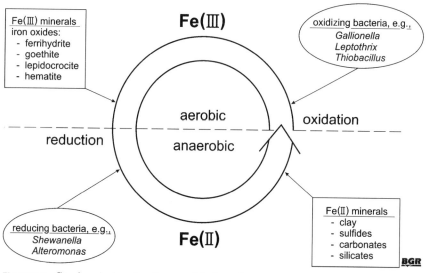

Figure 3.5 Geochemical cycle of iron and its key players.

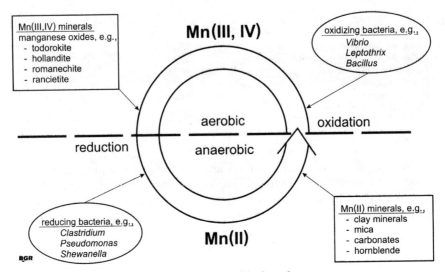

Figure 3.6 Geochemical cycle of manganese and its key players.

show the key minerals involved and also the influence of specialized bacteria on mobilization and precipitation (→ 3.2.6). The reduced bivalent iron species Fe^{2+} is referred to as ferrous iron, the oxic trivalent species Fe^{3+} as ferric iron. There are some important geochemical differences between iron and manganese which affect the formation of ochre (e.g., Stumm & Morgan 1996):

1. Manganese (Mn) is far less abundant than iron. On average, the earth's crust contains about 50 kg (110 lb) Fe per ton but only 0.95 kg (2.1 lb) Mn per ton (mass ratio Fe:Mn ≈ 50:1). Kölle (2001) calculated that in Germany, about 2,240 tons of iron and 251 tons of manganese are pumped (and treated) from 1,223 billion m³ (323,080 billion US gal) of groundwater each year (mass ratio Fe:Mn = 9:1).
2. Manganese can occur in three oxidation states in a normal environment (II, III, IV), as opposed to two for iron (II, III).
3. Oxidation of manganese(II) requires higher oxidation potentials (0.6–1.2 V) compared to iron(II) (0.0–0.5 V).

Depending on the redox conditions in the aquifer—or in the well—a spatial separation of iron and manganese is therefore not a surprise. Of the 70 ochre incrustations which we were able to investigate, 59 were dominantly made up of iron oxides (with very little manganese present), 9 of manganese oxides (with very little iron present), and only

2 samples (<3%) were of a mixed iron/manganese oxide type (Fig. 1-5). As expected from the geochemical abundance, iron ochre is more common than manganese ochre. If the water inside a well contains both dissolved iron and manganese but the redox potential—due to admixture of oxic water—rises only to <0.6 V, only iron is oxidized and precipitated. We found another example of this phenomenon at a well row in the Rhineland north of Cologne (Germany). One well—which is situated close to a river—was incrusted by manganese ochre while the others—farther away from the river—were clogged by iron ochre. This was of course an effect of different redox potentials caused by the different amounts of oxic bank filtrate water entering the wells.

For the sake of completeness, it should be noted that iron and manganese can also be mobilized as oxic aquatic species—under conditions of very low pH (pH << 3). Since such pH values hardly ever occur in nature, we can safely ignore them here. Nevertheless, low pH values are used for chemical well rehabilitations with acids ($\rightarrow$ 8.2).

In the real world redox zones are often less well defined than in textbooks. Figure 3-7 shows an example of the vertical distribution of some indicator species in groundwater of an aquifer system in northwestern Germany. It is obvious that the oxygen and nitrate zones are basically inseparable here (which is quite common). The oxic zone is underlaid by a distinct iron-rich zone, while the manganese zone is not developed. The concentrations of manganese are also much smaller, as expected from its lesser geochemical abundance. Sulfate and methane are not shown. Since acid inputs may occur with groundwater recharge and many redox reactions involve release or uptake of protons (H^+), changes in pH and hardness in the vertical dimension are also possible.

Measurement of the redox potential (Eh)—which can be performed onsite—is another inexpensive way to assess the redox conditions in an aquifer. Figure 3-8 shows the vertical distribution of such values from the same aquifer as in Fig. 3-7. It is still advisable to be cautious with the interpretation of such redox potential data. The measured potentials are most often mixed potentials stemming from different redox reactions, so the identification of redox zones is often ambiguous. Further, Eh measurements themselves are problematic. Redox electrodes can be quite sluggish in their response and are sensitive to the influence of the surroundings on the water, e.g., by air, during sampling.

Since most owners want their wells to be as productive as possible, wells are often screened over the whole depth of the aquifer (full penetration). This also helps to minimize head losses resulting from vertical flow components ($\rightarrow$ 2.2.3). From the geochemical point of view discussed above we can see immediately that this will lead to mixing of two or more hydrochemical zones (Fig. 3-9). The well can be considered a "hydrochemical short circuit," in which, e.g., dissolved ferrous iron and oxygen come into

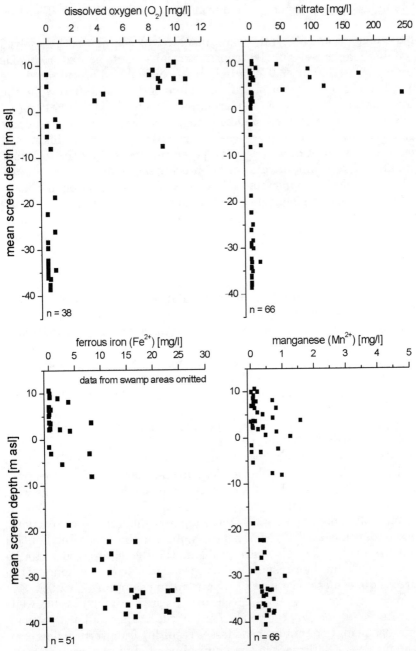

Figure 3.7 Vertical distribution of redox indicators in groundwater of the Bourtanger Moor area (near Haren, Ems, northwestern Germany). *After data by Houben (2000).*

Chemical Ageing Processes   65

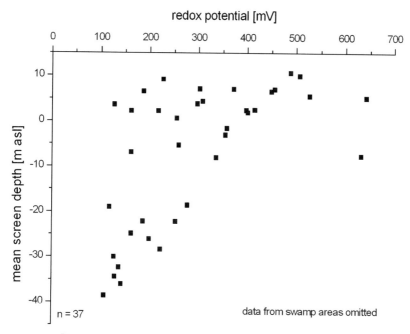

Figure 3.8 Vertical distribution of redox potential in groundwater of the Bourtanger Moor area (near Haren, Ems, northwestern Germany). *After data by Houben (2000).*

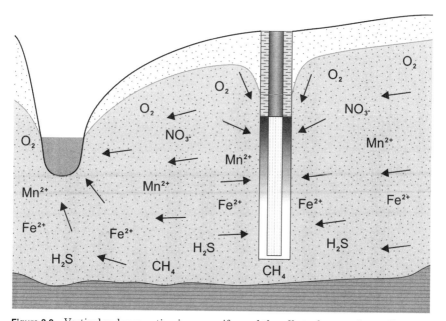

Figure 3.9 Vertical redox zonation in an aquifer and the effect of a pumping well.

contact. This of course leads to redox reactions aiming at the establishment of a new redox equilibrium. In this case this leads to the oxidation of ferrous iron to insoluble ferric iron and the formation of solid oxides which can plug the well. The same concept applies for the redox pairs manganese and oxygen, as well as for ferrous iron and sulfide (Figs. 3-4, 3-9). The turbulent or semiturbulent flow conditions inside and around the well are of course favorable for mixing incompatible waters and also facilitate degassing of solution-assisting gases.

The variations of hydrochemical parameters in the lateral dimension can also be quite large. Since the well is the end of all flow paths from its catchment area, it will receive all the different qualities present. Because of the inhomogeneities of matter input and soil or aquifer compositions, samples from neighboring wells may look completely different (e.g., Stuetz & McLaughlan 2003). The example presented in Fig. 3-10 shows the iron and manganese concentrations in water samples from a well row in central Germany near the Elbe River. The 10 shallow wells—each at an approximate distance of 50 m (164 ft) to each other—showed iron concentrations ranging from 7 to 40 mg/liter and manganese from about 1 to 3.5 mg/liter.

Temporal variations of iron and manganese concentrations in groundwater are also quite common. Figure 3-11 shows the fluctuation of

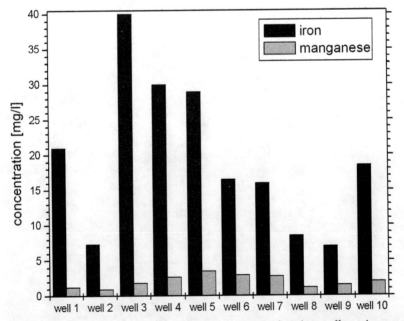

Figure 3.10 Variation of iron and manganese concentrations in a well row in central Germany [10 wells, distance between two wells is approximately 50 m (164 ft)].

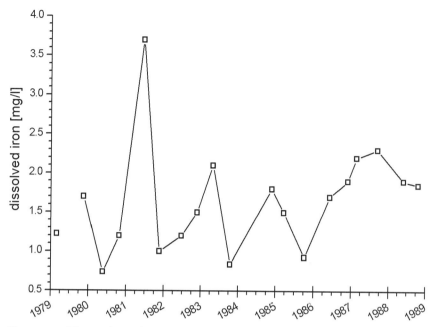

Figure 3.11 Fluctuations of dissolved iron concentrations in groundwater from a well near Krefeld (Germany).

dissolved iron concentrations in well water over some years. The variation is quite high, but the maximum and minimum of the curve cannot be correlated with particular periods of the year.

3.2.2 Oxidation of iron(II) and manganese(II)

Exposing dissolved ferrous iron to dissolved oxygen leads to the formation of ferric iron and its subsequent precipitation as iron oxide [Eq. (3.4)] (Stumm & Morgan 1996).

$$4\ Fe^{2+} + O_2 + 10\ H_2O \leftrightarrow 4\ Fe(OH)_3 \downarrow + 8\ H^+ \qquad (3.4)$$

Again, manganese behaves similarly to iron. Dissolved Mn^{2+} contained in reduced waters coming into contact with dissolved oxygen will experience oxidation to Mn^{3+} and Mn^{4+}, which will quickly precipitate as insoluble manganese oxides [Eqs. (3.5) and (3.6)].

$$4\ Mn^{2+} + O_2 + 6\ H_2O \Leftrightarrow 4\ MnOOH \downarrow + 8\ H^+ \qquad (3.5)$$

$$2\ Mn^{2+} + O_2 + 2\ H_2O \Leftrightarrow 2\ MnO_2 \downarrow + 4\ H^+ \qquad (3.6)$$

The kinetics of Eq. (3.4) have been studied by a variety of authors; see, e.g., Houben (2004) for a listing. For mildly acidic to neutral waters (pH > 5), a rate law of the type presented in Eq. (3.7) is commonly used:

$$r = k\ \{Fe^{2+}\} P_{O2} \{OH^-\}^2 \quad (3.7)$$

The reaction rate r describes the amount of mass converted from ferrous to ferric iron over a period of time (k is the reaction constant, which includes—among other factors—the influence of reaction temperature). Chemists describe the reaction as "first order" for the activity of ferrous iron and the partial pressure of dissolved oxygen. This means that there is a linear relationship between their activities and the reaction rate. Doubling the amount of iron or oxygen will double the reaction rate. The reaction is also "second order" for hydroxyl ion activity. The quadratic dependency thus makes the reaction strongly dependent on pH. Doubling the amount of OH^- ions will increase r by a factor of four! A rising pH value will therefore lead to strongly enhanced precipitation of ferric iron. A compilation of rate constants for dilute aqueous solutions from various authors is given by Houben (2004).

Instead of the partial pressure of oxygen, $P_{O2} = \{O_2\}$, the concentration of dissolved oxygen ($O_{2(aq)}$) can be calculated using Henry's law and the distribution coefficient K_H [Eq. (3.8)]:

$$K_H = \frac{[O_{2\ (aq)}]}{P_{O2}} \quad (3.8)$$

K_H is $1.26 \cdot \times 10^{-3}$ M/atm at 25°C, and it is $1.32 \cdot \times 10^{-3}$ M/atm at 10°C (Stumm & Morgan 1996). For all further calculations, a mean value of $k = 1.4 \cdot \times 10^{16}$ mol$^{-3}\cdot$min^{-1} is assumed.

Equation (3.7) can be transformed into the more convenient form of Eq. (3.9), as a function of pH rather than pOH, using the dissociation constant for water, $K_w^{\ 0} = (OH^-)(H^+) = 1 \cdot \times 10^{-14}$ mol^2/liter^2:

$$r = k_1\ \{Fe^{2+}\}(O_{2(aq)})\{H^+\}^{-2} \quad (3.9)$$

Although K_w is a function of temperature, Sung & Morgan (1980) showed that there is no measurable influence on the rate constant in the range 5–35°C. The final rate constant used here is then $k_1 = 1.4 \times 10^{-12}$ min^{-1}. It is a more or less lumped parameter that incorporates the original k, the Henry's law constant for oxygen dissolution, and the square of the dissociation constant for water.

The reaction half-life, $t_{1/2}$ for the first-order oxidation of ferrous iron at constant pH and oxygen concentration [Eq. (3.10)] can be derived by integration of Eq. (3.9):

$$t_{1/2} = \frac{0.693}{k_1[O_{2(aq)}]\{H^+\}^{-2}} \quad (3.10)$$

The oxidation of ferrous iron can be enhanced by trace quantities of catalysts, such as Co^{2+} and Cu^{2+}. Silica and anions that form complexes with Fe(III), such as phosphate, can also enhance reaction progress (Lowson 1982; Stumm & Morgan 1996). Specialized microorganisms can also have a pronounced influence ($\rightarrow$ 3.2.6). On the other hand, reaction rates are slowed down at higher ionic strengths (high salinities) and in the presence of anions such as sulfate and chloride (Kester et al. 1975; Sung & Morgan 1980). All in all, one has to keep in mind that in a natural system, variable amounts of catalysts, anions, microbes, and mineral coatings can be present at the same time. The definition of an exact rate law for such a system law remains at least difficult.

Applin and Zhao (1989) were the first to transfer the simple concept of Eq. (3.7) to the problem of iron encrustations in wells. Their findings underline the importance of pH and residence time on the amount of iron oxide that is deposited in a well. Figure 3-12 shows the strong effect of pH on the reaction rate and the half-life.

Thermodynamics suggests that nitrate, the second-strongest oxidant in regular groundwaters, is also capable of oxidizing ferrous to insoluble ferric iron [Eq. (3.11)]:

$$10\ Fe^{2+} + 2\ NO_3^- + 24\ H_2O \leftrightarrow 10\ Fe(OH)_3\downarrow + N_2 + 18\ H^+ \quad (3.11)$$

The purely chemical (abiological) reduction of nitrate by ferrous iron has been observed in the laboratory (Buresh & Moraghan 1976) but is less likely to play a major role in natural environments. Here, specialized nitrate-reducing microorganisms utilizing ferrous iron contribute

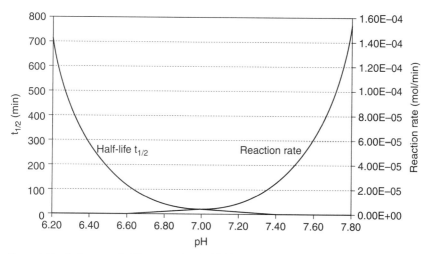

Figure 3.12 Influence of pH on reaction half-life and reaction rate for ferrous iron oxidation. *Modified after Houben (2004).*

significantly to the formation of ferric iron in the suboxic zone (Ehrenreich & Widdel 1994; Straub et al. 1996; → 3.2.6).

3.2.3 Auto-catalysis of iron(II) and manganese(II) oxidation

From water treatment research we know that clean gravel is not efficient for the removal of iron and manganese in gravel bed filters. Only when the gravel has a slight ochre cover will it perform as desired. Tamura et al. (1976) were able to provide an explanation for this. They showed that solid iron oxide has a catalytic effect on the oxidation of dissolved ferrous iron. They found a linear increase of the total rate constant with rising concentration of ferric oxide [Fe(III)] at pH 6.2. This is due to the sorption of ferrous iron onto the solid iron phase and its subsequent oxidation to ferric iron at the surface. Sorption of ferrous iron increases linearly with rising pH, probably due to the increasing negative surface charge of the oxide. Tamura et al. (1976) established an expanded rate law [Eq. (3.12)] that comprises two summands:

1. The reaction in solution described by Eq. (3.9) (homogeneous reaction)
2. The oxidation at the solid ferric oxide surface (heterogeneous reaction)

$$r = k_1 \{Fe^{2+}\}(O_{2\,(aq)})\{H^+\}^{-2} + k_2 \{Fe^{(III)}\}\{Fe^{2+}\}(O_2)\{H^+\}^{-1} \quad (3.12)$$

where k_2 is the product of the equilibrium constant K_{ads} for the adsorption of Fe^{2+} onto ferric oxide and the rate constant k_s of the oxidation at the surface. Tamura et al. (1976) give values of $K_{ads} = 1.40 \cdot \times 10^{-5}$ and $k_s = 1.217$ mol·min^{-1}. The rate constant k_2 for the heterogeneous reaction step is thus 1.71×10^{-5} mol·min^{-1}.

Equation (3.12) shows that the pH has a much more pronounced influence on the homogeneous part than on the heterogeneous part. The relative importance of the solid oxide surface is thus higher at lower pH when the homogeneous reaction alone is rather slow. The auto-catalytic effect gains measurable influence only at initial concentrations of ferrous iron above 3 mg/liter (Tamura et al. 1976). Otherwise the amount of precipitated iron oxides remains too small to provide enough catalytic surface area (Figs. 3-13 and 3-14).

Millero (1985) and Wehrli (1990) found that ferrous iron can interact more easily with dissolved oxygen when it is complexed with hydroxyl groups, which is much more likely at higher pH. In this case, the higher oxidation rates are a result of enhanced electron-transfer capacity. Since hydrous oxide surfaces act in a similar way, this also explains their catalytic behavior. This approach has been applied recently by Burke and Banwart (2002).

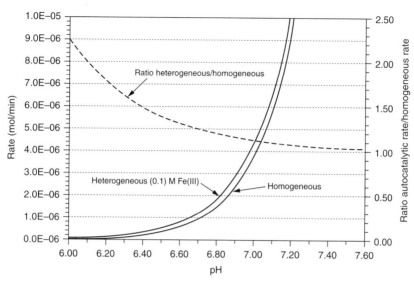

Figure 3.13 Influence of pH on reaction half-life and reaction rate for homogeneous and heterogeneous ferrous iron oxidation. *From Houben (2004).*

Figure 3-13 compares the initial rates and half-lives for the homogeneous [Eq. (3.9)] and heterogeneous [Eq. (3.12)] cases. The total effect of auto-catalysis is not very strong, even when the initial amount of ferric iron is as high as 1,000 mmol/liter. As expected from Eq. (3.12),

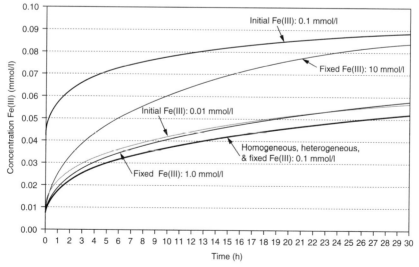

Figure 3.14 Reaction progress for ferrous iron oxidation in the presence of different amounts of solid iron oxide. *From Houben (2004).*

the relative importance of the auto-catalytic part of the reaction is more pronounced at lower pH, when the homogeneous reaction alone is rather slow.

In a natural environment the precipitating ferric iron oxide phases will quickly precipitate and adhere to the surface of mineral grains. Equation (3.12) predicts that this will occur preferentially on the surface of pre-existing iron oxides. Newly forming precipitates thus shield older ones. Concordantly, microscopic investigations of thin sections of encrustations revealed laminated textures resembling annular rings (→ 3.2.9). Because of this coating, only parts of the total amount of iron oxide can act as a catalyst. The catalysis is rather a function of the amount of available surface sites than of the total concentration of ferric iron. This leaves us with several possible approaches for modeling:

1. Homogeneous reaction only [Eq. (3.9)]
2. Heterogeneous reaction [Eq. (3.12)] with increasing catalytic efficiency over time
3. Heterogeneous reaction with a fixed amount of catalytic Fe(III) [the activity of Fe(III) in the right-hand product of Eq. (3.12) is constant; the catalytic surface remains constant due to continuous coating]
4. Heterogeneous reaction [Eq. (3.12)] with some initial Fe(III) present, e.g., natural iron oxide coating on grains

Approach 2 could lead to an overestimation of the rates for natural systems, because coating effects are not considered. Approaches 3 and 4 seem more realistic but suffer from the fact that catalytic surfaces are difficult to quantify. Neither addresses the fact that younger iron oxides (e.g., ferrihydrite) are much more reactive than older ones (e.g., goethite) (→ 3.2.7).

The different reactions were modeled using PHREEQC-2 (Parkhurst & Appelo 1999). A detailed description of the model setup can be found in Houben (2004). Figure 3-14 shows the results of several model runs, which include some fixed and initial activities of catalytic ferric iron. The differences between the purely homogeneous and heterogeneous reactions are negligible. The dramatic influences of initial and fixed ferric iron activities are clearly visible when comparing the reaction half-lives. Fixed amounts become important only at rather high concentrations, while initial amounts are dominant even at low concentrations because of their self-accelerating buildup of catalytic surfaces. An initial amount of 0.1 mmol/liter [~9 mg (3.17×10^{-4} oz) iron oxide] is sufficient to lower the half-time from 25 h to less than 1 h (Fig. 3-14). This explains why the gravel used for filter packs must be clean and white. A thin coating of natural ochre (the typical yellowish to reddish tarnish) will be enough to serve as a starting point for incrustation buildup.

3.2.4 Mass balancing of ochre buildup: example calculations

The following example calculations are based on a paper by Houben (2004). We will consider a well with realistic boundary conditions:

- Screen length 10 m (32.8 ft)
- Drilling diameter 800 mm (15 in)
- Casing diameter 400 mm (7.6 in)

This leaves a cylinder volume of 5.03 m^3– 1.26 m^3 = 3.76 m^3 (132 ft^3) for the gravel pack. The outer entrance surface area F of the cylinder is 25.1 m^2 (270 ft^2), and the inner area (screen) is 12.6 m^2 (136 ft^2). In order to calculate flow velocities, these numbers have to be corrected for porosity m and the area of well slots, respectively. We will further assume that precipitation occurs only in the gravel pack. The gravel pack is often simply dumped into the borehole from the surface, with no effort being made to solidify it. Therefore we can expect a rather loose packing and thus high porosities, probably in the range of 30–40%. With an average value of m = 35%, the total pore space should thus be 3.76 m^3 × 0.35 = 1.32 m^3 (46.6 ft^3). Flow is assumed to be laminar and not turbulent. At a pumping rate Q = 40 m^3/h (176 gal/min), the pore fluid is exchanged about 30 times per hour; average residence time then is 120 s (= 2 min). This time is higher than the half-lives calculated in Figs. 3.13 and 3.14. The flow velocity v_a in the screen is a function of pumping rate Q, area F, and porosity m and is given as $v_a = Q/F \cdot m$. This yields a velocity of 0.158 cm/s (5.18·× 10^{-3} ft/s) at the wall of the drill hole and of 0.315 cm/s (1.03·× 10^{-2} ft/s) at the screen.

Knowing the pore volume allows us to calculate the precipitating mineral mass that is required to clog it completely. The densities for goethite and ferrihydrite, 4.26 and 3.96 g/cm^3, respectively, cannot be used because the natural precipitates are highly voluminous due to their high porosity and high amount of adsorbed water. Volume corrections are therefore needed. The measured water contents of fresh iron oxides incrustations vary between 40 and 80 wt% The bulk density of a ferrihydrite well encrustation sample with a water content of 66%, collected during a rehabilitation of a drinking water well, was measured as 1.25 g/cm^3.

The pumped groundwater is assumed to contain 0.1 mmol/liter ferrous iron (≈ 5.6 mg/liter). During passage through the gravel pack it becomes saturated with respect to oxygen, as a result of turbulent flow inside the well and thorough mixing with air. At T = 10°C and pO_2 = 0.21 atm the dissolved oxygen concentration according to Henry's law is ≈ 0.28 mmol/liter (≈ 9.0 mg/liter). The total pore fluid volume of 1.32 m^3 (46.61 ft^3), thus initially contains 0.132 mol (= 7.37 g) iron. The half-lives of the homogeneous reaction at pH 7.0, 7.2, and 7.4 are then 17.9, 7.1, and

2.8 min, respectively (Figs. 3-12 and 3-13). Including the heterogeneous step further decreases the half-lives so that at pH 7.4 it almost equals the residence time. In this case a mass of 3.685 g Fe (= 5.86 g FeOOH) would precipitate per hour (141 g FeOOH per day, 51.33 kg/a). With a bulk density of 1.25 g/cm^3 it would thus take about 32 years to fill up the pore volume completely. Of course, the residence time becomes shorter with an increasing amount of precipitate—because of the decrease of porosity—but the pore spaces need not be completely filled to render the well useless, since the muddy precipitate tends to filter fine-grained and colloidal particles from the water ($\rightarrow$ 3.8).

The gravel pack is assumed to consist of spheres with a uniform grain size of 2 mm. Because of the loose packing, we will assume cubic arrangement of the grains. The spheres are covered by a monomolecular layer of iron oxide of infinitesimal thickness. A channel of 1 mm length then equals a surface area of $4 \times 0.25 \times 2$ mm $\times \pi = 6.28$ mm^2. Growth of encrustation is assumed to affect primarily the bottlenecks of the pores (Fig. 3-15, A, B). The remaining circular pore channel of 0.83 mm diameter and 1 mm length then presents a surface area of 2.60 mm^2, a decrease by a factor of ~2.4. Reestablishing the original monomolecular layer by the removal of the encrustation will lead to a regaining of hydraulic permeability but will increase the available auto-catalytic surface area and thus speed up ferrous iron oxidation. This explains why rehabilitated wells often show a very fast loss of hydraulic capacity shortly after the rehabilitation (Fig. 3-15).

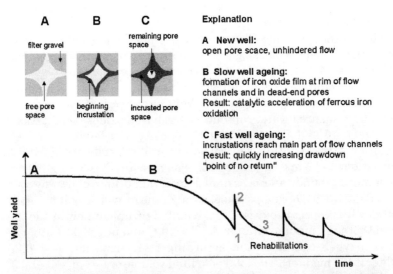

Figure 3.15 Influence of the catalytic activity of iron oxides present in the pore spaces of the gravel pack before and after rehabilitations. *From Houben (2004).*

3.2.5 Mineralogy of ochre

The analysis of the mineralogy and geochemistry of incrustations is a very useful tool to identify ageing processes and to plan successful rehabilitations (Brauckmann et al. 1990; Walter 1997; Houben et al. 1999; Houben 2003a). Hints for sampling and identification techniques can be found in Sec. 5.5.

The most common mineral phase in young iron oxide incrustations is ferrihydrite ($Fe_5HO_8 \cdot 4H_2O$) (Tuhela et al. 1992; Houben 2003a). Because of its very low crystallinity, it was identified as an independent mineral phase only as late as 1971. Prior to this discovery, young ochre gel was not regarded as a mineral but was often called "amorphous iron hydroxide" or "hydrous ferric oxide" (HFO). A tentative chemical formula of $Fe(OH)_3$ was proposed. Though both the formula and terms such as HFO are not precisely correct, they are still in common use. The problem in the identification of ferrihydrite is mainly its extremely small crystal size, which puts the latter practically in the range of nanoparticles (Banfield & Zhang 2001).

The most common mineral phase in older ochre incrustations is goethite (α-FeOOH). This mineral has been known since 1806 and is named after the German poet, author, and naturalist Johann Wolfgang von Goethe (1749–1832). Under the microscope, goethite is often recognizable by needle-shaped crystals, often arranged in a radial or spherical pattern.

Lepidocrocite (γ-FeOOH) sometimes occurs as a minor ochre component. The highest-crystalline type of natural iron oxide, hematite (α-Fe_2O_3) was not detected in any of the incrustations studied by the authors. Iron oxides are recognizable by their typical ochraceous reddish to brownish colors.

In mining areas which are often affected by sulfide oxidation and subsequently highly acidic drainage waters, dewatering wells may be plagued by the precipitation of schwertmannite ($Fe_{16}O_{16}(OH,SO_4)_{12-13} \cdot 10-12H_2O$), an iron(III) oxyhydroxysulfate common only in very acidic and sulfate-rich environments (Uhlmann & Arnold 2003).

Occasionally, the strongly magnetic mineral maghemite (γ-Fe_2O_3) is found in incrustations. It is not formed by the usual incrustation processes described in Sec. 3.2.2 but is a corrosion product of steel and iron ($\rightarrow$ 3.1).

Similar to iron ochre, manganese incrustations can contain mineral phases of both low ("amorphous") and high crystallinity. Since recrystallization can occur at relatively low temperatures, sample treatment should be performed with caution ($\rightarrow$ 5.5). The mineralogy of manganese ochre is much less understood than that of iron ochre. This is a result of the large variability of the chemical composition of minerals. Commonly found minerals are

Todorocite [$(Na,Ca,K,Ba,Sr)_{0.3-0.7}(Mn,Mg,Al)_6O_{12} \cdot 3.2-4.5H_2O$)]
Birnessite ($Na_4Mn_{14}O_{27} \cdot 9H_2O$)

These minerals are also found in deep-sea manganese nodules. Since manganese has two forms of oxidized species, both minerals with Mn(III) and Mn(IV) may occur, e.g., MnOOH or MnO_2, respectively. Manganese oxides are commonly black.

The attachment of iron and manganese oxides onto surfaces is the process that makes the whole issue so cumbersome for all well owners. Quartz grains—which are the most common constituent of gravel packs—usually have a very weak negative surface charge (up to -1 $\mu C/cm^2$ at pH 4–6, Scheidegger et al. 1994). In the same pH range, iron oxides have a positive surface charge of $+5$–10 $\mu C/cm^2$ (Scheidegger et al. 1994). Therefore it is easy to imagine how attracting electrostatic forces can promote the attachment of ochre to quartz grains. If no electrostatic attraction occurs, microbiology can fill the gap ($\rightarrow$ 3.2.6). Surfaces that seem smooth to the human eye are rough enough at the microscopic scale for microorganisms to find a hold. Therefore all screen and casing materials, including polished copper or stainless steel, will be affected by incrustation growth. Once a thin film of oxides has formed, it will act as catalyst for further oxidation processes ($\rightarrow$ 3.2.3) and attract the attachment of increasing amounts of oxides, as revealed by the "annular rings" visible under the microscope ($\rightarrow$ 3.2.9).

3.2.6 Microbiology of ochre formation

Iron is an essential nutrient for practically all bacteria, but some have a special affinity for it (and manganese). The geochemical cycles of iron and manganese (Figs. 3-5 and 3-6) are strongly influenced by specialized bacteria, which are thus called iron bacteria or iron-related bacteria (IRB) in the former case. While reducing bacteria mobilize iron and manganese from aquifer minerals, oxidation of the dissolved divalent iron and manganese by oxidizing bacteria causes the ochre problems of wells. Such bacteria are known to accelerate significantly the oxidation reactions described in Sec. 3.2.2 (Ralph & Stevenson 1995).

The IRB were first detected and described in the nineteenth century. The first book on iron bacteria, by Ellis (1919), reviewed some early important papers (the following references are all taken from Ellis). In 1843 Kutzing described iron-rich "living slimes," while the important IRB genus *Gallionella* was first described in 1897 by Migula. In 1904 Brown concluded that iron bacteria are the cause of the plugging of pipelines. The role of IRB for well ageing has been addressed in a growing number of papers and monographs ever since (e.g., Hässelbarth & Lüdemann 1967, 1972; Cullimore & McCann 1978, 1979; Barbic & Savic 1990; Tuhela et al. 1992, 1993, 1997; Ralph & Stevenson 1995; Tyrell & Howsam 1997; Cullimore 1989, 1990, 1993, 1999).

Since most groundwaters and well waters have a more or less neutral pH (pH 6–8), neutrophilic iron and manganese bacteria are most common. The probably most abundant types of IRB in ochre incrustations are *Gallionella* (especially *G. ferruginea*) and *Leptothrix* (especially *L. ochracea* and *L. discophorus*). The genus *Gallionella* was first described by Ehrenberg in 1836 and was investigated in much detail by Hanert (1968, 1973, 1974, 1981). It is easily recognized under the microscope by its twisted stalks, which resemble noodles (Figs. 3-16 and 3-17). *Leptothrix*, on the other hand, is easy to recognize by its hollow stems, and it can oxidize and deposit both iron and manganese. The features visible in Figs. 3-16 and 3-17 are not the bacteria themselves but rather the stems which the much smaller bean-shaped bacteria use to attach to the substrate. These stems are a good example of a directionally excreted extracellular biomass (→ 3.6). The bacterial cell can detach itself from its stem and float off to better breeding grounds.

Most authors agree that *Gallionella* is an autotrophic (chemolithotrophic) organism (Ehrlich 2002). This means that it takes its vital energy from the oxidation of dissolved ferrous iron to ferric iron [Eq. (3.13)]. Since the latter is practically insoluble in neutral waters, it will form iron oxides [Eq. (3.14)], which can cause incrustation of the bacteria, especially its stem. It is important to notice that *Gallionella* does not need organic carbon to synthesize its biomass. The carbon for this purpose is gained from dissolved inorganic carbon. The biomass in Eqs. (3.13) and (3.14) here is simplified as a primitive hydrocarbon ("CH$_2$O").

$$4\ Fe^{2+} + CO_2 + 4\ H^+ \Leftrightarrow 4\ Fe^{3+} + \text{``CH}_2\text{O''} \tag{3.13}$$

$$4\ Fe^{3+} + 8\ H_2O \Leftrightarrow 4\ FeOOH + 12\ H^+ \tag{3.14}$$

Figure 3.16 Iron bacteria under the scanning eelectron microscope: *Gallionella* = twisted stalks, *Leptothrix* = hollow stems. Photograph: Banfield & Zhang (2003), with permission.

Figure 3.17 Fragments of iron bacteria from an incrustation of a well in central Germany under the scanning electron microscope: left, *Leptothrix* (hollow stems); right, *Gallionella* (twisted stalks). *Photographs: Georg Houben & Martina Klinkenberg, BGR, Hannover, Germany.*

Iron- and manganese-oxidizing bacteria occur naturally in most aquifers and are well adapted to environments at the interface between the oxic and the reducing zone, where they find both oxygen and iron or manganese (Figs. 3-4 and 3-9). There, they digest the small amounts of iron and oxygen passing with the slow groundwater flow. The limited nutrient supply hence limits their numbers. Table 3-1 summarizes their ideal living conditions. Although most of the time they are attached to their substrate ("sessile"), they can detach themselves and are then transported passively with the groundwater flow. Once they reach a well they must suddenly feel they are in paradise: here large volumes of water flow through a small area, carrying cumulatively high amounts of iron and oxygen. This of course can lead to explosive reproduction and thus massive precipitation of ochre.

The stoichiometry of Eq. (3.13) shows that four parts (moles) of ferrous iron have to be oxidized to form one part of biomass ("CH_2O"). The conversion rate of iron is therefore quite high. In other words: to produce 1 g of biomass, the bacteria need to precipitate 11.2 g of iron oxide (FeOOH).

The energy gain from the oxidation of ferrous iron is rather low, probably in the range of 27–42 kJ (6.5–10 kcal) per mole (Ehrlich 2002). Assuming it takes 120 kJ of energy to assimilate 1 mole of biomass (Ehrlich 2002), 2.86–4.44 moles of dissolved iron need to be oxidized. This results in a production of about 160–248 g (0.35–0.55 lb) ferric iron (= 305–474 g (0.67–1.04 lb) $Fe(OH)_3$). This is only valid assuming 100% efficiency for the energy conversion from iron to biomass. In reality, the efficiency of iron bacteria (e.g., *Thiobacillus ferrooxidans*) to "produce" biomass from iron oxidation is much lower, ranging from

TABLE 3.1 Ideal Living Conditions for Iron Bacteria Such as *Gallionella ferruginea* According to Various Authors

	Reference			
	Hanert (1968, 1974, 1981)	Hässelbarth & Lüdemann (1967)	Driscoll (1989)	Summary
pH	6.0–7.6	5.4–7.2	6.0–7.6	± Neutral
Redox potential (mV)	200–320	−10 ± 20	200–300	Slightly oxic
Dissolved oxygen (mg/liter)	0.1–1.0	> 0.005	0.1–1.0	Slightly oxic
Dissolved iron(II) (mg/liter)	ND*	0.2–1.2	1.0–25	Normal
Dissolved CO$_2$ (mg/liter)	(Present)	(Present)	>20	Normal

ND* = no data available.

30% to 3% (Ehrlich 2002). In this case the amount of iron oxide produced would be even 3 to 33 times higher than calculated above.

The unfavorable iron-to-biomass ratio explains why the iron bacteria develop such a ravenous appetite for iron. They simply need to take up vast amounts of iron to produce their minute amounts of biomass. For good reason, Cullimore & McCann (1979) coined the term "rusty little monsters" for them.

The ratio between the iron oxide mass and the biomass in such ochre incrustations should therefore be at least about 10:1. Our own geochemical investigations revealed that the ratio for young incrustations is in the range 10:1 to 20:1 (Fig. 3-18; refer to App. B for how to read box-whisker plots). This is probably due to an additional chemical oxidation of ferrous iron and the attachment of iron oxide colloids to the ochre. In old, highly crystalline incrustations, we found 35 to 60 times more iron oxide than biomass. The probable cause is the hardening of the incrustation caused by recrystallization. This induces a strong decrease in permeability and thus nutrient supply to the bacteria, which then float away or die off.

The ochre biofilms present in wells often contain more than one genus of IRB. Unlike *Gallionella*, the other iron bacteria are probably not completely autotrophic (Chapelle 1993; Ehrlich 2002). Many are heterotrophic, which means that they require oxygen as an electron acceptor and a source of organic carbon to synthesise biomass. Genera such as *Leptothrix* and *Siderococcus* probably belong to this type. They can utilize organic substances such as humic acids (Ehrlich 2002). The negatively charged (anionic) humic acids (here simplified as "CHO$^-$") can form complexes with dissolved ferrous iron

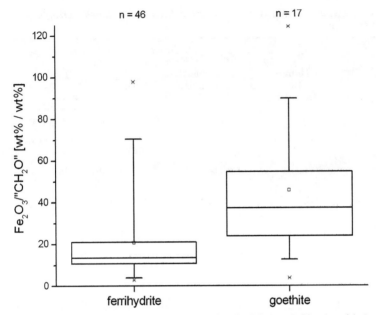

Figure 3.18 Iron to biomass ratio in young (ferrihydrite) and older (goethite) iron oxide incrustations.

[Eq. (3.15a)], which partially protects the latter from oxidation. If heterotrophic bacteria nibble away the protective humic acid anion [Eq. (3.15b)], the solitary ferrous iron is now much more prone to oxidation [Eq. (3.15c)].

$$Fe^{2+} + 2 \text{ "CHO}^-\text{"} \Leftrightarrow (Fe^{2+}(\text{"CHO}^-\text{"})_2) \quad (3.15a)$$

$$(Fe^{2+}(\text{"CHO}^-\text{"})_2) + 2\,O_2 + 2\,H^+ \Leftrightarrow Fe^{2+} + 2\,CO_2 + 2\,H_2O \quad (3.15b)$$

$$4\,Fe^{2+} + O_2 + 6\,H_2O \Leftrightarrow 4\,FeOOH + 8\,H^+ \quad (3.15c)$$

The oxidation of manganese(II) to insoluble manganese(III, IV) is also mediated by specialized bacteria, e.g., *Leptothrix discophorus*. Whether the energy obtained from this reaction is used directly in the bacterial metabolism is yet unclear.

Both iron and manganese bacteria cannot survive for a long time in water fully exposed to the atmosphere. Since they are also not pathogenic (cannot live inside the human body), they are not at all relevant from a hygienic point of view. The rare occurrences in tap water should therefore not be of concern, except for the persistent and ugly stains on clothes which the ochre causes.

Like most other bacteria, IRB are capable of the production of extracellular biomass. The composition and use of this "slime" is discussed in more detail in Sec. 3.6. The sheaths (tubes) and stalks of IRB are basically directionally excreted extracellular biomass. Some IRB, such as *Siderocapsa*, produce a slime capsule. Since the slime has a negative (acidic) surface charge, it acts as a "magnet" for all positively charged particles (e.g., Ehrlich 2002). Iron and manganese oxides have such a positive charge and therefore adhere strongly to the bacteria. Figure 3-19 shows iron-oxidizing bacteria (nitrate reducers) heavily encrusted by tiny iron oxide crystals.

Nitrate-reducing microorganisms—which are common in the suboxic zone—can also oxidize dissolved ferrous iron. They can thus contribute significantly to the formation of ferric iron—even when oxygen has been depleted (Ehrenreich & Widdel 1994; Straub et al. 1996). The stoichiometry of the dominant equation (3.11) predicts that the mass of iron oxide formed per mass of reduced nitrate is quite high [~1,061 mg Fe(OH)$_3$ per 124 mg of nitrate, or 8.56:1]. This easily explains the massive incrustation of such bacteria as visible in Fig. 3-19. Some of these organisms grow better when an organic substrate is available. Nitrogen gas is the most common product of such nitrate reduction [Eq. (3.11)].

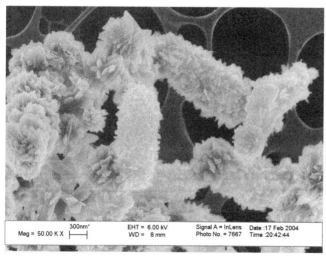

Figure 3.19 Scanning electron microscope exposure of nitrate-reducing bacteria (beta-proteobacteria, similar to the *Acidovorax* sp. strain G8B1) heavily covered by tiny iron oxide crystals. *Photograph: courtesy of Andreas Kappler, Centre of Applied Geosciences, University of Tübingen, Germany, taken at the California Institute of Technology, Pasadena, CA, USA.*

Microorganisms can also act indirectly on the formation of ochre. Any changes to pH and oxygen content caused by their metabolism will result in changes to the chemical reactions leading to ochre formation. Walter (1997) found that ochre biofilms indeed increase the pH of the water, possibly by consumption of (organic) acids or metabolic uptake of dissolved CO_2. As we have seen in Sec. 3.2.2, increases in pH will accelerate the oxidation rate of ferrous iron.

3.2.7 Ageing (recrystallization) of iron and manganese oxides

Practical experience has shown that old ochre incrustations are much more difficult to remove than younger ones. This phenomenon is based on the maturation of the iron and manganese oxide minerals. The first precipitates are phases which require only low crystallization energy. At the same time, their mineral grains are usually very fine-grained and often form gel-like ("amorphous") masses.

Ferrihydrite (→ 3.2.5) is such a fine-grained phase of low crystallinity. From a thermodynamic point of view, such phases are not stable (metastable) and will eventually recrystallize to more stable phases with lower surface energies. This process usually involves the dissolution of small mineral grains and the reprecipitation of coarser grains. The effect is quite common in nature and is not restricted to ochre minerals. It was first described by Nobel laureate Wilhelm Ostwald and is therefore often called the Ostwald sequence (Langmuir & Whittemore 1971; Morse & Casey 1988). The presence of ferrous iron accelerates the recrystallization sequence from ferrihydrite to iron oxides of higher order (Pedersen et al. 2005).

The low-crystallinity iron oxide phases ferrihydrite and lepidocrocite recrystallize to the thermodynamically more stable phases goethite and finally hematite (Murray 1979). The latter phase has so far not been detected in well incrustations. Ageing involves the release of water from the crystal lattice. A very simple ageing sequence for iron oxides is shown in Fig. 3-20

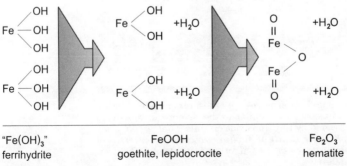

Figure 3.20 Simplified ageing sequence of iron oxides.

as a stepwise process of water release. Similar ageing sequences are also known for the manganese oxides (Hem et al. 1982; Hem & Lind 1983).

The mineralogical ageing sequences described so far induce a drastic decrease of surface area, an increase in particle size, and a decrease in chemical solubility (Fig. 3-21). This can be envisaged as a rearrangement of several small crystallites to fewer bigger ones (Fig. 3-22). The newly formed crystalline phases are much less soluble, much harder, and therefore much more resistant to mechanical and chemical attacks. Goethite has a hardness of 5.0–5.5 on the Mohs' scale. Fully crystallized goethite incrustations recovered during redrilling resemble red concrete ($\rightarrow$ 3.9). For iron oxides, the reactive surface area decreases from median values of about 170 m^2/g (1,829 ft^2/g) for ferrihydrite to about 75 m^2/g (807 ft^2/g) for goethite (Fig. 3-23). The surface area can be considered to be equivalent to the number of surface sites that dissolving chemicals can attack ($\rightarrow$ 8.2). The decrease in surface area, i.e., reactivity, and the hardening therefore explain why old incrustations are so difficult to remove.

The kinetics of the ageing and recrystallization of synthetic iron oxides was investigated by Schwertmann and Murad (1983) in long-term experiments. They found a strong acceleration of recrystallization with higher pH. Some reaction half-times are given in Table 3-2. In the ideal case, the transformation from ferrihydrite to goethite at pH 7 could be completed in about 1 year. The presence of certain inhibitors such as

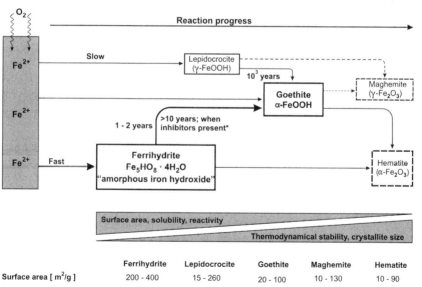

Figure 3.21 Ageing sequence ("Ostwald ripening") of iron oxides and effects on mineralogy and mineral properties.

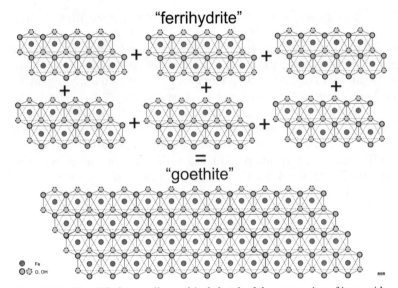

Figure 3.22 Simplified crystallographical sketch of the coarsening of iron oxide grains during ageing ("Ostwald ripening").

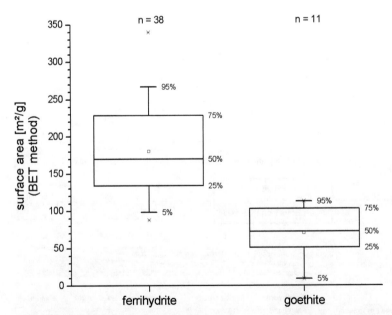

Figure 3.23 Box-whisker plot of the surface area of young (ferrihydrite) and older (goethite) iron oxide well incrustations.

TABLE 3.2 Kinetics of the Recrystallization of Iron Oxides

	Unit	pH				
		5	6	7	8	9
Reaction constant K	(d^{-1})	3.38×10^{-3}	4.20×10^{-3}	6.30×10^{-3}	10.00×10^{-3}	21.65×10^{-3}
Half-life $t_{1/2}$	(d)	205	165	110	68	32

After Schwertmann & Murad (1983).

phosphate, silicate, and organics—which are quite common in natural groundwaters—can significantly slow down ageing processes. Real ochre incrustations will probably take a few years or decades to convert completely to more crystalline phases.

The ageing of ochre minerals is not only of scientific interest. The loss of solubility and the hardening of the ochre strongly influence the success of rehabilitations. Figure 3-24 shows the different degrees of dissolution obtained in experiments for incrustation samples of increasing crystallinity. Rehabilitations need to be performed as early as possible and while the incrustations are still soft and soluble. One must not

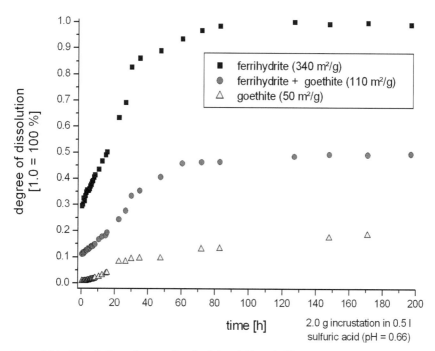

Figure 3.24 Dissolution of young (ferrihydrite-rich) and older (goethite-rich) iron oxide well incrustation samples in sulfuric acid. (Values in parentheses are the surface area of the mineral samples.)

wait and hope the problem will fade away. It will not do so, and it will actually worsen over time.

3.2.8 Interactions of ochre minerals with trace elements

Iron and manganese oxides have some remarkable chemical properties which affect the mobility and the availability of trace elements in nature (Cornell & Schwertmann 2003). The most important process is the sorption of trace elements onto the surfaces of the ochre minerals. We can envisage this surface as a collection of functional groups consisting of, e.g., iron hydroxide complexes (Figs. 3-22 and 3-25). Negative ($\equiv$FeO$^-$), positive ($\equiv$FeOH$_2^+$), and neutral ($\equiv$FeOH) functional groups can exist at the same time. This causes the interesting fact that iron oxides can sorb both anions and cations at the same time (Fig. 3-25). The hydroxyl group can be detached as a whole, which allows exchange versus a negatively charged ligand (anion) from the surrounding solution [Eq. (3.16)].

$$\equiv\text{FeOH} + \text{L}^- \leftrightarrow \equiv\text{FeL} + \text{OH}^- \quad (3.16)$$

If only a proton (H$^+$) is detached instead of the whole hydroxyl group, a positively charged ligand (cation, M^{z+}) can be sorbed [Eq. (3.17)].

$$\equiv\text{FeOH} + \text{M}^{z+} \leftrightarrow \equiv\text{FeOM}^{(z-1)} + \text{H}^+ \quad (3.17)$$

Which type of surface charge dominates and which type of ligand will thus be sorbed predominantly is strongly influenced by the pH of the surrounding solution (Cornell & Schwertmann 2003). At a pH of around 9.0, the surface has a nearly neutral charge. At pH > 9.0, negative charges gain

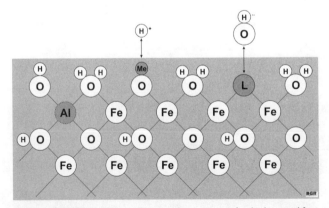

Figure 3.25 Mechanisms of trace element uptake in iron oxides. Me = cationic metal, L = anionic ligand, Al = substituted aluminum in crystal lattice.

increasing importance, favoring cation sorption. At pH values < 9.0, which includes the common range of groundwater pH, positive charges dominate, favoring anion uptake. Therefore, in most natural environments, iron oxides will sorb more anions than cations (Figs. 3-26 and 3-27).

Ions with small radii and high charges, such as phosphate (PO_4^{3-}), arsenate (AsO_4^{3-}), and heavy metals, have the highest sorption affinities for iron oxides. Regular groundwater anions such as sulfate and bicarbonate are less affected. At pH 7, the sorption affinity for goethite is (Cornell & Schwertmann 2003)

$$SeO^{3-} > AsO_4^{3-} = PO_4^{3-} > SiO_4^{4-} \gg Cl^- > F^-$$

Furthermore, iron and manganese oxides are capable of including cations having ionic radii similar to those of Fe(III) and Mn(III; IV), respectively, into their crystal lattice (isomorphous substitution, Fig. 3-25). These include, e.g., aluminum, chromium, cobalt, nickel, copper, and zinc. Maximum substitutions of Mn^{III}, Co^{III}, Ni^{II}, Cu^{II}, and Zn^{II} in goethite are 0.15, 0.10, 0.06, 0.05, and 0.07 mol·mol^{-1}, respectively (Cornell & Schwertmann 2003).

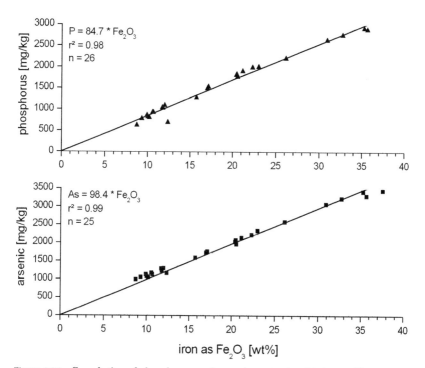

Figure 3.26 Correlation of phosphorus and arsenic contents with iron oxide contents of well incrustations from Schwabach, Germany. *Modified after data by Houben (2003a)*

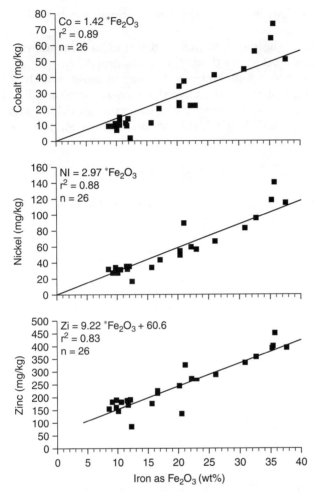

Figure 3.27 Correlation of heavy metal contents with iron oxide contents of well incrustations from Schwabach, Germany. *Modified after data by Houben (2003a).*

Some manganese oxide minerals have so-called tunnel structures. A surplus of negative surface charges is then compensated for by the uptake of cations of large ionic radii, such as barium and lead, into the tunnels. These cations are too large to be incorporated into the regular crystal lattice. Known examples are the minerals

Hollandite $BaMn_8O_{16}$
Todorocite $(Na,Ca,K,Ba,Sr)_{0.3-0.7}(Mn,Mg,Al)_6O_{12} \cdot 3.2 - 4.5H_2O$
Romanechite ("psilomelane") $(Ba,H_2O)_2Mn_5O_{10}$

Ochre incrustations are known to contain significant amounts of trace elements (Walter 1997; Houben 2003a). Figures 3-26 and 3-27 show the elevated trace element contents in well incrustations discussed in this book for anionic species and some selected heavy metals. As expected from the neutral pH of the well water, far larger amounts of anions are sorbed than cationic metals. The linear correlation between the iron oxide content and the trace element concentrations gives evidence that the latter are sorbed to the ochre and do not occur as separate phases.

What can we infer from these consideration for practice?

- Concentrations of arsenic and heavy metals in incrustations can exceed legal limits for soils and solid waste (Figs. 3-26 and 3-27). Disposal of such incrustations must follow the applicable legally required procedures. Young incrustations deposited at landfills may undergo recrystallization processes ($\rightarrow$ 3.2.7). This could lead to a remobilization of trace elements.

- Adsorbed phosphate can severely diminish the dissolution rates of iron oxides (Borggaard 1992) by blocking surface sites and thus negatively affect chemical rehabilitations.

- Adsorbed phosphate slows down the ageing of iron oxides ($\rightarrow$ 3.2.7) because of the blockage of surface sites otherwise used to "click" small crystallites together.

- Uptake of dissolved trace elements by iron and manganese oxides incrustations can deplete water from these elements and influence the quality of samples (Walter 1997; Houben 2003a).

The last topic is worth some further consideration. In order to investigate this effect, example calculations were carried out for a well field near Schwabach in southern Germany (Houben 2001, 2002, 2003a, Table 3-3). The concentrations in the well water are given in the second column of Table 3-3. Using a pumping rate of 140,000 m^3/a (4,943,400 ft^3/a), the annual mass flux for some species was calculated (third column). The uptake in 1 kg of incrustation can be extrapolated using the linear trends shown in Figs. 3-26 and 3.27 to 100% Fe$_2$O$_3$ (fourth column). The fifth and sixth columns of Table 3-3 show the percentages of the total mass removed from the water by an annual precipitation of 1 or 10 kg (2.2 or 22.0 lb) of Fe$_2$O$_3$, respectively. The influence on the concentrations of dissolved heavy metals and arsenic in water samples is almost negligible, but in the case of phosphorus a significant amount is removed—at least in strongly incrusted wells. Water analyses from such wells should thus be treated with caution.

TABLE 3.3 Calculated Changes of Water Quality as a Result of Iron Oxide Incrustation for a Water Well in Schwabach, Southern Germany

	Dissolved conc. (mg/liter)	Annual mass flux at $Q = $ 140,000 m³/a (g/a)	Content in incrustation (g/kg)	Error caused by 1 kg Fe_2O_3 (%)	Error caused by 0 kg Fe_2O_3 (%)
Phosphorus	0.006	840	9.84	1.17	11.7
Arsenic	0.050	7,000	8.47	0.12	1.2
Zinc	0.008	1,120	0.98	0.09	0.9
Nickel	<0.01*	1,400	0.30	0.02	0.2
Cobalt	<0.01*	1,400	0.14	0.01	0.1
Barium	0.191	26,740	10.73	0.04	0.4

*Measured values below detection limit; detection limit was used as concentration for calculations.
From Houben (2003a).

Ageing (recrystallization) of ochre minerals (→ 3.2.7) involves a decrease in surface area and thus in the number of reactive surface sites. Comparing the trace element concentrations of young and aged incrustation samples shows clearly that sorbed phosphate and arsenic, both negatively charged anionic species, must have been lost during ageing (Fig. 3-28). We envisage that the anions present at the crystals' edges are expelled during reassembly (coarsening) of the small ferrihydrite to larger goethite crystals (Fig. 3-29).

For the heavy metals, the story is somewhat different. Figure 3-30 shows how barium experiences a decrease of sorbed concentration (mobilization) during ageing (lead—not shown in Fig. 3-30—behaves similarily). The comparatively large ionic radii of barium and lead probably prevent their substitution into the crystal lattice. Barium might also be affected

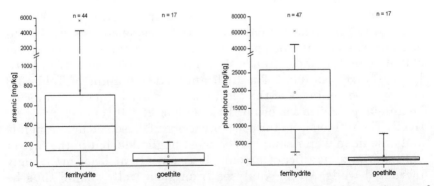

Figure 3.28 Consequences of iron oxide ageing on the contents of anionic constituents of well incrustations.

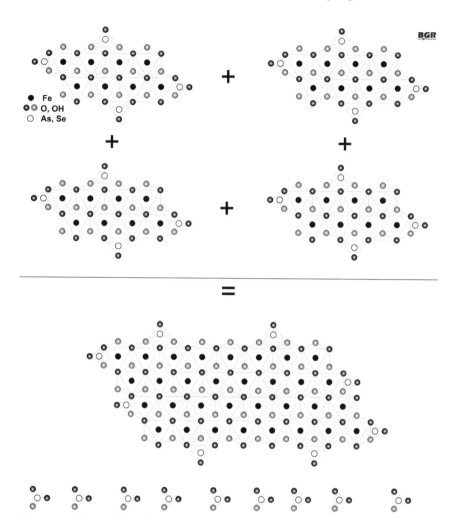

Figure 3.29 Iron oxide ageing as coarsening process and the simultaneous release of sorbed anions.

by a collapse of the tunnel structures of associated manganese oxides during ageing. The concentrations of other heavy metals such as copper, nickel, cobalt, and zinc (latter not shown) do not decrease during ageing but rather show a slight increase (Fig. 3-30). This is probably due to the uptake of the metal ions into the crystal lattice of the ochre minerals. Any changes in the number of surface sites therefore do not affect them: they simply stay in place during ageing. Figure 3-31 compares the ionic radii of some of the relevant metals. The radius of nickel(II) is 0.69 Å and thus is quite close to that of iron(III), which is 0.64 Å. The larger barium(II) cation with its 1.34 Å radius will not fit into the crystal lattice. This is

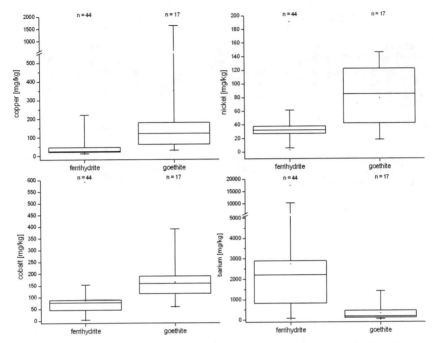

Figure 3.30 Consequences of iron oxide ageing on the trace metal contents of well incrustations from Germany.

in good agreement with experimental results by Ford et al. (1997), Martinez and McBride (1998, 1999), and Martinez et al. (1999). They found that thermal aging of ferrihydrite leads to mobilization of lead. On the other hand zinc, copper, cobalt, and nickel contents increase with increasing crystallization. The experiments by Ford et al. (1997) and by Martinez and McBride (1998) showed that the concentrations of zinc, copper, and nickel in the solid phase increase with proceeding thermal transformation (ferrihydrite to goethite). Similar behavior of heavy

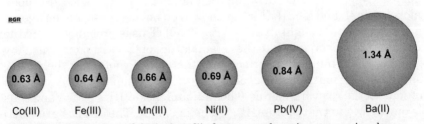

Figure 3.31 Comparison of the ionic radii of some metal species common in ochre.

metals was reported from the field for aging red bed sediments by Zielinski et al. (1983).

3.2.9 Structure and texture of ochre incrustations

Samples of hardened incrustations can be recovered during redrilling of abandoned wells. Their internal structure can then be investigated by microscopy on polished sections or thin sections. Brittle samples can be stabilized by resin impregnation. The authors were able to perform such studies on several ochre samples and also on some carbonate incrustations.

Internal laminations resembling annular tree rings are the most striking feature found (Fig. 3-32). The laminae usually have thicknesses from 0.2 mm up to 2 mm. They are probably caused by variations in the pumping regime (discharge rate, etc.), the natural hydrochemistry, and possibly seasonal influences. Quite often the differences of their translucence are caused by varying contents of incorporated silicate (quartz) grains. Whether the laminae actually represent 1 year or other time periods still needs to be investigated.

The ochre minerals are usually too fine-grained and closely matted to be distinguished individually. A common feature in old incrustations is the occurrence of radial goethite spheres and hemispheres (Fig. 3-32). They contain goethite needles which crystallized radially and/or spherically from a common seed point.

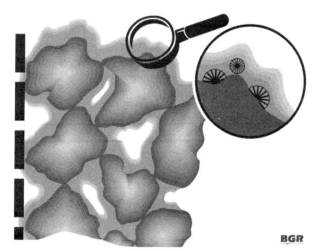

Figure 3.32 Internal structure of ochre incrustations: "annular rings" and radial goethite crystals.

3.3 Carbonate Incrustations (Scale)

Carbonate incrustations—commonly referred to as scale—naturally require a carbonate-rich environment to form. This does not need to be a limestone aquifer, as any gravel or sandstone aquifer with several weight percent of carbonate detritus is usually sufficient.

Unlike the formation of ochre and sulfides, mixing of waters is not the main cause of carbonate precipitation. Mixing of two waters, both in equilibrium with calcium carbonate, will lead—counterintuitively—to undersaturated water which can dissolve carbonates. This phenomenon is known as *Mischungskorrosion* (mixing corrosion) and was detected as early as the beginning of the twentieth century (Tillmanns & Heublein 1912). Precipitation can occur only when the water is saturated or oversaturated with respect to carbonates (see Sec. 5.8 for a detailed discussion of the calculation of saturation indices).

The "carbonate equilibrium" is discussed in much detail in many hydrochemistry textbooks (e.g., Stumm & Morgan 1996; Appelo & Postma 2005). The solubility of carbonates in water depends strongly on the amount of available dissolved carbon dioxide (CO_2). According to Eq. (3.18), carbon dioxide enters the water from the atmosphere—or soil gas—and causes dissolution of solid carbonates. The maximum amount of soluble carbonate is then defined by the equilibrium of CO_2 dissolution in water. Carbon dioxide gas is a component of atmospheric air (0.035 vol%) but can be enriched in soil gas (up to 7 vol%) due to microbial degradation of biomass.

$$CaCO_3 + CO_{2(g)} + H_2O \Leftrightarrow Ca^{2+} + 2\,HCO_3^- \tag{3.18}$$

Waters that are rich in CO_2 will thus dissolve more carbonate, while its removal turns the water into an oversaturated state—which can lead to precipitation of inorganic carbon (carbonates). Therefore all processes which decrease the solubility of gases in water will enhance the precipitation of scale. Gas solubility in water is a function of (listed according to importance)

- Pressure
- Temperature
- Agitation
- Salt content of the water

Although the last factor can be safely ignored for practically all cases of concern to us (high concentrations of dissolved salts decrease the solubility of gases in water), the first three must be considered.

The solubility of nonreactive gases in dilute solutions can be described according to the Henry-Dalton law [Eq. (3.19)]:

$$\lambda = K_H \cdot p \qquad (3.19)$$

where λ = solubility, M/L^3
p = partial pressure of the dissolving gas in the gas phase, $M \cdot L^{-1} \cdot T^{-2}$
K_H = Henry constant of the gas (temperature-dependent), $L^{-2} \cdot T^2$

Therefore, with constant K, solubility increases with increasing pressure (up to some equilibrium value). Any decline in pressure will thus release carbon dioxide from the water and facilitate precipitation of solid carbonates. The decline in pressure experienced by water entering a well is probably much too small to be of any influence. On the other hand, the high flow velocities and possible resulting turbulences in the gravel pack, the screen slots, and the well interior strongly agitate the water, which will cause some degassing. This is a common observation we all know from everyday life: if we open a soda bottle that has been shaken vigorously, the gas will escape at once—leaving us soaking wet.

The temperature dependence of the gas solubility λ in a solvent can be expressed by [Eq. (3.20)]:

$$\lambda = \alpha \,(1 - T/273) \qquad (3.20)$$

where T = temperature, T
α = solubility constant of the gas, $1/T$

Values for both the reaction constants K and α can be found in many hydrogeochemistry books or databases (as cited above).

Accordingly, gas solubility decreases with increasing temperature. Figure 3-33 shows the solubility of atmospheric CO_2 in water as a function of temperature. Under turbulent flow regimes—as they may occur around or inside wells—some of the kinetic energy that is "lost" can be transformed into heat, which should slightly raise the water temperature. The net effect of this is probably small. Another, more obvious heat source is the submersible pump, which is cooled by the flowing water.

Counterintuitively, scales are often found in zones of high flow velocities, where one would expect erosion to limit precipitation. Laboratory experiments by Zeppenfeld (2005) have shown that the nucleation rate of calcium carbonate crystals actually increases with increasing flow rate. This can be explained by an increased resupply of the crystal-building dissolved ions to the growing mineral surface.

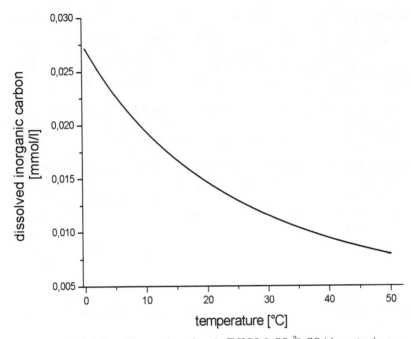

Figure 3.33 Solubility of inorganic carbon (= $\sum HCO_3^-$, CO_3^{2-}, CO_2) in water in contact with the atmosphere (partial pressure $pCO_2 = 10^{-3.5}$) as a function of temperature (calculated using PHREEQC).

Several groups of microorganisms are known to interfere directly or indirectly into the carbonate equilibrium and cause the precipitation of solid carbonates (Ehrlich 2002). Many of those require light for photosynthesis and are thus unlikely to be found in wells (e.g., algae, cyanobacteria). Sulfate-reducing bacteria (SRB) are known to produce carbonates as by-products ($\rightarrow$ 3.5). In many cases carbonate-incrusted wells do not show sulfate-reducing environments at all, and carbonate scales are often sulfide-free. Therefore the role of SRB or other microorganisms should not be overestimated. The authors of this book found no references in the literature to the presence or activity of carbonate-precipitating organisms in water wells.

The most common carbonate mineral phases in well incrustations are calcite and aragonite (both $CaCO_3$), and to a lesser extent dolomite [$CaMg(CO_3)_2$]. Most carbonate minerals have very short-lived meta-stable precursor phases, which are the first precipitates from solution but in turn quickly convert to the common carbonates (Gabrielli et al. 1999; Tlili et al. 2003). Aragonite itself is also meta-stable and will eventually—but on a much slower time scale than the precursor crystals—transform into calcite. The light orange-colored iron carbonate (siderite, $FeCO_3$) is restricted to anoxic redox environments and the presence of Fe^{2+} but with a simultaneous absence of sulfide (otherwise iron sulfides would form; → 3.5). Carbonates usually contain insignificant amounts of trace elements because of their low sorption capacity and because most metals do not fit into the crystal lattice. Zinc is a remarkable exception to this rule.

Pure calcium carbonate is snow-white, but in most cases natural samples contain some trace amounts of, e.g., iron (oxide), which can cause reddish colors. Color alone is therefore not a good feature by which to identify carbonate incrustations. Carbonates will react with acid to form CO_2. Therefore calcium carbonate is readily identified by the strong "fizzing" when it is exposed to a few drops of dilute acid. The reaction is somewhat less intense for dolomite. Siderite will hardly "fizz" at all.

Carbonate incrustations recovered from wells often show distinct "annular rings" of variable thicknesses and coloring. They are probably due to variations in chemical and flow conditions of the well.

The calcite crystals found in well scale are often quite large (millimeter sized), which points to fast growth rates. Incrustations often have a thickness of several centimeters and are very dense and hard (Fig. 3-34).

Figure 3.34 Calcium carbonate incrustation recovered from a 6-in (15-cm) water well near Königslutter (Germany). The scale bar is 5 cm (2 in) long. (<1 wt%). *Photograph: Andrea Weitze, BGR.*

3.4 Aluminum Hydroxide Incrustations

Aluminum hydroxides are probably among the rarest types of well incrustations. They were described in Australia by McLaughlan et al. (1993) and McLaughlan (1996) and in Germany by Baudisch (1989). Their occurence is restricted to certain geochemical boundary conditions:

- High acid input
- Steep vertical pH gradient
- Aluminum-rich soil material
- Low pH buffering capacity of soil

Acid input can stem from the mineralization of soil biomass and manure (especially its nitrogen [N(III)] content). More drastic acid input is often related to acidic rain water ("acid rain") from the combustion of sulfur-rich fossil fuels. Most soils have a natural buffering capacity—mostly due to carbonates—which will keep the pH up, although not infinitely. In sandy soils with low carbonate contents, e.g., podzols, the carbonate buffering capacity can be overpowered quite easily. If the pH then drops to <4.2, the acid attacks aluminum-containing minerals such as gibbsite [Eq. (3.21)] and clay minerals [Eq. (3.22), here: kaolinite]:

$$Al(OH)_3 + 3\ H^+ \Leftrightarrow Al^{3+}_{(aq)} + 3\ H_2O \quad (3.21)$$

$$Al_2Si_2O_5(OH)_4 + 6\ H^+ \Leftrightarrow 2\ Al^{3+} + 2\ H_4SiO_4 + H_2O \quad (3.22)$$

The released Al^{3+} will to a certain extent be sorbed onto the soil matrix but will eventually migrate downward. Figure 3-35 shows the vertical distribution of pH and dissolved Al^{3+} in the groundwater of an area in Germany with soils of low buffering capacity that have been affected by "acid rain." Obviously, this has led to the development of a vertical pH gradient with low pH and high aluminum concentrations close to the surface and reverse conditions deeper in the aquifer.

If the acidic and aluminum-enriched water percolates downward, it encounters deeper groundwater of higher pH. The resulting admixture causes an overall rise in pH and thus a reprecipitation of aluminum hydroxide. A well screened over both the acidic and the neutral zone will be affected strongly by this process (Fig. 3-36). Involvement of microorganisms in these reactions is unlikely.

Pure aluminum hydroxide has a white color and can therefore easily be confused with calcium carbonate. Contrary to carbonates, aluminum hydroxides will not "fizz" when exposed to a few drops of dilute acid.

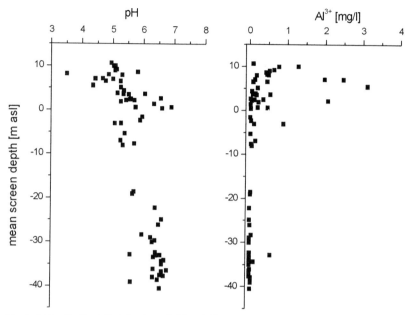

Figure 3.35 Vertical distribution of pH and dissolved aluminum concentrations in the Bourtanger Moor area (near Haren, Ems, northwestern Germany). *Modified after data by Houben (2000).*

3.5 Metal Sulfide Incrustations

During chemical rehabilitations with acids, a strong smell of "rotten eggs" (hydrogen sulfide, H_2S) often arises from the well. The conclusion which is often drawn from this is that the well is incrusted by metal sulfides. In most cases this is probably not entirely true. The human nose is very sensitive to H_2S and is able to detect trace amounts (0.025–1 ppm), which can arise from the small amounts of metal sulfides (<1 wt%) present in most ochres (→ 3.2). Pure sulfide incrustations will produce very high concentrations of gaseous H_2S. Unfortunately, such high concentrations (>200 ppm) quickly overpower the human nose, so no smell is detected.

Pure sulfide samples were not detected during our mineralogical studies of well incrustations. Well screens overlapping the sulfide (H_2S) zone and the upper or lower iron zones in Fig. 3-4 will lead to incrustations dominated by reduced mineral phases. Such overlaps therefore tend to occur in wells which reach deep down into the reducing parts of the aquifer. In oxygen-free reducing environments, sulfate-reducing bacteria utilize (dissolved) organic carbon to break down sulfate to sulfide while producing inorganic carbon [Eq. (3.23)]:

$$2 \text{ "CH}_2\text{O"} + SO_4^{2-} \Leftrightarrow H_2S + 2 \, HCO_3^- \qquad (3.23)$$

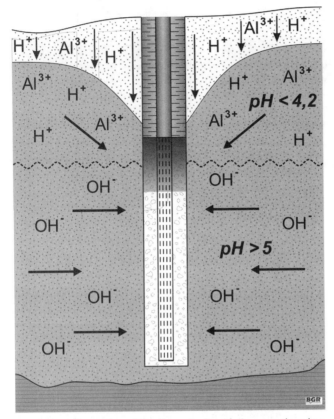

Figure 3.36 Formation of aluminum hydroxide incrustations in a well at a site with acidified soil.

The reaction proceeds rather slowly, with reaction half-lives in the range of 100 years (Böttcher et al. 1989). The dissolved sulfides H_2S und HS^- originating from Eq. (3.23) react with dissolved ferrous iron ($Fe_{(aq)}^{2+}$) to form iron monosulfides and native sulfur [Eq. (3.24)]. Because of their solubility in acids, the former are called "acid-volatile sulfur" (AVS).

$$2\ FeOOH + 3\ HS^- \Leftrightarrow \text{"FeS"} + S^0 + H_2O + 3\ OH^- \qquad (3.24)$$

Amorphous FeS, mackinawite ($FeS_{0.9}$), and greigite (Fe_3S_4) were identified as the most common monosulfide phases in the reduced incrustations investigated by the authors. Monosulfides are black colored (the Black Sea takes its name from their presence) and can thus easily be confused with manganese oxides. Exposure to a few drops of dilute acid will cause the characteristic smell of rotten eggs. Under the scanning electron microscope they show typical spheroidal or framboidal shapes. Native sulfur (S^0 or S_8) was also found in some incrustations.

Monosulfides slowly react with native sulfur to more stable phases such as marcasite and pyrite (both FeS_2) [Eq. (3.25)] (Berner 1970, 1984; Schoonen & Barnes 1991a, 1991b; Wilkin & Barnes 1996; Oles & Houben 1998). In most incrustations studied by the authors, this process had not proceeded far, because most of the sulfide was still present as AVS.

$$FeS + S^0 \Leftrightarrow FeS_2 \qquad (3.25)$$

Iron sulfides can incorporate significant amounts of chalcophilic (= "sulfur-friendly") trace metals into their crystal lattice. In order of affinity, these are cobalt, nickel, mercury, arsenic, lead, and zinc (Belzile & Lebel 1986; Morse & Arakaki 1993; Huerta-Diaz et al. 1998; Morse & Luther 1999).

Since the reduction of sulfate [Eq. (3.23)] and the genesis of the monosulfides [Eq. (3.24)] produces bicarbonate and raises the pH, this is likely to lead to a precipitation of carbonates (Ehrlich 2002). The investigated reduced incrustations indeed contained significant amounts of carbonate, mostly in the form of calcite ($CaCO_3$). The iron carbonate siderite ($FeCO_3$) is sometimes found as well. It is—similar to sulfides—an indicator of a reducing environment, because it can only be formed in the presence of reduced iron (Fe^{2+}). On the other hand, it can form only in the absence of sulfides; its co-formation with the AVS phases is therefore not possible. Some samples which contain both AVS and siderite therefore represent changing geochemical environments. The additional—and quite common—presence of iron oxides (ferrihydrite) in reduced incrustations is also a strong indicator for such temporal variations. Local geochemical variations inside a biofilm may also be a possible explanation.

As mentioned above, SRB are key players in the genesis of sulfide incrustations. Sulfate-reducing bacteria include several genera, including *Desulfovibrio, Desulfobacterium, Desulfococcus, Desulfomonas,* and more. Unlike iron bacteria, SRB cannot tolerate oxygen and are thus called obligate anaerobic bacteria. Therefore they often "hide" underneath other more oxygen-tolerant bacteria in biofilms. Scratching open ochre crusts will therefore often reveal a black sulfide core. Almost 2,300 years ago, Aristotle was the first to give evidence of the presence of sulfur bacteria in wells. He found whitish (bio)films inside wells which turned black after some time. What he saw were probably sulfur bacteria (e.g., *Beggiatoa alba*) which formed black iron sulfide after some time. The presence of SRB in modern wells was confirmed by van Beek and van der Kooij (1982).

Care has to be taken not to confuse reduced incrustations with microbially induced anaerobic corrosion products (→ 3.1.3). SRB play an important role in both cases, and the reaction products are also very similar.

3.6 Bioclogging

Clogging caused by microorganisms (bioclogging) can occur through several different mechanisms:

- by bacterial cells
- by bacterial exopolymers ("slime")
- by metabolic production of gas bubbles (CO_2, CH_4, N_2, H_2S)
- by metabolic mineral precipitates ($\rightarrow$ 3.2, 3.5)
- by accumulation of suspended detritus on bacterial surfaces

In this chapter we will consider only microbial deposits which are not connected to the buildup of mineral mass or the collection of detritus. The commonly applied term "biofouling" is somewhat ambiguous due to its indiscriminate use, which sometimes includes iron bacteria ($\rightarrow$ 3.2.6).

The growth of bacterial biomass can reduce the porosity and thus the permeability of porous media (Vandevivere & Baveye 1992; Baveye et al. 1998; Thullner et al. 2002a, 2002b). Wells with nutrient-rich water are often affected by the buildup of massive microbial slime deposits. They are easily identified through camera inspections. Water qualities in favor of slime buildup occur with bank filtrate wells and with pump-and-treat schemes aimed at the recovery of polluted groundwater (Hijnen & van der Kooij 1992; Taylor et al. 1997). Slimes are also among the most common ageing causes for infiltration wells ($\rightarrow$ 4.7). The most important nutrients that foster microbial growth are phosphorus (as phosphate), nitrogen (as nitrate), and carbon (especially when available as easily degradable compounds, e.g., sugars). Sulfur and iron are the main trace nutrients for specialist microorganisms.

The main cause for slime buildup in wells is the presence and growth of diverse bacterial populations with a large variety of different bacteria types (Taylor et al. 1997). Heterotrophic bacteria which can utilize organic carbon as their main energy source are very common. Aerobic heterotrophs include *Pseudomonas, Flavobacterium, Nocardia,* and *Citrobacter,* whereas the anaerobic heterotrophs include *Actinomyces, Bacteriodes, Bacillus,* and *Agrobacterium.* Quite often the latter are accompanied by sulfate-reducing (e.g., *Desulfovibrio,* $\rightarrow$ 3.5) and/or sulfur-oxidizing (e.g., *Thiobacillus*) bacteria. Iron bacteria such as the reducing *Shewanella* and the oxidizing *Gallionella* and *Crenothrix* may also occur ($\rightarrow$ 3.2.1; 3.2.6). As we see from the bacteria types listed above, both aerobic and anaerobic bacterial groups may be present at the same time. This indicates the presence of varying geochemical microenvironments within the bacterial community. Such a community is called a biofilm. While oxygen-tolerant or even oxygen-utilizing bacteria may settle on the parts of the biofilm that are exposed to groundwater flow, reducing bacteria can "hide" underneath. Symbiotic relationships among different bacteria may develop occasionally.

Biofilms contain not only bacterial cells but also extracellular biomass (hydrous gel, i.e., "slime") which the bacteria secrete. This consists mainly of polysaccharides, but includes some proteins, lipids, and nucleic acids. The extracellular biomass is often summarized under the term exopolysaccharides (EPS). EPS protect the bacteria from attack, foster attachment to their substrate ($\rightarrow$ 3.2.6), help them to survive starvation and drought periods, prevent loss of extracellular enzymes, and are used for communication and exchange with other bacteria. Their volume can be substantially larger than the volume of the bacterial cells alone.

Biofilms in wells may also be a shelter for certain types of fungi genera such as *Aspergillis*, *Candida,* and *Penicillum* (Taylor et al. 1997). Their numbers are three to five orders of magnitude lower than those of the bacteria. Sometimes higher developed organisms, e.g., eukaryotes, use biofilms as their feeding ground. Figure 3-37 summarizes the stages of development of a typical biofilm.

Under favorable conditions the development of a biofilm can occur within weeks. As we all know, bacteria can multiply very quickly by cell division. An example calculation can show the potential. Huisman & Olsthoorn

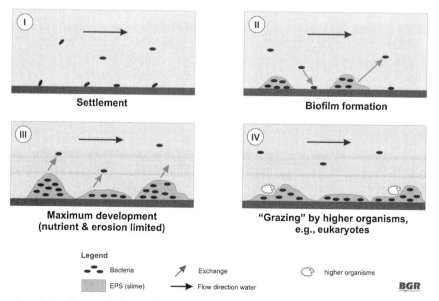

Figure 3.37 Stages of biofilm formation.

(1983) present the following equation (3.26) to calculate the buildup of bacteria in infiltration wells:

$$n = \frac{v_e \cdot n_0 \cdot T_d}{0.693 \cdot (e^{0.693 \cdot t/T_d} - 1)} \qquad (3.26)$$

where n = number of bacteria at time t
n_0 = initial number of bacteria
t = time, T
T_d = doubling time for bacterial population, T
v_e = entrance velocity, L/T

Assuming the following input data,

$n_0 = 100/cm^3 = 10^8/m^3$

$t = 7$ days

$T_d = 0.3$ day

$v_e = 1$ m/h $= 24$ m/day (78.7 ft/day)

we obtain a total number of 1.09×10^{16} bacteria after 7 days. If we further assume that one bacterium has a volume of 2 µm^3 = 2×10^{-18} m^3, the total volume of the bacteria after 7 days is

$$(1.09 \times 10^{16}) \times (2 \times 10^{-18}) \text{ m}^3 = 0.022 \text{ m}^3 = 21{,}800 \text{ cm}^3 \text{ (0.75 ft}^3\text{)}$$

With a gravel pack porosity of 45% this equals a biomass thickness of 4 cm (0.01 ft). Figure 3-38 shows the dramatic consequences of bacterial growth. Luckily this is only a theoretical concept. Bacterial growth is not unlimited, because of the limited availability of nutrients (and additionally erosion by flowing water). Figure 3-39 shows a more realistic population growth curve.

3.7 Special Incrustation Types

Other types of incrustations may occur under specific hydrogeological and hydrochemical conditions. Because these are usually not found in wells used for public and industrial water supply, they will be discussed here only briefly.

Incrustations from geothermal wells and hot springs often contain ochre as well as carbonates, but some more "exotic" mineral phases may also occur, e.g.,

- Gypsum ($CaSO_4 \cdot 2H_2O$)
- Barite ($BaSO_4$)

- Celestite ($SrSO_4$)
- Iron sulfate ($FeSO_4$)
- Silicates (e.g., amorphous quartz)
- (Heavy) metal sulfides

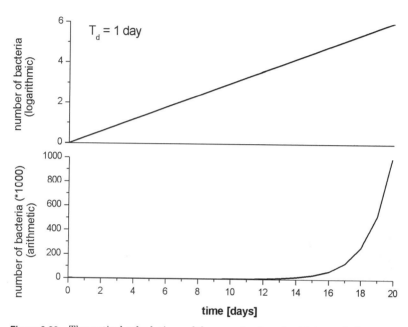

Figure 3.38 Theoretical calculations of the growth of a microbial population.

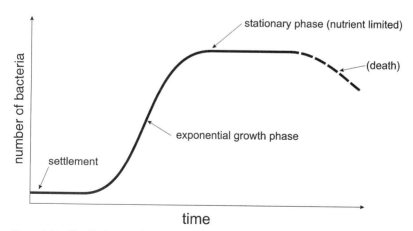

Figure 3.39 Realistic growth curve for a bacterial population.

Their occurrence is a major problem for the operation of geothermal plants and has therefore received much attention in the literature (Rothbaum et al. 1979; Sanyal et al. 1985; Yokoyama et al. 1988; Criaud et al. 1989; Gallup 1989; Akaku 1990; Gallup & Reiff 1991; Manceau et al. 1995; Thomas et al. 1992; Brown et al. 1995; Hinman & Lindstrom 1996; Gallup 1998; Drieux et al. 2002). Similar incrustation types were also detected in oil wells (Saunders & Rowan 1990; Kalfayan 2001).

Field drains (horizontal drainage pipes), which are commonly used to dewater bogs and swamps, are very often encrusted by iron oxides (Trafford et al. 1973; Ivarson & Sojak 1978; Süsser & Schwertmann 1983; Wheatley et al. 1984; Abeliovich 1985; Stuyt & Oosten 1987). Their mineralogy is basically the same as described in Sec. 3.2.5. Groundwater recharge trenches suffer from the same problems (Warner et al. 1994).

Acid mine drainage (AMD), caused by the oxidation of solid metal sulfides, mostly pyrite, in mines and mine tailings results in very acidic waters which are rich in sulfate and ferrous iron. Upon contact with air these waters can precipitate several mineral phases which can clog drainage systems. These minerals include iron oxides as described in Sec. 3.2 (Ferris et al. 1989; Singh et al. 1999) and jarosites [e.g., $KFe_3(SO_4)_2(OH)_6$, potassium iron sulfate hydroxide]. In sulfate-rich waters with a pH value of less than 3, schwertmannite [$Fe_{16}O_{16}(OH)_{12}(SO_4)_2$] forms instead of ferrihydrite (Uhlmann & Arnold 2003). Schwertmannite is meta-stable and will eventually recrystallize to goethite (Cornell & Schwertmann 2003).

Incrustations often hamper the operation of landfill leachate collection systems (Mitchell et al. 1993; Brune et al. 1994; Fleming et al. 1999; Rowe et al. 2000). Common incrustation components are iron oxides and sulfides as well as carbonates and gypsum. Calcite ($CaCO_3$) is the most common carbonate phase, although the mineral vaterite (also $CaCO_3$) may also occur.

The buildup of the minerals struvite $(NH_4)Mg(PO_4)\cdot 6(H_2O)$ and sometimes vivianite [$Fe_3(PO_4)_2\cdot 8H_2O$] is a common problem in sewers and wastewater treatment plants (Parsons & Doyle 2004). As their chemical composition indicates, these minerals are made up largely of typical wastewater nutrients such as ammonium and phosphate. They occur under anaerobic conditions when the levels of soluble ammonia (NH_3-N) and phosphate (PO_4-P) are elevated due to the decomposition of activated sludge.

3.8 Passively Incorporated Components of Incrustations

The geochemical analysis of incrustations from water wells often reveals significant portions of silicon and occasionally aluminum. This should not be interpreted as result of the in-situ formation of silicates or alumosilicates in the incrustations (such processes require more extreme environments, e.g., high temperatures and salinities). The silicon content stems from passively incorporated silicate minerals such as quartz (SiO_2), clay minerals [e.g., kaolinite, $Al_2Si_2O_5(OH)_4$], and feldspar (e.g., $KAlSi_3O_8$). The origin of the silicates is mostly located in the aquifer, although quartz may also stem from the gravel pack. From there they are mobilized through the groundwater flow and attach themselves, e.g., to the sticky, gel-like ochre. Microscopic and mineralogical investigations (thin sections) clearly show the detrital origin of this material.

3.9 Spatial and Temporal Distribution of Incrustations

Many camera inspections show that the spatial incrustation distribution inside the well is inhomogeneous (→ 2.2.3). Incrustations tend to be more abundant in the upper parts of the screen than in the lower. In addition, often only one-half of the filter tube is strongly encrusted, while the other half is only moderately affected (Fig. 3-40).

One possible way of studying the vertical distribution of incrustations is the monitoring of material removed during rehabilitations performed in vertical sections. Figure 3-41 shows the findings from a chemical rehabilitation of a well in Krefeld (Germany). The well interior was brushed before the application of the chemicals, which in turn means that all of the removed incrustations came from the annulus. To quantify the amounts, the concentration of dissolved iron was measured in the water pumped from each screen section after the application of the chemicals. The water was sampled with a high temporal resolution (5 min) and later analyzed

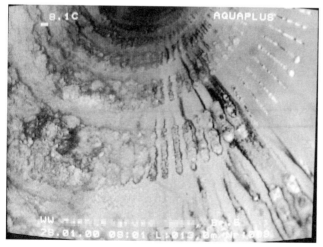

Figure 3.40 Spatially inhomogeneous distribution of ochre inside a water well. *Photograph: Aquaplus, Kronach, Germany.*

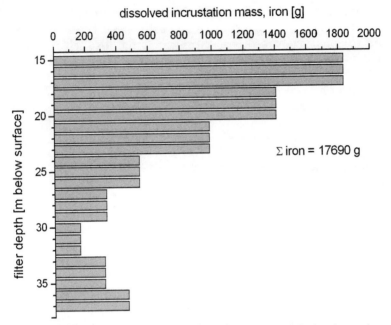

Figure 3.41 Vertical distribution of iron oxide mass removed during the section-wise chemical rehabilitation of a water well in Krefeld (Germany). *Modified after Houben (2006).*

in the laboratory. Multiplication of the measured concentrations with the volume of water pumped during the sampling step—calculated from the measured pumping rate—yielded the removed mass per section. The distribution of incrustations strongly resembles the relative inflow rates from the model calculations presented in Sec. 2.2.3. The slightly elevated degree of incrustation at the bottom of the filter is likely to be an effect of partial penetration of the well causing locally elevated inflow ($\rightarrow$ 2.2.3).

In order to assess both the vertical and horizontal distribution of incrustations, two core drillings through the annulus of an abandoned well were performed. The selected well had been abandoned due to severe buildup of ochre and belonged to the Wildeshausen water works (northern Germany) of the Oldenburgisch-Ostfriesischer Wasserverband (OOWV). Details of this study can be found in Houben (2006). One core drilling was performed in the direction facing the natural groundwater flow (upgradient), while the other was drilled 180° into the opposite direction (downgradient). The former drilling was cored from 6 to 100 m (20 to 330 ft), the latter from 42 to 100 m (138 to 330 ft). Due to technical constraints, the drilling had to be kept a few centimeters away from the well casing.

Samples were taken from the undisturbed inner part of the cores and analyzed for their incrustation content as described by Houben (2006).

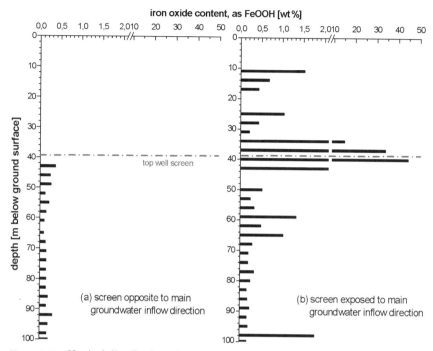

Figure 3.42 Vertical distribution of incrustation (iron oxide, as FeOOH) in drill cores from the annulus of an abandoned water well in Wildeshausen (Germany). *Modified after Houben (2006).*

Figure 3-42 shows the vertical distribution of iron oxide contents calculated from the measured concentration. From the opened drill cores it was already evident that the cores from the section opposite the natural groundwater inflow were only very slightly incrusted (Fig. 3-42a). The material was usually loose and showed only traces of ochre. Since drilling had to be kept a few centimeters away from the well tubing, the maximum incrustation thickness at this side was expected not to exceed a few centimeters. On the other hand, cores from the opposite side showed distinct red coloring. Especially in the upper sections of the filter, the pore spaces of the entire core were completely filled with ochre (Fig. 3-42b) and the gravel grains were basically "glued" together. Again, these results confirm the findings from the camera inspections and the numerical flow models (→ 2.2.3), which predict such behavior.

Apparently the Wildeshausen well contained a substantial amount of incrustation in the annulus above the well screen (Fig. 3-42b). During well completion only a small (2-m, 6.56-ft) clay seal was installed several meters above the top of the screen, while the remaining space was backfilled with coarse material. For this reason, a lot of near-surface groundwater descended to the well through the annulus and caused these deposits. Similar observations were made in eastern Germany on excavated shallow wells (Riempp 1964).

The inhomogeneous spatial distribution of well incrustations reflects the flow field superimposed by the well onto the natural groundwater flow regime. The upper screen sections and the section facing the natural groundwater flow direction receive most of the total inflow. This involves elevated flow velocities which might be responsible for increased mixing and degassing of carbon dioxide. In addition, the high amount of water stemming from the oxygenated upper parts of the water column provides a high oxidation capacity. All these factors favor the enhanced formation of mineral incrustations. The spatial distribution of incrustations in both lateral and cross-sectional view is summarized in Fig. 3-43.

The thickness of incrustations outside the screen is another point of interest. Incrustation samples recovered from screen fragments during

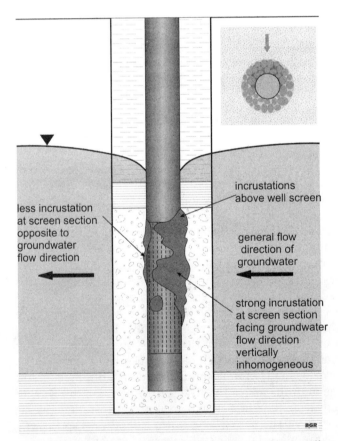

Figure 3.43 Spatial distribution of incrustations in water wells. *From Houben (2006).*

Chemical Ageing Processes 111

Figure 3.44 Incrustation (ochre + enclosed gravel) on a piece of the 2–in (1 cm) observation well recovered from the annulus during redrilling of a well. *Photograph: Andrea Weitze, BGR.*

the redrilling of wells often have thicknesses of a few centimeters (Fig. 3-44). They represent the oldest and therefore hardest incrustations, since the incrustations start to grow from the screen surface toward the gravel pack. Often they are still firmly attached to the screen (Fig. 3-45).

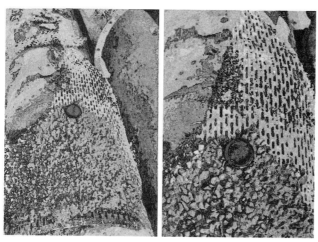

Figure 3.45 Incrustation (ochre + enclosed gravel) on a well screen fragment recovered from the annulus during redrilling of a well. *Photograph: Christoph Treskatis.*

Outer parts are younger and thus less well crystalliszed (→ 3.2.7). During redrilling, they are probably detached so the resulting sample is thinner than the overall incrustation, e.g., as shown in Fig. 3-43.

The core drilling through the annulus of an abandoned well as presented above gave another indication for the thickness of the incrustations. Since the 10-cm (2-in)-wide cores—at least at the upgradient side—were completely filled with incrusted material, more incrustations had to be expected on the outside. Since the drilling had to be kept a few centimeters away from the tubing for technical reasons, the drill core was also a few centimeters away from the well tubing. The total thickness of the incrustations is thus more than 10–15 cm (0.3–0.5 ft) (Fig. 3-46)! Thicknesses of several decimeters are probably not unusual.

The only way to finally determine the thickness of incrustations would be to excavate the well and perform a postmortem study of the annulus and its incrustation distribution. Although this sounds practicably

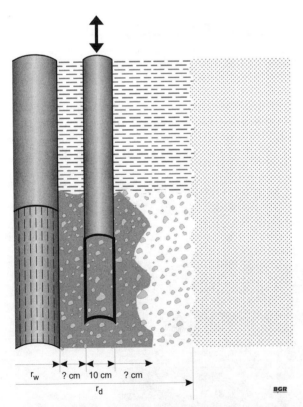

Figure 3.46 Thickness of incrusted gravel pack inferred from core drilling through the annulus of an abandoned water well in Wildeshausen, Northern Germany (r_w = radius of well, r_d = radius of drillhole).

impossible, it has actually been carried out (Etschel 2004). During the 1960s, the dewatering wells of the lignite open pit mine Wackersdorf (Bavaria, Germany) were seriously affected by well ageing and yield losses. It was therefore decided to drill several new wells applying different drilling techniques (with and without drilling fluid and additives) to test which technique would lead to the best performance. The wells were excavated after a certain period of time by the proceeding mining front, followed by a careful examination of the incrustation buildup (Fig. 3-47). It turned out that under the prevailing conditions—very high dissolved iron concentrations—all wells incrusted very quickly and the pore space of the gravel pack was completely filled by ochre. Incrustation thicknesses of several decimeters should therefore be expected in old or heavily incrusted wells.

Due to the redox zonation described in Sec. 3.2.1 and the inhomogeneous flow distribution in screens as described in Sec. 2.2.3, incrustation growth starts at the top of the screen. If incrustation proceeds to a total clogging of the upper screen parts, the incrustations will gradually grow downward (Fig. 3-48). The subsequent increase in flow resistance will of course steepen the hydraulic gradient and the drawdown. In the worst case the drawdown can reach the screen. This is usually a "kiss of death" for the well because it involves a massive influx of oxygen and a high degree of turbulence, which results in massive ochre buildup.

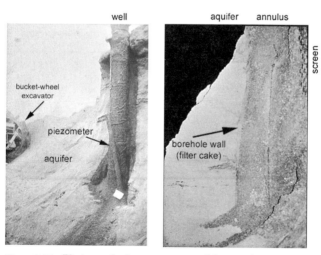

Figure 3.47 Photographs from an excavated dewatering well at the Wackersdorf lignite open-pit mine (Bavaria, Germany). The annulus is completely incrusted with ochre. Note the influence of the filter cake at the former borehole wall and the extrusion of incrusted gravel through holes in the filter cake. *Photograph: E & M Bohrgesellschaft mbH, Hof/Saale.*

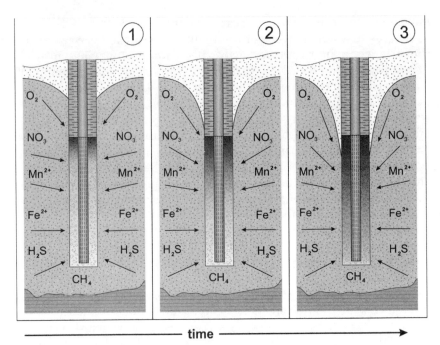

Figure 3.48 Temporal development of incrustation growth in a well.

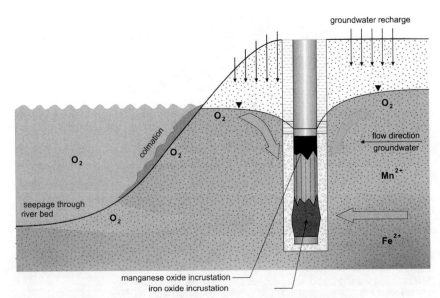

Figure 3.49 Spatial distribution of incrustations in a bank filtrate well.

The spatial distribution of incrustations in wells pumping bank filtrate can deviate significantly from the patterns depicted in Figs. 3-43 and 3-48. Here, the deeper iron-rich groundwater (from the "land side") encounters the oxic bank filtrate water at deeper screen sections. This is due to the fact that the term "bank filtration" is somewhat misleading. Most of the river water exfiltrates through the permeable river bed, which is often cleared of colmation by the prevailing higher flow velocities. The banks, on the other hand, are often clogged by plants and fines deposited by the slower water flow (Fig. 3-49). Mixing oxic bank filtrate with reduced groundwater in the deeper screen sections will of course lead to precipitation of ochre in those locations. Sometimes the distribution of redox potentials over the screen can be high enough to cause two distinct incrustation zones, one upper, manganese oxide zone and one deeper, iron oxide zone (Fig. 3-49). Camera inspections and depth-related mass balancing of chemical rehabilitations (see Fig. 5-15) performed on such wells gave proof of the pattern shown in Fig. 3-49.

The uneven spatial distribution must be considered when planning rehabilitation measures. Wasting working hours—or even chemicals—on only lightly incrusted screen sections should be avoided. The time and chemicals saved there can be applied with much more efficiency on heavily incrusted sections, e.g., through repeated treatment.

Chapter 4

Mechanical Causes of Well Ageing

4.1 Suffossion and Colmation

4.1.1 Mobilization and transport of particles in the subsurface

Flowing groundwater can transport dissolved as well as suspended matter. While the dissolved constituents will be transported wherever the water flows, suspended particle transport is a function of the particle size and of the width of the pore spaces (or pore necks) and sometimes also of electrostatic interaction between particles and matrix. Figure 4-1 gives an overview of the particle sizes of matter transported in groundwater.

Particulate matter often causes turbidity but can be removed by sedimentation. Colloids cannot be removed by sedimentation because of their small particle sizes and the resulting low settlement velocities of $<10^{-2}$ m/s (0.0328 ft/s), which effectively keeps them suspended. Colloid contents of up to $>10^6/cm^3$ are possible in surface waters. They may be colorless or they may add a certain color to the water. Colloids may comprise the following phases (or combinations thereof):

- Silicate particles: clays, highly polymerized silica
- Carbonates (e.g., $CaCO_3$)
- Humic substances: macromolecular humic acids, biological detritus
- Iron(III) and manganese(III, IV) oxides
- Aluminum hydroxides
- Sulfides and polysulfides (under anoxic redox conditions)
- Microorganisms: bacteria, viruses, fungi

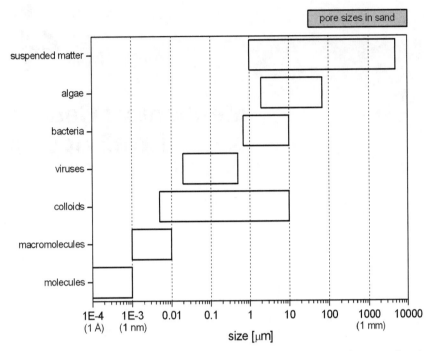

Figure 4.1 Size ranges of matter transported in groundwater compared to pore sizes in sand. *After data of multiple authors.*

Colloids are constantly created through physical and chemical processes in the aquifer, e.g., by rock weathering or erosion. Simultaneously, they are also constantly removed by attachment to surfaces, mechanical filtration, and coagulation. Therefore we assume a dynamic equilibrium between them and their surroundings. The electrostatic interaction of colloids is mainly a function of the amount of charge present in their surroundings. Coagulation to larger particles—which may settle out—is for instance promoted by waters of elevated ionic strength (= higher dissolved salt content). Because of their large surface area, many nonsilicate colloids are chemically very active, e.g., through sorption. This makes them prime carriers for the migration of trace metals, radionuclides, and microorganism in the subsurface.

The process of erosion and transport of particles in the subsurface is called suffossion (sometimes incorrectly spelled "suffusion"). As can be easily imagined, the flow velocity of the passing water has a major influence on its erosional force. It exerts both drag and lift forces (F_d, F_l), which depend on the flow velocity and the shape of the grains (Fig. 4-2).

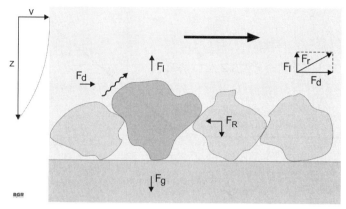

Figure 4.2 Forces acting on grains induced by flowing water (v = velocity, z = elevation above ground, F = force, d = drag, l = lift, g = gravitation, r = resulting, R = resistance).

We all know the concept of drag (coefficient) from our cars. Smooth shapes produce less drag. In order to mobilize the aquifer grains, the resulting shear forces F_r must exceed the gravitational forces F_g (weight) and the forces exerted by the surrounding grains F_R (weight, cohesion, e.g., through electrostatic attraction). Experiments by Gruesbeck and Collins (1982) showed that a critical flow velocity has to be exceeded before particles become entrained.

The velocities needed to mobilize grains from river beds was measured by Hjulström (1935) in a classic paper. He did not find a linear relationship between the flow velocity and the grain diameter of the moved particles. A linear dependency was observed only for coarser grains (silt to sand). Finer particles require much higher flow velocities for erosion than was originally expected. The explanation for this is the electrostatic attraction between fine-grained particles, especially clays. A graph strikingly similar to Hjulström's plot was developed experimentally by Muckenthaler (1989) for subsurface erosion (Fig. 4-3). Again, fine-grained particles require more energy to be mobilized, because of the adhesive forces between them.

4.1.2 Mechanical filtration processes

Basic filtration theory. The basic processes of filtration are shown in Fig. 4-4. The reader is referred to the review papers by McDowell-Boyer et al. (1986) and Ryan and Elimelech (1996) for a more detailed discussion of subsurface particle transport and retention. The two processes on the left side of Fig. 4-4 are mechanical and depend solely on the grain sizes of the particles and the pore size of the medium. Bridging is

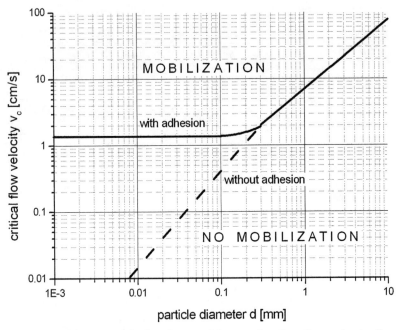

Figure 4.3 Mobilization of subsurface particles as a function of groundwater flow velocity. *Redrawn after Muckenthaler (1989).*

also influenced by the rate and velocity of flow. The process described on the right side of Fig. 4-4 includes interactions between the medium and particles caused by physical and chemical properties of the water, medium, and particles.

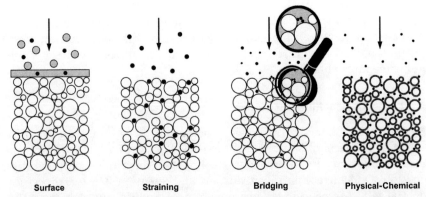

Figure 4.4 Filtration processes in porous media. *Redrawn and modified after McDowell-Boyer et al. (1986).*

The first filtration mechanism is surface or cake filtration. Cake filtration occurs when a solution containing particles with sizes larger than the pore necks is injected into a porous medium. The particles will immediately be deposited at the surface because the pores are too small for the particles to pass through. The particles will accumulate on top of the porous medium, forming a layer called the filter cake. When a thick filter cake has formed, the permeability of the porous medium is substantially decreased. The impermeable cake can then also retain smaller particles which would otherwise pass the filter.

If the average suspended particle size is smaller than the average grain (or pore neck) size of the medium, the particle suspension will enter and pass through the medium. Depending on the grain and particle size distribution, the larger particles will eventually be trapped in the smaller pore necks, because the particles are too large to pass through. This mechanism is known as straining or size exclusion. Hydrodynamic bridging is a special type of straining in which single particles are small enough to pass through a pore. Only when two or more particles arrive at a pore throat at the same instant do they form a particle bridge and block the pore. As a result, other small particles can accumulate on the bridge and pores can be fully blocked. Bridging is promoted by high flow rates, because this increases the probability of several particles appearing at a pore neck at the same time.

Particles much smaller than the pores of the porous medium can only be retained when the attracting forces between the medium and particles dominate over repulsing forces. In addition to the medium and particle diameters, these processes are strongly influenced by the physical and chemical properties of the medium, the particles, and the surrounding water (Fig. 4-5). The contact between particle and medium can be effected by

- Gravitational settling (for larger particles), described by Stokes's law
- Brownian motion (diffusion), for very small particles
- Simple collision (interception), depending on the ratio of particle to medium grain size
- Attracting electrostatic forces between the particle and medium caused by surface charges and London/van der Waals forces (dipole–dipole attraction)

Filtration theory for gravel packs. As we can easily imagine, the spacing of the pores and the grain sizes of both the moving particles and the filter medium are the main parameters defining which and how many particles will pass through a gravel pack. If too many grains pass, the well may suffer from abrasion and sedimentation (→ 4.2, 4.3). If too many are retained, the

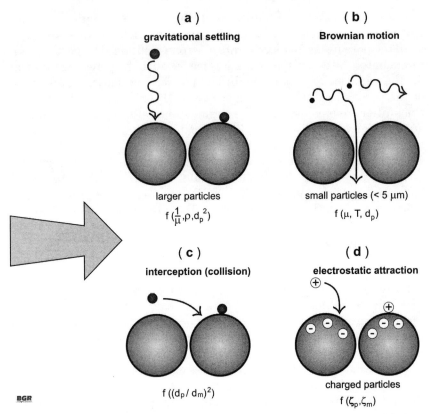

Figure 4.5 Contact processes leading to mechanical-chemical filtration of particles in a porous medium (μ = viscosity of water, ρ = density of water, T = temperature, d_p, d_m = diameter of particle, medium, ζ_p, ζ_m, = zeta potential (surface charge) of particle, medium).

pores will be blocked, causing a decrease in hydraulic conductivity. A filter must therefore fulfill two opposing requirements: it must be

- Fine enough to hold back formation particles
- Coarse enough to induce only low hydraulic head (energy) losses and still allow desanding

For this reason, the selection of the filter is always a compromise.

If we consider the pore spaces in a hypothetical gravel pack with uniform spheres from a geometric point of view, we can easily see that packing has a strong influence on filtration behavior (Fig. 4-6). Cubic packing—which we can expect only in very loose, unconsolidated sediments—implies very high porosities and large pore sizes. Filling the pores with smaller particles—as depicted in Fig. 4-6—will lead to drastic reduction of porosity, from 48%, to 12.5% and thus permeability. The

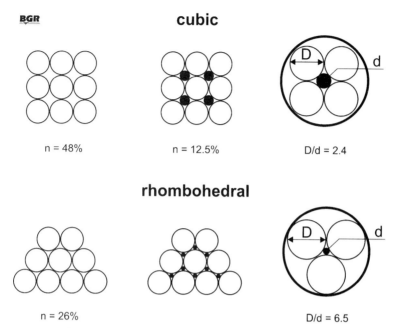

Figure 4.6 Geometric arrangement of grains and resulting effect on porosity (n) and mechanical filtration properties (n = porosity, D = diameter of granular filter, d = diameter of particle).

porosity and pore sizes of a rhombohedral packing—which we can assume to be more realistic—are much smaller. In that case, a ratio between medium and particle diameter of just below 6.5 will be sufficient to prevent particles from entering into the gravel pack.

Of course, the geometric considerations presented so far significantly differ from real-world observations. In reality, grains—and thus pores—are not spherical. Furthermore, grain sizes do not tend to be uniformly distributed. Poorly sorted sediments with a high percentage of small grains are difficult to control. The filtration of aquifer-borne particles through gravel packs was studied in detail by Saucier (1974) and by Gruesbeck and Collins (1982). Saucier found that the filtration properties can be assessed by comparing the grain size at 50% of the cumulative sieve curve (d_{50}) of the gravel pack and the grain size of the particles. Figure 4-7 shows four distinct zones. In the first zone (left) the gravel pack is too fine and will retain all particles, e.g., as filter cake. Energy will be wasted because of decreased permeability. If the ratio is higher than 6, permeability will be reduced because particles now enter the gravel pack and become wedged in the pores (Fig. 4-8). This type of filter plugging is called "inner colmation" for gravel packs. At ratios higher than approximately 12, the sand grains will start to pass through

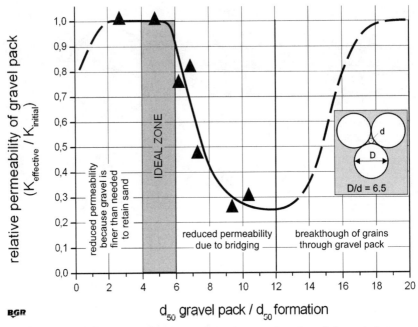

Figure 4.7 Permeability and particle mobilization as a function of the ratio between grain size of gravel pack and formation. *Modified after data by Saucier (1974).*

the filter and cause sand intake (breakthrough). Only the narrow field at ratios between 4 and 6 represents a good compromise between excessive permeability reduction and the ingression of particles. It is noteworthy that these values are still close to the factors calculated from the

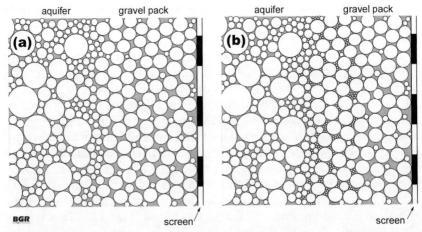

Figure 4.8 (a) Initial state of a gravel pack and (b) ingression of particles leading to plugging (inner colmation).

simplified and purely geometric considerations presented above. The common methods for determining the grain size of gravel packs are all based on grain size distribution curves and all give similar ranges of ratios between aquifer and gravel pack granulometry (DVGW W 113; U.S. Environmental Protection Agency EPA-570/9-75-001; U.S. Army Corps of Engineers TI 814-01).

4.1.3 Filtration as a cause of mechanical well clogging

Even when the gravel pack is ideally designed, mechanical well clogging can still occur. As explained above, mechanical clogging is related to particle migration and retention in the medium. As a result of pumping-induced strongly increased flow velocities, particles will be mobilized and transported toward the well. The closer the water gets to the well, the higher velocities it will have, and therefore more particles will be mobilized. Close to the well is a region within the aquifer where large velocities and more particles are present. This may result in a blocking of pores near the borehole wall (Fig. 4-9). This blocking process consists of two stages: the first occurs while drilling the well and the second stage occurs after completion, when the well starts producing.

Most wells are drilled using a drilling mud. This mud consists of very fine-grained swelling clay (bentonite) or organic material (e.g., carboxymethylcellulose, CMC). Even when plain water is used instead of mud, clay, and silt particles from the aquifer will be suspended in the drilling fluid. Although many drillers will probably disagree, in many cases remains of the drilling fluid (respectively, the filter cake) will be left attached to the borehole wall (Fig. 4-10a). This skin will have thicknesses of few millimeters to a few centimeters. Because of its very low permeability, it constitutes an efficient flow barrier and can strongly reduce hydraulic conductivities around the well (van Everdingen 1953; Hurst 1953; Hurst et al. 1969). Permeability decreases to less than 10% of the original value have been observed (Howsam & Hollamby 1990). During drilling, this permeability decrease is a positive effect because it keeps circulation losses low. If the (bentonite) cake is not removed during well completion and development, it causes high entrance losses and a low well yield. In addition, fines transported from the aquifer matrix will not be able to pass. Their accumulation will decrease the permeability even further (Fig. 4-10b). This process is sometimes called "outer colmation," to distinguish it from the somewhat rarer "inner colmation," which affects only the gravel pack. Possible remedies will be discussed in Sec. 8.2.7. Outer colmation is a classical example of surface filtration (Fig. 4-4). Skin layers with higher permeability than the aquifer may also occur, e.g., when flow-impeding material around the borehole wall is removed, e.g., by desanding or acidization.

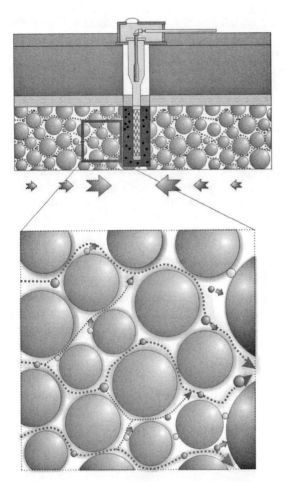

Figure 4.9 Schematic representation of water movement in an unconsolidated aquifer. Close to the well, high velocities cause more particles to be mobilized toward the well. Near the borehole wall, the pores are blocked by particles (indicated by the box). *Figure courtesy of Bert-Rik de Zwart, KIWA, The Netherlands).*

Remains of filter cakes have often been found in the field. One example is shown in Fig. 4-11, which clearly shows incomplete removal of the drilling mud after well completion.

Remains of drilling mud have also been found in downhole samples from water supply wells at four different locations in the Netherlands (Timmer et al. 2000, 2003). Figure 4-12 shows a microscopic thin section along the borehole wall of a clogged well. Two different types of particle accumulations can be distinguished. The two types differ in color, composition,

Mechanical Causes of Well Ageing   127

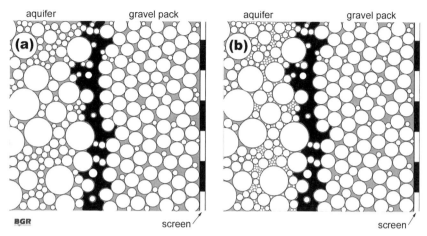

Figure 4.10 (a) Low-permeability remains of drilling fluid (black) at the former borehole wall and (b) the collection of fine particles behind this interface (outer colmation).

layer orientation, and size. Morphological considerations revealed that the deposit close to the gravel pack in Fig 4-12 was deposited first and infiltrated from top to bottom into the formation. This result is associated with the drilling mud, in which both chemical and natural additives were used. After the deposition of this layer, fine-grained particles from the aquifer accumulated on top of the remains of the drilling mud. This image thus illustrates the effect of "outer colmation."

Figure 4.11 Annulus of a well excavated at the Wackersdorf lignite open-pit mine (Bavaria, Germany). The mud cake at the former borehole wall is still visible in the form of a thin clayey layer around the gravel pack, although no bentonite or CMC was added during drilling. The gravel pack is heavily incrusted by iron oxides
Photograph: E & M Bohr Gesellschaft mbH, Hof / Saale, Germany.

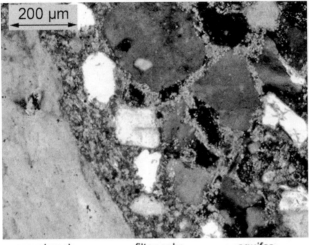

Figure 4.12 Thin section from a downhole sample through the annulus of a clogged well from the Ritskebos well field in Noord Bergum (Friesland, Netherlands). Visible are a filter cake (at the interface between gravel pack and aquifer) and clogged aquifer pores to the right. The large grain of light color at the left edge is a quartz grain from the gravel pack. The smaller grains to the right are aquifer material. *Photograph courtesy of Toine Jongmans, Wageningen University, The Netherlands).*

Experience from the Netherlands has shown that water supply wells will clog even after complete removal of the drilling mud. In such wells, pores are usually blocked by particles in a radius of 5–10 cm (2–4 in) behind the borehole wall. The sizes of the involved particles range between 1 and 20 µm, of which the smaller particles are most abundant. The particles consist of organic material, limestone fragments, and clay. They originate directly from the aquifer, confirming the mobilization of particles by erosive force due to the high velocities induced by pumping.

The main process for pore blocking in the above-described case is bridging. This process occurs under specific subsurface conditions. Bridging may cause plugging even when the size of the individual particles is sufficiently small to pass the filter—as opposed to straining, in which a single particle cannot pass through. Bridging occurs when several particles arrive at the bottleneck of a pore at the same time. They are wedged together and block the pore throat while the rest of the pore remains clear (Fig. 4-13).

The formation of particle bridges can be compared to traffic near and on a roundabout during rush hour. At these crowded times, the roundabout will congest due to all the traffic. While approaching the roundabout, traffic moves more slowly because entering cars have to yield to

Mechanical Causes of Well Ageing   129

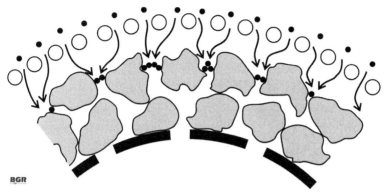

Figure 4.13 Bridging and straining of particles (black) during passage through a gravel pack.

the traffic already on the roundabout. This congestion also causes a traffic jam in the streets approaching the roundabout. It is the same with particles approaching the borehole wall (Fig. 4-13). Bridging is promoted by high flow velocities, which increase the probability of several particles entering the pore throat at the same time. Bridging depends not only on grain and pore sizes of the medium and the particles but also on their concentration per unit of time. Intense pumping, which is a common technique in desanding and rehabilitation, can therefore worsen the situation if the flow direction is not reversed from time to time to destroy the bridges (Fig. 4-14).

In order to validate bridge formation around wells, downhole samples have been studied in the Netherlands. Figure 4-15 shows a thin section of a sample taken at 5 cm inside the aquifer. It can be clearly seen that the accumulation of particles starts as a bridge in the smallest pore throats.

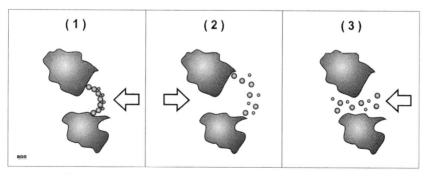

Figure 4.14 Buildup (1), destruction (2), and removal (3) of bridging by reversal of flow direction.

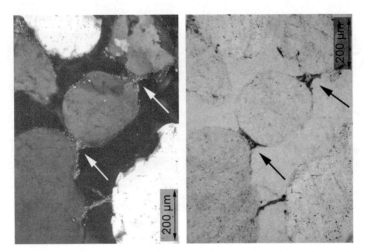

Figure 4.15 Thin-section photograph of bridging of pore throats by particles (arrows). Sample collected from the formation of a water supply well from the Ritskebos well field in Noord Bergum (Friesland, The Netherlands) by sideway core drilling from the well interior. Scale: 0.6 × 0.8 mm. *Photograph courtesy of Toine Jongmans, Wageningen University, The Netherlands.*

In order to investigate further the most important factors in bridge formation, laboratory experiments have been carried out at the Technical University of Delft (The Netherlands). The most important parameters for mechanical clogging are

- Concentration of particles
- Flow velocity
- Ratio between the size of the particles and aquifer grains (aspect ratio)
- Particle properties; shape, surface charge, and roughness
- Flow geometry; radial or linear flow

Measurements in the Dutch wellfield Noord Bergum indicate that there is a strong correlation between the flow rate and the number of mobilized particles (Prins 2004). High particle concentrations are an indicator of high clogging rates. For lower flow rates, fewer particles are mobilized, so clogging rates are lower (Fig. 4-16). Wells with high particle concentrations (hundreds per milliliter) have a higher probability of clogging than those with very low concentrations (<15 particles per milliliter). It is interesting to note that the flow velocities at the borehole wall in Fig. 4-16 are exactly in the recommended range (→ 2.3).

At the same well in Noord Bergum, water-level recordings showed an evident relation between the operating intervals of the submersible

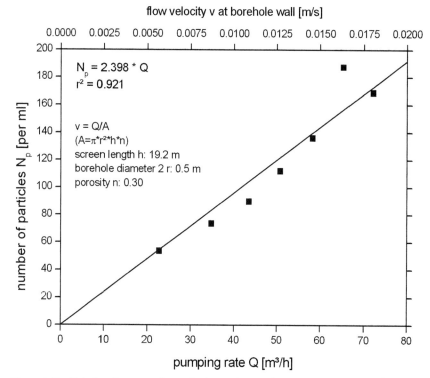

Figure 4.16 Relation between flow velocity and particle concentration of Well 50A in Noord Bergum, The Netherlands. N_p = number of particles, r = correlation coefficient, Q = flow rate, A = area of borehole wall cylinder, n = porosity. *After data by courtesy of Ate Oosterhof, Vitens Friesland, The Netherlands.*

pump and entrance losses (Fig. 4-17). Water-level recorders measured the water level in the well. The difference between the static and dynamic (operating) water levels allowed the calculation of the degree of well deterioration. During times when the pump was turned on and off frequently, the well showed hardly any decline in yield (Fig. 4-17, phase 2). When the pump was operated less frequently (Fig. 4-17, phases 1 and 3), the well started to clog. This behavior was also found in other wells in the southern part of the Netherlands (Limburg province). Probably, when the pump is turned on, the shockwave—in combination with the acceleration of the water—breaks some of the bridges and removes particles from the borehole wall. Column experiments by Gruesbeck and Collins (1982) point in the same direction: changes in flow rate caused short bursts of fines in the effluent. This mobilization was probably caused by the formation of turbulent vortices which increased the viscous forces acting on fines and enhanced their entrainment. The turbulence decayed quickly as a result of the fluid viscosity.

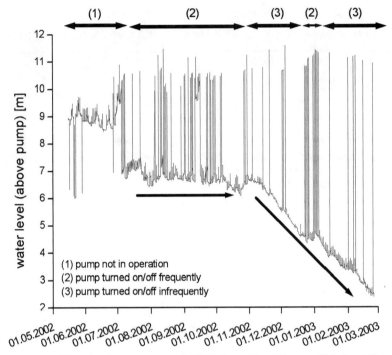

Figure 4.17 Water level measurements in Well 50A at Noord Bergum (Friesland, The Netherlands). The difference in water level is an indicator of well capacity. The arrows show that the drawdown remains constant during frequent pump operation and that well capacity decreases significantly when pump operation is less frequent. *After data by courtesy of Ate Oosterhof, Vitens Friesland, The Netherlands.*

From a practical point of view, both flow rate and frequency of pump operation can be used to control mechanical clogging. However, thoughtful design for well operation is required.

4.2 Abrasion

Abrasion can be defined as the wearing off of material from the surface of a solid by the mechanical action of moving solid particles suspended in a fluid. In our case, pumps, tubing, and screen are the fixed solid components which are acted on by solid particles suspended in flowing groundwater. The passing particles exert a force on the surface of the well material. By definition,

$$\text{Force} = \text{mass} \times \text{acceleration}$$

Therefore dense particles at high flow velocities exert large forces on surfaces they encounter. The following properties of the abrading particle are therefore of influence:

- Mass
- Velocity (acceleration)
- Shape (sharp pointy edges will incur more damage than well-rounded grains because the force is transmitted through a small area)
- Number of particles per unit of time ("constant dripping wears away the stone")
- Hardness contrast between particle and surface

In most cases, sand grains—commonly almost entirely quartz (SiO_2)—are the main cause of abrasion. Because of its high hardness (7 on the Mohs scale), it can cause damage to all materials of lower hardness. While steels have varying hardnesses between 5.0 and 8.5, copper—which has sometimes been used in well construction—with a hardness of 2.5–3.0, is much more prone to abrasion damage. Other materials of low hardness are stoneware, polyvinylchloride (PVC), and wood-based casings and screens. Abrasion can also damage corrosion-inhibiting passive layers or coatings (e.g., polyethylene on steel). This allows easier access for corrosive agents and thus accelerates corrosion significantly (Fig. 4-18). Screen slots and screen slot bridges are especially prone to abrasion damage (Fig. 4-19). Widened slots or broken bridges may allow the intrusion of gravel pack material, which potentially endangers the pump and the structural integrity of the whole well.

Nevertheless there is at least one good thing to say about abrasion: it prevents the attachment of incrustations and biofilms.

A special type of mechanical damage can occur during well maintenance. While tripping in or out of the pump and riser pipe—or rehabilitation equipment or geophysical probes—it must be ensured that nothing slams into the well wall or is dragged along it. Damaged coatings, holes, and leaks may be the result of careless handling. Careful operation and the use of flexible spacers and centralizers help to minimize the problem.

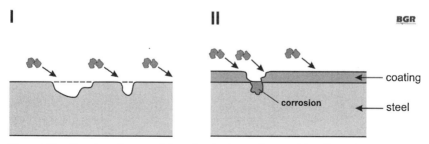

Figure 4.18 Abrasion of a material surface by impacting particles (I), and abrasion of a coated surface and its implications (II).

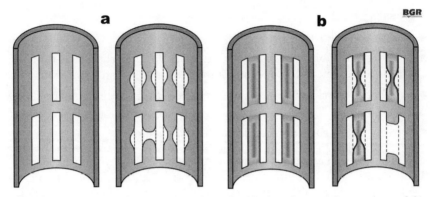

Figure 4.19 Abrasion effects on screen slots: (a) widening of screen slot opening and (b) destruction of screen slot bridges.

4.3 Sand Intake

Particles passing the interface between aquifer and gravel pack will plug the annulus (→ 4.1.3) or enter the well (sand intake). There they can either drop out of suspension and form a bottom sediment in the well sump or enter the pump and subsequently the pipelines. Wells with elevated particle content suffer from various symptoms (Fig. 4-20):

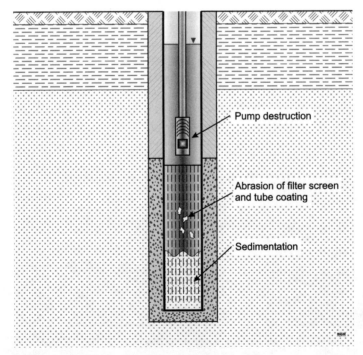

Figure 4.20 Problems caused by sand intake in wells.

- Abrasion of screen slots and screen slot bridges (Fig. 4-19)
- Destruction of protective surface layers (Fig. 4-18)
- Increased wear of the pump impellers
- Filling of well sump and screen, potentially leading to a decrease in well yield
- Ground settlement and resulting stresses on the well if substantial amounts of grains are removed from the aquifer

Causes of elevated sand intake include:

- Inappropriate dimensioning of gravel pack granulometry and screen slot size
- Partial blockage of screen slots (e.g., by incrustations), leading to elevated flow velocity in the remaining open slots
- Gravel pack designed too thin (Fig. 4-21a)
- Tubing not centered in the borehole (Fig. 4-21b)
- Detachment of sand pockets from the borehole wall during insertion of gravel pack (Figs. 4-21c, 4-22)
- Small fine sand or silt layers not detected during exploration and drilling (Fig. 4-23)
- Insufficient well development (desanding)
- Pumping rate too high (high flow velocity)
- Leaky casing seals
- Corrosion holes
- Widening of screen slots and destruction of screen slot bridges caused by abrasion and/or corrosion (Figs. 4-18, 4-19)

The first four causes can be avoided from the beginning by proper planning, site investigation, and well development (→ 2); the last two

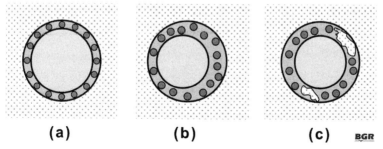

Figure 4.21 Some common causes of sand intake in wells. (a) gravel pack too thin; (b) screen not centered in the borehole (gravel pack too thin at one side); (c) sand pockets from aquifer slumped into gravel pack during insertion.

aquifer annulus screen sand pocket aquifer

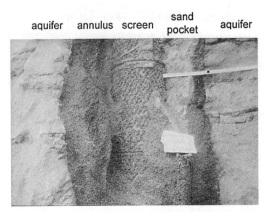

Figure 4.22 Sand pocket slumped into the gravel pack during well completion. Excavated well of the Wackersdorf (Bavaria, Germany) lignite open-pit mine. *Photograph: E & M Bohrgesellschaft mbH, Hof/Saale, Germany.*

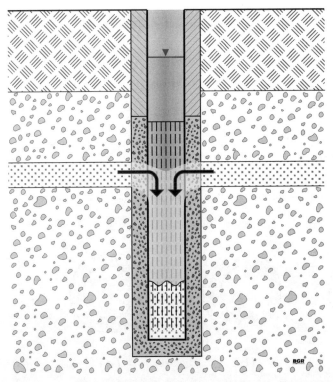

Figure 4.23 Sand intake into a well from a small, fine sand layer not detected during exploration.

Mechanical Causes of Well Ageing

causes will require reconstruction measures (→ 9). Literature surveys show that most authors agree that inappropriate dimensioning of gravel pack and screen slots as well as insufficient well development are the most common causes (e.g., Petersen et al. 1955) of increased sand intake. Of course, poorly sorted sediments are much more prone to sand intake than are well-sorted sediments. Sand intake is the main ageing process in 7% of the ageing wells in Germany (→ 1).

Sand intake is common in both unconsolidated (sand, gravel) and (semi-)consolidated (sandstone, dolomitic limestone) aquifers. Table 4-1 gives an overview of causes and effects of sand intake.

TABLE 4.1 Causes and Effects of Sand Intake in Water Wells

Cause	Phenomena and effects
Gravel pack granulometry not adapted to aquifer (also for multiple and preglued gravel packs)	*Sand intake* (gravel pack too coarse) *Colmation* of former borehole wall *Sedimentation* in well sump
Insufficient thickness of gravel pack	*Sand intake* *Sedimentation* in well sump
Faulty well completion (gravel pack not present around full screen circumference or does not cover full screen length)	*Sand intake* (due to suffossion) *Sedimentation* in well sump
Screen slot geometry or screen type not adapted to aquifer granulometry; faulty gravel pack granulometry	*Sand intake* *Colmation* of gravel pack (fines cannot pass slots) *Sedimentation* in well sump
Lacking or insufficient desanding	*Sand intake and/or turbidity* *Colmation* of gravel pack (fines cannot pass slots) *Colmation* of former borehole wall (if gravel pack is too fine) *Sedimentation* in well sump
Excessive pumping and increased entrance velocities (→ 2.2)	*Sand intake and/or turbidity* *Sedimentation* in well sump
Excessive pumping and increased entrance velocities during start-up of the pump	*Sand intake and/or turbidity* *Sedimentation* in well sump
Layers of silt, fine sand, or loose deposits in cleavages, etc., not detected and not sealed off during exploration and well drilling	*Sand intake and/or turbidity* *Sedimentation* in well sump
(Inappropriate) rehabilitation techniques	*Settling* of the annulus *Sand intake and/or turbidity*, e.g., through destruction of filter gauze, preglued gravel pack, multiple gravel packs

After DVGW W 130.

Particle mobilization and turbidity can occur over long periods of time, sometimes even without any measurable decrease in well yield. This is common in fine-grained consolidated aquifers such as the only partially consolidated Bunter Sandstone in Europe. The particles stem from fine deposits present in cleavages or from the eroding aquifer matrix. High pumping rates and entrance velocities definitely aggravate the problem. Figure 4-24 shows turbidity and pumping rates as a function of time of a well screened in the karstic limestone of the Attendorn syncline (Sauerland region, Germany). In this case, a reduction of the pumping rate and thus the friction force of the groundwater flow was sufficient to lower the turbidity to values in accordance with drinking water regulations.

Sand intake due to insufficient or not executed well development (desanding) can be eliminated by retroactive desanding—if an adequately dimensioned gravel pack was installed (Fig. 4-25). Desanding aims at the removal of fines from the gravel pack and the near-field of the aquifer to enhance flow and minimize energy losses. Depth-specific techniques (packered) with repeated reversal of flow direction should be preferred (→ 4.1). Monitoring the amount of removed material during operation is essential.

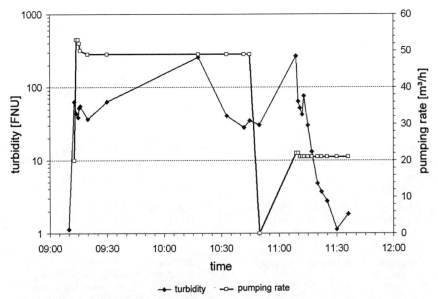

Figure 4.24 Turbidity and pumping rate of a well screened in karstic limestone of the Attendorn syncline (Sauerland region, Germany).

Mechanical Causes of Well Ageing

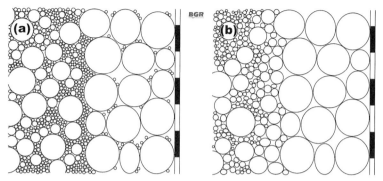

Figure 4.25 Texture of the near-field of a well (a) after drilling and (b) after desanding.

Countermeasures have to be assessed and selected for each individual case:

- Desanding at regular intervals (with flow reversal)
- Reconstruction of the gravel pack (→ 9)
- Reconstruction of the screen and casing (→ 9)
- Installation of a suction flow control device (→ 11)
- Decrease in the pumping rate

4.4 Damage by Plant Roots

A rare type of mechanical well ageing is the ingrowth of plant roots, especially from trees with deep and strong roots, into the screens of shallow wells (Fig. 4-26). Roots reaching the annulus can distort its texture and can cause serious damage to screen and screen slots, at least in wells with weaker material (PVC, wood, stoneware). We all know that roots—even those of small plants—produce enough pressure to crack pavement easily. A PVC well screen is therefore also no obstacle for a growing root.

Schneider (1983) describes a case in which a shallow well with a stoneware screen was affected by the roots of a neighboring (10-m, 33-ft) poplar tree. The roots were cut from the well interior, but this yielded only short-lived improvement. Even cutting down the whole tree proved to be insufficient. Finally, the whole tree stem had to be dug out. The authors cannot give general recommendations on how far from a well trees should be allowed. At least 10 m (33 ft) is usually a good estimate for regular central European conditions, but elsewhere in the world there may be trees with deeper rooting depths.

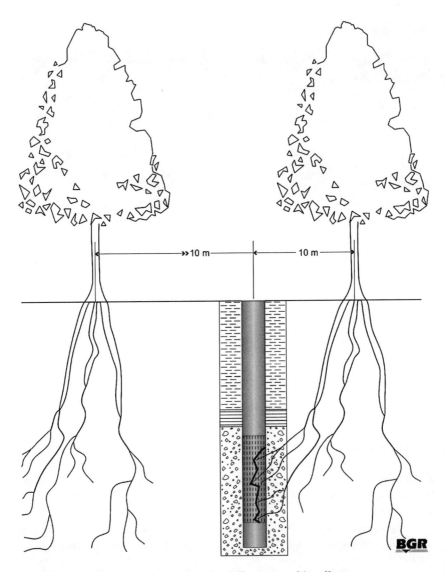

Figure 4.26 Ingrowth of tree roots into the well screen and its effects.

4.5 Ground Movement

4.5.1 Soil settling and subsidence

Vertical movement (settling) of constructions may occur due to their own weight or an imposed working load. In general, damage is not caused by the settling itself, but rather by differential settling of different components. Loads can be both static and dynamic. The latter comprise, e.g., the agitations or impulses caused by machinery or flowing water.

The removal of aquifer and gravel pack grains through flowing water can change the texture and thereby the load distribution.

Drawdown and recovery cycles of the water table around an operating well alter the distribution of pressure and loads acting on a well. Lowering the water table always implies a decrease of the buoyant force of the groundwater. Especially fine-grained layers (silt, clay, peat) are prone to react to this by compaction. In a well screened in hardly compressible gravel overlain by a clay layer, such compaction causes strong differential stresses: while the screen remains fixed, the casing in the clay layer is exposed to a strong, vertically directed load. This can lead to buckling of the casing and finally to fracture and collapse of the well. While the regular operation of wells hardly ever causes such problems, dewatering and rising water tables on a regional scale, e.g., in mining districts or bank infiltration zones, can be a problem.

Mining poses another source of ground subsidence through the collapse of abandoned underground excavations. The authors are aware of at least two cases in which wells collapsed due to mining-related land subsidence. Well owners who want to claim compensation are advised to install vertical ground motion detectors. Otherwise the legal argument can be difficult. Wells in areas with known land subsidence should be equipped with ductile and sturdy screen material (e.g., steel).

Well yield can be seriously affected by settling of the annular filling material and surrounding aquifer strata. In the following, we will therefore focus on the near-field of wells. Materials that can be affected by settling include the annular fillings (gravel pack, filter sand, clay seals, and cements). According to DIN 4019, we can distinguish the following settling types for the annuli of wells:

- *Initial settling* based on elastic, reversible deformation of the medium, instantaneous at load changes (only in wells with plastic fill material, e.g., clay)
- *Primary settling* or compression settlement (consolidation), through expulsion of pore water (Fig. 4-27) and subsequent compression of the granular matrix (e.g., during insertion of the annular fill material)
- *Secondary settling* or creep settling, based on slow rearrangement of the texture

Settling of annular fill material can be calculated according to DIN 4019 using Eq. (4.1). The load distribution over the irregular-shaped borehole wall is included in the settling coefficient. The stress distribution can be assessed according to Fig. 4-28 for each horizontal layer of the fill material:

$$s = \frac{\sigma_0 \cdot b \cdot f}{E_m} \qquad (4.1)$$

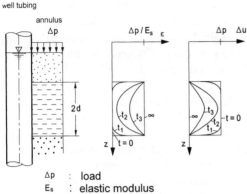

Δp	:	load
E_s	:	elastic modulus
ε	:	compression
Δu	:	pore water over pressure
2d	:	thickness of individual clay layer
t	:	time increment
z	:	depth

Figure 4.27 Consolidation of the annulus through expulsion of pore water from clayey sealing material induced by settling of the underlying gravel pack. *Modified after Zilch et al. (2002).*

where s = settling, L

σ_0 = mean soil pressure at the base of the gravel pack, F/L^2

b = length of the base area of the fill, L

f = settling coefficient according to DIN 4019, part 1 [contains integration of stresses over depth (Fig. 4-28)], *dimensionless*

E_M = mean compressibility of the fill material, F/L^2

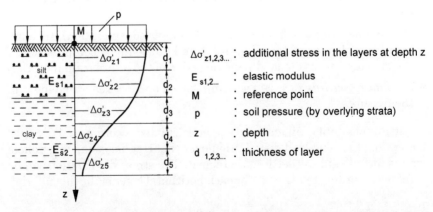

Figure 4.28 Stress distribution in clayey annular filling material. *Modified after Zilch et al. (2002).*

The transfer of stresses between grains is the main process in the deformation of the gravel pack. Compressibility of unconsolidated rocks can be measured by compression tests. Primary settling is influenced mainly by the load imposed by the overlying fill, the type of insertion (e.g., grouting pipes), and the shape of the borehole wall. Irregular backfilling (bridging), excentric positioning of the tubing, and insufficient well development can cause differential settling in the annulus.

The settling of the gravel pack can induce shear stresses on the overlying annular seal. This may lead to the development of microfractures and thus secondary pathways which may compromise sealing function.

Geophysical borehole investigations (→ 5.3) are useful tools to measure the bulk density and the hydraulic integrity of annular seals and gravel packs. Packer flow meters and neutron–neutron measurements are among the most useful methods.

Settling of gravel packs and the subsequent decrease in permeability (and well yield) have been detected in the field after

- Desanding
- Rehabilitation (Berger 1997)
- Unstable gravel packs (Herth & Arndts 1994)

Methods for quantification include the recording of flow rates and their spatial distribution over the screen (flow meter) and the pressure head differences between pumping and observation well in the annulus, measured before and after application.

Desanding operations with low packer distances and high pumping rates can significantly modify the gravel pack texture by removing large amounts of fine grains. This can cause settling of the remaining grains and thus higher compactness.

Although many contractors will deny this, on rare occasions rehabilitation has been known to decrease well yield rather than increase it. Berger (1997) describes such a case, where geophysical logging showed higher entrance losses and lower permeabilities after a mechanical rehabilitation (impulse method) (Table 4-2). Settling and compaction of the gravel pack and insufficient removal of incrustations were the causes.

Practical experience has shown that applying the following recommendations will minimize gravel pack settlement:

- Employment of pipes and pump for gravel pack insertion. This reduces the height of fall and possible breakup of colliding grains. In addition, this will prevent damage to the casing coating. Quick insertion also avoids a gradation of the gravel during fall.
- Application of surge blocks in the screen during insertion, to enhance primary settling

TABLE 4.2 Head Differences Between the Well and an Observation Well Installed Inside the Annulus Due to Settling and Compaction after the Application of a Mechanical Rehabilitation Method

	Pumping rate		
Head difference [m]	5 [m^3/h]	10 [m^3/h]	15 [m^3/h]
Before rehabilitation [m]	0.72	2.23	5.47
After mechanical rehabilitation [m]	0.79	2.24	5.22

After data by Berger (1997).

- Installation of protective casing ("lost casing") in easily erodable formations
- Constant volume control of inserted goods and sounding of actual filling height at several locations in the annulus
- Reduction of fluid density by exchanging drilling fluid with water prior to insertion of the gravel pack (this enhances gravel pack insertion and facilitates even distribution)

4.5.2 Earthquakes

Earthquakes cause differential acceleration and possibly resonance frequencies in soils and buildings (→ 7.5.1), leading to possible damage. Wells are no exception to this rule. Deep wells equipped with low-strength or brittle casing or screen material are especially at risk. The authors know of at least one case in the Lower Rhine basin where the 1995 Roermond earthquake (magnitude ~5.9) was responsible for shearing off the casing of a deep well. This led to a full collapse of the well and the subsequent burial of the pump. The well was located close to a tectonically active fault zone (Rur rim fault). Wells in mining districts and earthquake-prone regions should therefore be equipped with sturdier casing and screen material.

4.6 Vandalism

Vandalism is a serious problem that affects most technical outdoor installations, especially in remote areas. Analogous to Murphy's law, we could write for wells and observations wells:

- If something sticks out from the ground surface, they will try to break it.
- If they see an open hole in the ground (or one that can be opened easily), they will throw things into it (Fig. 4-29).
- If they find a piece of wall, they will spraypaint something onto it.

Mechanical Causes of Well Ageing 145

Figure 4.29 Example of vandalism: rocks thrown into an observation well. *From Nolte et al. (2004).*

Preventive measures include:

- Make everything as sturdy and unbreakable as possible.
- Seal and secure everything tightly.
- Make the cap housing of observation wells level with the ground surface.
- Fence off the well site.
- Install a (fake) camera.
- Put up a "high voltage" sign (this is not really legal but. . .).

4.7 Ageing Processes in Infiltration Wells

Infiltration wells are used to discard excess water and to counteract negative effects of dewatering schemes, e.g., for groundwater-dependent biosystems, and salt water intrusion (Bouwer 2002). In recent years, infiltration wells have received renewed attention as part of aquifer storage and recovery (ASR) schemes (Pyne 1995). They suffer from a variety of ageing processes. A multitude of papers (e.g., Rebhun & Schwarz 1968; Vecchioli 1970; Ehrlich et al. 1973; Oberdorfer & Peterson 1985; Bichara 1986; van Beek et al. 1998; Rinck-Pfeiffer et al. 2000; 2002; Moorman et al. 2002; Buik & Willemsen 2002) and monographs (Olsthoorn 1982; Huisman & Olsthoorn 1983) have been devoted to the topic. Ageing processes include:

- Biofouling (→ 3.6)
- Mineral incrustation (→ 3.2)

- Mechanical plugging (→ 4.1)
- Entrapment of air bubbles
- Combinations of the above processes

Although the first three processes listed above also occur in pumping wells and are also discussed in detail elsewhere in this book (→ 3, 4.1), entrapment of air bubbles is unique to infiltration wells. Gas bubbles may occur under the following circumstances:

- Degassing of dissolved air in pipes or pump when pressure falls below atmospheric pressure
- Entrainment of air bubbles during free fall of water
- Mixture of warm and cold water; solubility of air in water decreases with increasing temperature
- Microbial mineralization of biomass to, e.g., nitrogen gas in the annulus or the aquifer

At flow velocities of the infiltration water greater than 0.3–0.4 m/s (1.0–1.3 ft/s), water carries air bubbles downwards. Air entrapment can be identified when the water inside the well "foams" after operation. Countermeasures include:

- Positioning the end of the supply pipe below static water level to prevent entrapment of air by falling water
- Installing an orifice at the end of the supply pipe to increase pressure in the pipe

Biofouling is one of the most common phenomena in injection wells, especially when the injected water contains dissolved and/or suspended organics. Water repumped from injection wells often shows much higher bacterial counts than the injected water (Rebhun & Schwarz 1968; Vecchioli 1972). This clearly demonstrates the growth of bacterial colonies in the near-field of the well using the nutrients of the injection water. Massive slimes often develop (Ehrlich et al. 1973). Bioclogging is usually more widespread in space than mechanical clogging and is accentuated by slow flow rates. The polysaccharide content of the injection water can be used as a prediction tool: concentrations of less than 7 mg/liter already lead to severe clogging (Rinck-Pfeifer et al. 2000, 2002). On the other hand, bacteria can break up mechanical clogging caused by organic matter through mineralization. Oberdorfer & Peterson (1985) noticed that their wells clogged at first but improved after some time after the establishment of heterotrophic bacteria. Acid generated during the mineralization may also improve the permeability by dissolving calcite (Oberdorfer & Peterson 1985; Rinck-Pfeifer et al. 2000, 2002). Gas bubbles generated during the breakup of biomass may counteract these permeability increases.

Huisman & Olsthoorn (1983) found that mechanical colmation of infiltration wells can occur in the gravel pack, at the borehole wall, and in the surrounding formation. The additional head loss caused by the buildup of a filter cake is a function of the infiltration rate and the particle content of the infiltrated water (Huisman & Olsthoorn 1983):

$$s_c c = \frac{k - k'}{k \cdot k' \cdot (n - n') \cdot \rho d} \cdot \frac{Q_0 \cdot c \cdot t}{4 \cdot \pi^2 \cdot r_0^2 \cdot m^2} \quad (4.2)$$

where s_c = additional head loss, L
k = permeability of aquifer, L/T
k' = permeability of filter cake, L/T
$n - n'$ = change in porosity through colmation,
ρ_d = density of filter cake, M/L^3
Q_0 = infiltration rate, L^3/T
c = gravimetric particle content of water, M/L^3
t = time, T
r_0 = radius of well, L
m = thickness of aquifer, L

Assuming that the first fraction of Eq. (4.2) is a constant α and with the entrance velocity $v_e = Q_0/2 \cdot \pi \cdot r_0 \cdot m$, we can simplify the equation to

$$s_c = \alpha \cdot v_e^2 \cdot c \cdot t \quad (4.3)$$

A common measure to assess the clogging potential of infiltration wells is the membrane filter index (MFI) (Olsthoorn 1982) [Eq. (4.4)]. The MFI is determined experimentally in the laboratory and describes the slope of the graph of the inverse of the flow rate against the amount of water passing a 0.45-μm filter under constant pressure and standard conditions.

$$MFI = \frac{\mu_d}{2 \cdot p \cdot A_f^2} \cdot \frac{c}{k'} \quad (4.4)$$

where MFI = membrane filter index
μ_d = dynamic viscosity of water, $M \cdot L^{-1} \cdot T^{-1}$
p = pressure loss, $M \cdot L^{-1} \cdot T^{-2}$

A_f = area of filter, L^2
c = gravimetric particle content of water, $M \cdot L^{-3}$
k' = permeability of filter cake, $L \cdot T^{-1}$

Clogging rates calculated from MFI values are often overestimated, probably because the filtration through a 0.45-μm filter cannot be compared to filtration in real-world pores. Therefore, Buik & Willemsen (2002) proposed a correction term for the MFI for natural aquifers.

The development of additional head losses in infiltration wells will proceed according to the underlying processes (Fig. 4-30). Head losses due purely to air entrapment will be constant when the equilibrium of gas solubility in water is reached. Biological clogging will also become constant because the nutrient supply is limited. Only the mechanical clogging by suspended matter could—at least in theory—increase linearly, due to a continuous buildup of the filter cake (Huisman & Olsthoorn 1983; Pyne 1995).

Bichara (1986) summarizes the parameters that control the speed of clogging for infiltration wells as follows:

- Type of suspended particles
- Size of suspended particles
- Shape of suspended particles

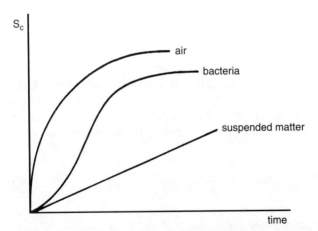

Figure 4.30 Development of additional head loss s_c in infiltration wells due to different ageing processes. *After Huisman & Olsthoorn (1983).*

- Concentration of suspended particles
- Chemical compatibility of injected water and groundwater
- Recharge rate
- Temperature contrast between injected water and groundwater

According to Bichara (1986), the development of the injection flow rate is a function of the injected particle mass (particle concentration multiplied by the recharge rate) divided by the open area of the sand face (borehole wall) (Fig. 4-31). Bichara found that wells with gravel packs can take 10–15 times higher amounts of suspended solids before they clog than can wells without a gravel pack. Doubling the gravel-pack thickness also doubles the time before the well clogs.

Delineation of indicator values to distinguish clogging from non-clogging injection wells is a difficult task. The concentrations of suspended solids, turbidity, and organic carbon are probably the best

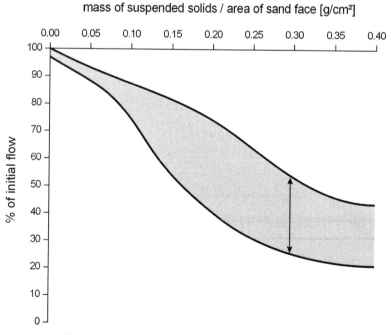

Figure 4.31 Flow rate of an injection well as a function of the ratio of the injected particle mass to the open screen area. The shaded field and the arrow indicate the scattering of experimental data obtained using different boundary conditions (e.g., screen types). *Modified after data by Bichara (1986).*

TABLE 4.3 Guidelines for the Detection of Clogging in Recharge Wells

	Recharge water		Redevelopment		
Clogging	TSS [mg/liter]	Turbidity [NTU]	TOC [mg/liter]	Pumping	Surging/jetting
Slight	<10	<5	<10	Frequent	Monthly
Notable	10 < TSS < 20		10 < TOC < 25	Daily	Weekly
Severe	>20		>25	Daily	On demand

TSS = total suspended solids, TOC = total organic carbon.
From Perez-Paricio (2001).

indicators. Some values derived by Perez-Paricio (2001) are presented in Table 4-3.

4.8 Ageing Processes in Open Borehole Wells

Many wells in consolidated formations are not screened. The water flows directly from the aquifer into the well interior. Sometimes a lost casing is cemented in the upper part to prevent loosened parts of the rock or soil material from falling into the well. Wells in consolidated formations suffer from the same ageing processes as wells in unconsolidated aquifers. These include mineral incrustations (→ 3), bioclogging (→ 3.6), and sand intake (→ 4.3). The latter, of course, applies only to semiconsolidated formations.

In consolidated rocks, the drillhole will cut through the cleavages (joints) and schistosity of the natural tectonic stress field. This can lead to the detachment of rock blocks along these joints and result in an

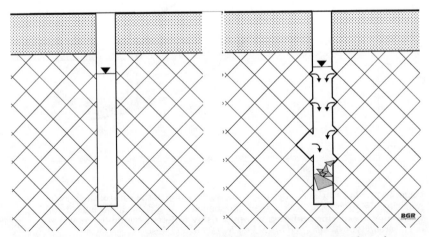

Figure 4.32 Ageing of open-borehole wells: detachment of rock fragments along cleavages.

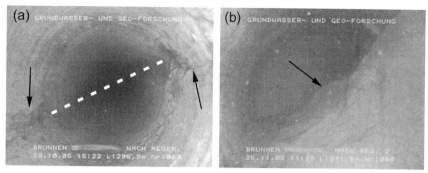

Figure 4.33 Photographs of open-borehole wells: (a) irregular borehole due to detachment of rock (arrows) along a fracture (dashed line); (b) partially blocked borehole. *Photographs: Jürgen Wagner, Grundwasser-und Geo-Forschung, Neunkirchen, Germany.*

irregular-shaped drillhole (Figs. 4-32, 4-33a). This process may continue—albeit more slowly—after construction and during operation. Destabilized rocks may fall into the well (Fig. 4-33b) and damage pumps and riser pipes, and also render their replacement difficult or impossible.

Chapter

5

Identification of Ageing Processes and Performance Assessment of Wells and Well Rehabilitation

5.1 Camera Inspections

Visual inspection of a well interior allows the appraisal of the constructional condition of the well and the amount of incrustations present. Modern submersible well cameras have developed a long way from the single-shot black-and-white photo cameras of the 1950s (Plumley 2000). Nowadays, cameras are able to view both directly downwards (axial) and sideways (radial), the latter sometimes over an angle of 360°. To do so, some are equipped with two separate fixed zoom lenses (Fig. 5-1); others have a single pivoted zoom lens. An additional light source must be included in the camera. Since modern cameras allow full color coverage, incrustation types can at be least tentatively narrowed down. Inspections both above and below the water table are possible. Since turbidity of the water can severely diminish the quality of the pictures, it may be necessary to let the well rest a couple of hours prior to inspection to allow sedimentation. Sometimes turbidity is caused by light incrustation fluffs which will not settle. In this case it may be necessary to pump the well prior to or during inspection.

During camera inspection, the following constructional and ageing-related features should be assessed:

- Signs of corrosion
- Mechanical damage

154 Chapter Five

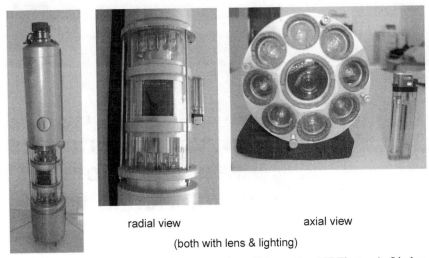

radial view   axial view
(both with lens & lighting)

Figure 5.1 Modern submersible color camera for well inspection (JT Electronic, Lindau, Germany), with axial and radial views. *Photograph: Cleanwells, Rottweil, Germany.*

- Status of casing connections
- Amount of sedimentation in the well sump
- Amount and type of incrustation
- Spatial distribution of incrustations

The camera images can be viewed on TV screens on-site. Storing these recordings on VHS tapes or DVD is now state of the art. Printouts of single shots for reports are possible. Additional information such as depth and water temperature can be continuously recorded and displayed on screen. Text frames on location, date, and special features are also possible (Fig. 5-2). The camera inspection equipment, i.e., camera, winch, electronics, screens, and storage media, fits into a single utility vehicle. Maximum depths of up to 1,200 m (4,000 ft) are possible, though most systems are limited to depths of a few hundred meters.

One has to be aware that camera inspections can only assess the interior of wells, and most of the problems are located outside. Yet, as was shown in Sec. 3.9, the distribution of incrustations inside the well usually mirrors the conditions behind the screen. Therefore the results of camera inspections are useful tools in the planning of rehabilitation. Geophysical methods are recommended if additional information on the status of the annulus is needed (→ 5.3).

If the well interior is heavily incrusted, constructional failures such as corrosion, open connections, and damage may be hidden from view. An additional camera inspection after the rehabilitation is therefore a good option.

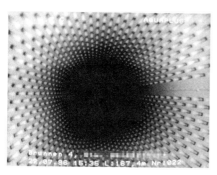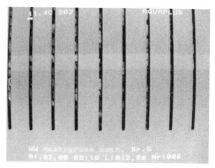

Figure 5.2 Views from a camera inspection of a new well in axial (left) and radial (right) views. Note the text frames with information about well name, date, time, and recording depth. *Photographs: Aquaplus, Kronach, Germany.*

5.2 Step-Discharge Tests (Well Performance Tests)

Step-discharge tests are the most common technique for quantifying the performance (hydraulic yield) of a well. Realization and evaluation are described in detail by Krusemann and De Ridder (1994) and by Batu (2004). They should include at least two, better three to five, different pumping rates. The transient development of drawdown has to be monitored for each rate in intervals that allow good coverage of the drawdown curve. Measurements have to be performed until the water levels remain practically constant (steady state). The duration of a single pumping stage may be between 1 and 24 h.

The intervals of measurements for the individual pumping stages according to DVGW W 111 are

- Pumping time up to 10 min: every 1 min
- Pumping time up to 60 min: every 5 min
- Pumping time up to 3 h: every 10 min
- Pumping time up to 5 h: every 30 min
- Pumping time of >5 h: every 60 min
- Recovery: 2/3 of the pumped time with corresponding measuring intervals

A practical step-discharge test might proceed as follows:

- Turn off the well (and neighboring wells).
- Read the water meter.
- Measure water level recovery (reading every 10 min).
- After the water level has stabilized, turn on the pump at about 1/3 the regular pumping rate

- Measure transient water levels in the well and in neighboring observation wells (if present) at intervals as given above until the water levels stabilize.
- Read the water meter at regular intervals, especially immediately before changing pumping rates.
- Increase the pumping rate to 2/3 the regular rate; measure transient water levels in the well and neighboring observation wells (if present) at intervals as given above until the water levels stabilize.
- Read the water meter at regular intervals, especially immediately before changing pumping rates.
- Increase the pumping rate to the regular rate; measure transient water levels in the well and neighboring observation wells (if present) at intervals as given above until the water levels stabilize.
- Read the water meter at regular intervals, especially immediately before stopping the pump.
- Turn off the pump and measure transient water levels during recovery until drawdown stabilizes.

To ensure full comparability, all tests on a given well should always follow the same procedure, including pumping rates and intervals.

The steady-state drawdown [difference between initial (static) and lowest pumping water level] of each step and its pumping rate (obtained from reading the water meter and time) are used in the calculation of the specific yield Q_s of the well [Eq. (5.1)]:

$$Q_s = \frac{Q_n}{s_n} \qquad (5.1)$$

where Q_n = pumping rate of step n, L^3/T
s_n = maximum, steady-state drawdown of pumping step n, L

The specific yield Q_s and the performance curve (Fig. 5-3) can be used to assess the current hydraulic well performance. The latter is obtained by plotting the drawdown at the end of each pumping phase (y axis) against the pumping rate (x axis). The data points for the final drawdown can usually be connected using a linear function of the type

$$s = BQ \qquad (5.2)$$

where s = steady-state drawdown at step n, L
Q = pumping rate of step n, L^3/T
B = constant, aquifer head loss (laminar flow), T/L^2
BQ = laminar aquifer loss term (after Jacob see e.g., Batu 1998), L

Identification of Ageing Processes and Performance

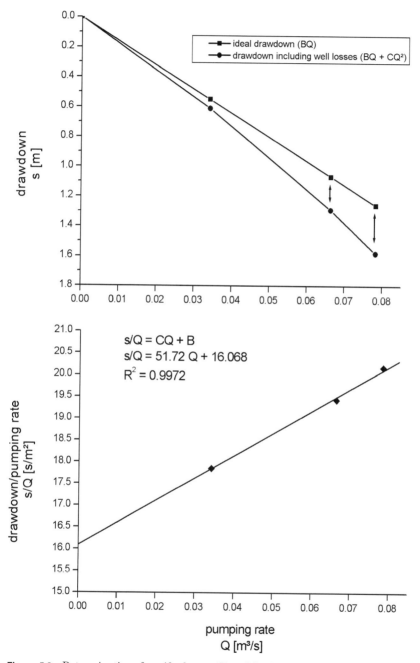

Figure 5.3 Determination of aquifer losses (B) and "turbulent" losses (C) from step-drawdown tests of a well in the Lower Rhine area (Germany).

At high pumping rates, turbulent flow may cause deviations from the linear dependency of Eq. (5.2). Jacob suggests that drawdown then is the sum of a first-order laminar flow component caused by aquifer head losses and a second-order (turbulent) component caused by well entrance losses [Eq. (5.3)]:

$$s = BQ + CQ^2 \qquad (5.3)$$

where C = constant, well-head loss (turbulent flow), "well loss coefficient," T^2/L^5
CQ^2 = turbulent well loss (after Jacob), L

Figure 5-3 shows the procedure. A plot of drawdown versus pumping rate yields a nonlinear curve (upper diagram, gray curve, circular symbols). Plotting s/Q versus Q (lower diagram) allows one to read values for B and C, allowing the construction of the "ideal" linear drawdown–pumping rate graph (upper diagram, black curve, rectangles). Comparing this to the real nonlinear curve gives an impression of the degree of "turbulent losses" (indicated by the arrows). In the case presented in Fig. 5-3, "turbulent losses" gain significance at pumping rates >0.03 m³/s (≈ 100 m³/h).

This concept is an oversimplification, of course, because BQ already includes major portions of the well losses and CQ^2 can include some aquifer losses (Driscoll 1989). Nevertheless, the concept is quite useful to evaluate the magnitude of turbulent losses. Dividing Eq. (5.3) by Q yields the linear equation (5.4):

$$s/Q = B + CQ \qquad (5.4)$$

If we plot s/Q versus Q, the resulting graph is a straight line with a slope of C and an intercept with the y axis of B, which can be read from the graph. The relative degree of turbulent flow L_p can be estimated using Eq. (5.5), although care should be taken due to the restrictions mentioned above.

$$L_P = \frac{BQ}{BQ + CQ} \cdot 100 \qquad (5.5)$$

Walton (1962) tried to classify wells of different ageing stages according to their well-loss coefficients C (Table 5-1). Care has to be taken in applying this concept, because many wells may start with high well losses due to erroneous dimensioning of gravel pack and screen slots.

Repeated well tests allow an assessment of the development of well yield over time and the efficiency of well rehabilitation (Fig. 5-4). The

TABLE 5.1 Well-Loss Coefficient C and Its Implications for Well Performance

Well condition	Well-loss coefficient C	
	s^2/m^5	s^2/ft^5
Properly developed and designed	<1,900	<5
Moderate deterioration	1,900–3,800	5–10
Severe clogging	>3,800	>10
Difficult to rehabilitate to original capacity	>15,200	>40

From Walton (1962).

best reference, against which all well tests during the operational life of the well should be plotted, is of course the well test performed immediately after construction and development. This is very helpful to determine when to rehabilitate (→ 5.8) and how successful a rehabilitation has been (→ 10.3).

In many cases, short well tests with one to three steps and a duration of 1–3 h are sufficient. Pumping rates should not exceed those applied during regular well operation. Both the pumping rates and the drawdown have to be monitored (Fig. 5-4). In order to assess the success of a rehabilitation, a yield test has to be performed before the rehabilitation. The differences between the drawdown/time and the drawdown/pumping rate curves prior to and after the rehabilitation show the yield gain obtained. Intermediate pumping tests between mechanical and chemical rehabilitation steps are useful in the quantification of the relative contribution of each step (Fig. 5-5). Short step-discharge tests should be performed on a regular basis, e.g., twice per year, to constantly monitor the development of the hydraulic well performance (→ 2.5).

Most commonly, the final drawdown is measured at a constant pumping rate and compared to prior measurements (Figs. 5-5, 5-6). Measuring the pumping rate required to reach a certain constant drawdown is also possible but is rarely done. The calculation of the well-yield coefficient has proven to be very useful. It is simply defined as the applied discharge rate divided by the steady-state drawdown obtained. Units are therefore $m^3/h \cdot m$ or m^2/h ($ft^3/h \cdot ft$ or ft^2/h). This parameter enables us to compare the yield at different pumping rates (Fig. 5-4). The use of relative yields has proven to be even more useful. If we know the well yield immediately after construction and set this value to 100%, we can easily compare the actual yield and yields obtained by rehabilitation. A short example shows why this method is useful:

160 Chapter Five

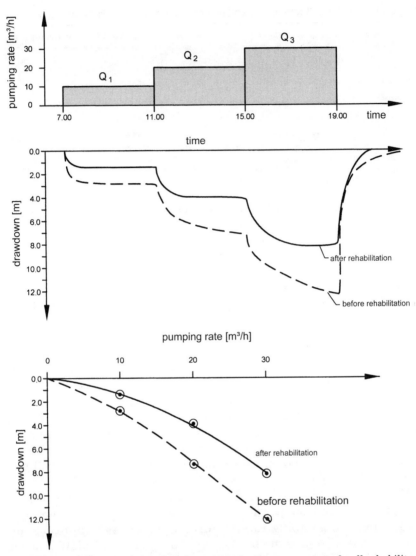

Figure 5.4 Application of step-drawdown tests for the assessment of well rehabilitation. *Modified after DVGW W 111.*

After a rehabilitation, a well has a yield of 2 m³/(h·m), compared to 1 m³/(h·m) before. The yield gain is therefore an impressive 100%. However, if we take into account that the well had a yield of 10 m³/(h·m) at the time of its construction, we can easily see that the rehabilitation has only improved well yield to 20% of the original value. In this case we should consider abandoning the well.

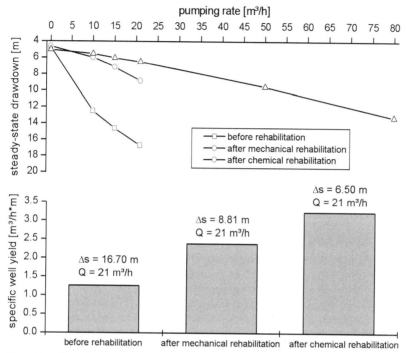

Figure 5.5 Drawdown versus pumping rate and development of specific well yield during a combined hydromechanical and chemical (reductant) rehabilitation of a well in the Lower Rhine area (Germany). The chemical rehabilitation step provides 43% of the total yield gain. *After data by Robert Plängsken, Neukirchen-Vluyn, Germany.*

The performance of a well can also be calculated using the empirical "wire-to-water ratio" (Driscoll 1989). This ratio describes the energy consumption of the pump as a function of pumping rate and pressure head [Eq. (5.6)].

$$WW = \frac{Q \cdot h \cdot 0.65}{3.956 \cdot P_I} \quad (5.6)$$

where WW = wire-to-water ratio
Q = pumping rate [liter/min]
P_I = energy consumption [kW]
h = total pressure head [m]

The total pressure includes all pressure losses in riser pipes, fittings, and pipelines. Considering the discharge-dependent energy demand of the pump and its (total) efficiency factor, Eq. (5.6) can be

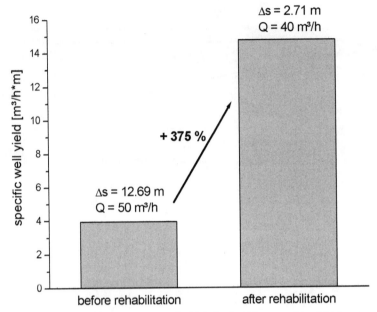

Figure 5.6 Absolute increase of well yield obtained by a combined hydromechanical and chemical (reductant) rehabilitation of a well in the Lower Rhine area (Germany). *After data by Robert Plängsken, Neukirchen-Vluyn, Germany.*

used to calculate the energy demand to overcome a given pressure head. An example is given in Sec. 6.3.

The total cost of pumping can be calculated by multiplying energy demand by cost per kilowatt. Increasing power demand can have several causes:

- Increase of pressure head at the suction side of the pump, e.g., caused by increasing entrance losses or a decreasing water table
- Increase of pressure head at the pressure side of the pump, e.g., caused by pressure losses in riser pipes and pipelines through deposition of scale
- Wear-and-tear defects to the pump: housing, impeller, etc., e.g., caused by mineral deposition, abrasion, cavitation

The development of pressure head and pumping costs—at constant pumping rates—can be used as indicators for well ageing, similar to the well yield tests described above. Increasing heads and energy demand to pump the same amount of water are a good indicator that ageing processes are affecting the pump and possibly the well.

Another simple and helpful indirect indicator of well and pump ageing is to mesasure the temperature at the pump motor. Incrustations around the pump act as an effective thermal seal, diminishing the cooling effect of the flowing water.

5.3 Borehole Geophysics

Borehole geophysical investigations can yield important information on the constructional state and the degree of ageing of a well. Comparing geophysical investigations before and after a rehabilitation can be used to verify decreases or increases in well yield. The parameters that can be assessed by borehole geophysical methods include:

- Geological formations behind the screen and annulus if the drilling diameter is not much larger than the casing diameter
- Physical and chemical parameters of groundwater
- Type and status of casing and screen:
    - Thickness/corrosion/leaks
    - Mechanical damage
    - Tightness of connections and seals
- Type and status of annular filling materials
    - Presence and position of annular seals and cementation
    - Presence and thickness of gravel pack
    - Accumulation of particles or incrustations in gravel pack
- Position and status of screen
- Flow in the near-field of the well, distribution of flow over the screen
- Sedimentation in the well sump

A detailed description of borehole geophysical methods and their application in hydrogeology can be found in Rider (1996). Table 5-2 summarizes the most important geophysical methods and their application in the performance assessment of wells and their rehabilitation and reconstruction. Borehole geophysical methods have proven to be useful and nondestructive tools in the exploration of well and formation characteristics. Nevertheless, the additional cost has so far precluded its widespread application in the water-well industry.

The sensitivity of geophysical methods is subject to a variety of geometric and petrophysical constraints. Because some influences may overlap and complicate interpretation, it is advisable to use a combination of several methods.

One also has to be aware that many geophysical methods require specialist training, experience, and special tools for both implementation and evaluation. The usual drilling company may not be able to perform these tasks. Specialized companies—often from the oil industry business—have to be consulted.

TABLE 5.2 Borehole Geophysical Methods Used for Well Construction, Operation, and Rehabilitation

Application	Aim	Method combination	Comments
Control of constructional state	■ Overall state ■ Tightness of tubing connections ■ Position of screen ■ Flow distribution in screen ■ Permeability in the near-field of the well ■ Type and position of annular filling material ■ Position and tightness of annular seal ■ Position, density of gravel pack and content of fines ■ Corrosion	TV FEL RGG.D or GG.D SGL or GR FLOW FWPACK MAL NN GDT	Smallest probe diameter: 22 mm (<1 in); for >100 mm (4 in) diameter, segmented measuring techniques are recommended (RGG.D, SGL)
Prior to rehabilitation	■ Overall state ■ Flow distribution in screen ■ Permeability in the near-field of the well ■ Corrosion of tubing ■ Colmation, fines content in gravel pack	TV FLOW FWPACK EMDS SGL or GR RGG.D or GG.D CAL SAL/TEMP	Important: to select proper rehabilitation scheme and to avoid collapse of well during rehabilitation
After rehabilitation	■ Overall state ■ Flow distribution in screen ■ Permeability in the near-field of the well ■ Colmation, fines content in gravel pack	TV FLOW SGL or GR RGG.D or GG.D FWPACK	Quality control of rehabilitation
Prior to decommissioning	Presence of annular seals and tightness of annular seals	NN SGL or GR RGG.D or GG.D CAL GDT	Extremely important when well penetrates more than one aquifer
Prior to reconstruction, e.g., retrofitting of annular seal	■ Presence and tightness of annular seals (360°)	SGL or GR RGG.D or GG.D NN	Quality control of reconstruction schemes

Abbreviations: CAL = caliper log (mechanical of tubing); EMDS = electromagnetic wall thickness (steel tubes); FEL = focused electro log (specific electrical resistance); FLOW = flow meter; FWPACK = packer flow meter; GDT = gas-dynamic tracer; GG.D = density measurement using gamma-gamma log; GR = gamma ray (natural gamma radiation); IL = induction log (inductive measurement of electric conductivity); MAL = magnetic log; NN = neutron–neutron log; RGG.D = annulus scanner (360° gamma–gamma measurement); SAL/TEMP = salinity/ temperature; SGL = modified GR, simultaneous measurements in segments; TV = submersible television camera.

In deep wells—for all practical purposes, wells at depths between >25 and >100 m (80 to 300 ft)—the well axis may not be vertical. Plumbness and straightness can be assessed by different methods. In heavily tilted wells, the tubing is exposed to mechanical stresses. In some cases this leads to leaky connections. Groundwater and fines may seep through these openings. All equipment lowered into or extracted from these wells is threatened by the possibility of getting stuck. Geophysical probes are very expensive and sometimes contain radioactive material. It therefore must be ensured that the probes not only enter into the well but are also retrieved. The use of a "dummy" probe before installing the "real" probe is always a good idea, to make sure the well is penetrable. Recovery of the probe may be difficult and may damage both probe and well.

In wells with very turbid waters, in which camera inspections are inapplicable, acoustical methods may serve as substitute ("borehole televiewer"). In this case the interior is scanned by ultrasonic waves and an image is then calculated from the reflections. This method works only below the water level (Fricke & Schön 1999).

A look "beyond" casing and screen is possible either through a measurement of the natural signals of the surrounding formation and the annular fill, or the attenuation of induced signals in the annulus or the surrounding formation. Possible signals are

- Induced acoustic waves (usually ultrasonic)
- Induced electrical currents
- Induced magnetic signals
- Natural magnetic signals
- Natural radioactive radiation [gamma (γ) radiation]
- Induced radioactive radiation [gamma (γ) or alpha (α) radiation]

The Focused Electro Log (FEL) is a screening method to validate the tightness of connections in nonconductive tubing material (e.g., PVC). Connections showing high resistances in the log can be assumed to be tight, since the flow of an electric current indicates the flow of water. Figure 5-7 shows a measured FEL. Connections are clearly visible as a regular pattern of small zones of decreased electric resistance. Presumed leaks are thus marked with an "L?" in Fig. 5-7. Packer tests or tracer methods (Fig. 5-8) can be used to quantify the leakage rate.

Electromagnetic measurement of material thickness (EMDS)—a method often applied in the oil industry—is useful to investigate the effects of corrosion on metal casings. Corroded wells with low thickness might collapse, e.g., during rehabilitation.

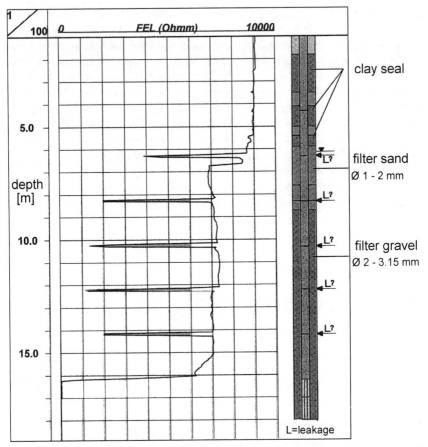

Figure 5.7 Focused Electro Log (FEL). Casing connections are visible as peaks of decreased electric resistance. Presumed leaks are marked as "L?." *Modified after Baumann and Tholen (2002).*

The quality of the annulus cementation is often measured using the Cement Bond Log (CBL)—again a method common in the oil industry. The measured parameter is the attenuation of induced seismic or ultrasonic waves in the annulus. In well-cemented wells, attenuation is high because a lot of the induced energy is transferred through the dense cement into the formation.

Verification of the presence, position, and homogeneity of an annular seal can be important to assess the degree of separation of different aquifers and the exclusion of contaminated surface water. If the sealing materials have a higher gamma-ray signal than the surrounding geologic material, passive measurements of the gamma radiation are useful. In some cases the bentonites used for sealing are artificially enriched by monazite—a

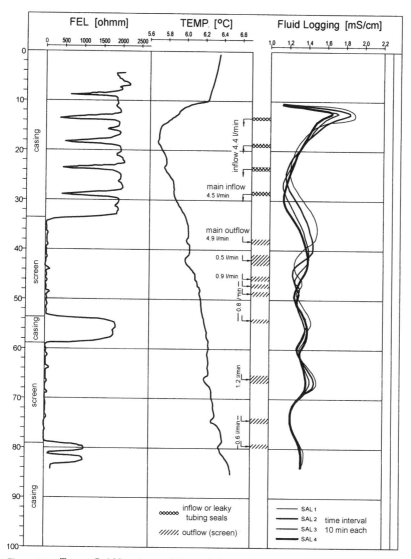

Figure 5.8 Tracer fluid logging and focused electro log (FEL) used for the detection of leaky casing connections. *Modified after Lux (1997).*

mineral with a known high gamma radiation. In this case detection is quite easy although many natural bentonites cannot be distinguished from the surrounding formations. Sometimes magnetite is added to the bentonite to increase the weight and speed up sedimentation in the annulus. Because this mineral is ferromagnetic, such seals can be detected by their high magnetic susceptibility in magnetic logs. This method cannot be applied in areas with high natural contents of ferromagnetic minerals. Luckily, this occurrence is quite rare, because of the geologically rare abundance of iron ores. If the sealing material contains no markers such as monazite or magnetite and seals cannot be distinguished from the geologic background, density measurements are required.

A proof of the tightness of an annular seal can be obtained by performing a "gas-dynamic tracer test" (GDT). A gas, preferably nitrogen, is injected into the well screen below a packer. The gas then rises into the gravel pack and collects under the annular seal. From there it slowly dissipates sideways into the aquifer. The gas bubble constitutes a temporary zone of very low density which should be easily detectable by appropriate methods. The Neutron Gamma Log (sometimes also called the Neutron–Neutron Log, NN) is a useful method for this. The NN probe emits fast neutrons from a radioactive source. These interact with hydrogen that is present in water molecules. This releases gamma rays, which can be measured in the probe. After subtraction of the natural gamma radiation from the geologic background, the water content and thus the porosity can be calculated. Through an appropriate application of probe intervals, the penetration depth of the irradiation can be limited approximately to the annulus. This diminishes effects induced by the geologic background. For the GDT, recurrent neutron–neutron measurements monitor the decrease of gas under the seal. A good seal should retain the gas for several hours. If the annular seal is leaky, gas penetrates into it and immediately rises upwards.

Clogging of the filter pack through incrustations or mechanical colmation by fines are the main causes of well ageing (→ 3, 4). Plugged pore spaces increase the bulk material density, because pore water is exchanged for mineral mass. Geophysical borehole measurements are suitable to detect such plugged pore spaces. If the pores are plugged by fine particles from the adjacent aquifer, these can be detected by their elevated natural gamma-ray signature. The clay minerals associated with the fines are usually enriched in radioactive potassium (^{40}K), which is a source of gamma rays. Such clay mineral accumulations are therefore easy to detect in Natural Gamma Ray Logs (GR).

Figure 5-9 shows the results of a Segmented Gamma Log (SGL) measurement in a well before and after a rehabilitation (Baumann & Tholen 2002). This method is also based on a Natural Gamma Ray Log, in the present case measured in three sections, from 0° to 120°, from 120° to 240°, and from 240° to 360°, enabling a full 360° "view" of the well. The measurements show the intact clay seal at about 10.0 m (33 ft). In the "view" before the rehabilitation, some darker shading is visible in the upper parts of the gravel pack [especially between 15 and 18 m (50 and 60 ft)]. This indicates elevated densities caused by pore spaces plugged by fines. Notably, the plugging is more pronounced on one side of the well (secs. 0–120° and 240–360°). The same plot after the rehabilitation (Fig. 5-9, right) shows

Identification of Ageing Processes and Performance 169

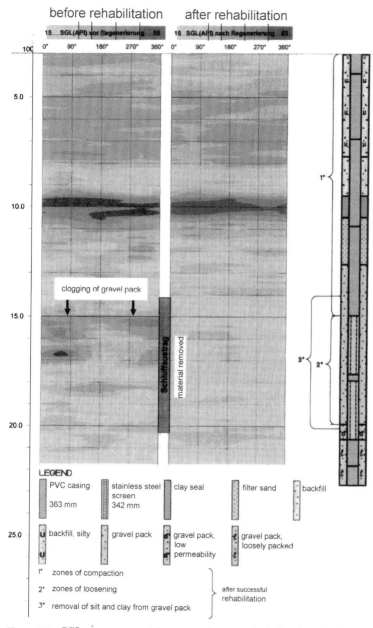

Figure 5.9 SGL measurement (segmented gamma log) showing the density distribution in a well before and after rehabilitation. *Modified after Baumann and Tholen (2002).*

much lighter shading in the same region. This indicates lower densities and thus successful removal of the fines. Together with well performance tests, the success of the rehabilitation could thus be shown.

The bulk density of annular filling material can be determined by the Gamma-Gamma Log (GG). This is useful to detect missing material, compaction, rearrangement, and bridging in the annulus. This method uses the absorption of gamma rays induced by a radioactive source by minerals. The absorption is a measure of the electron density of the mineral and thus its mass density. Incrustation minerals tend to increase the total density.

The emplacement of the gravel pack in deep boreholes (>200 m, 660 ft) is always a difficult task. The oil industry often uses a method termed "gravel pack" for monitoring. A Gamma-Gamma probe is installed in the screen and the density is measured at short intervals. The inserted gravel expels the water into the annulus and increases the density. This method is much more accurate than the usually employed plumb techniques.

Flow meter investigations are a well-known, relatively inexpensive, and simple method to measure water inflow into the screen and its spatial distribution over the screen. Many drilling companies offer them as standard service. The tool is simply a propeller (with the propeller axis aligned to the well axis) which is lowered or lifted through the screen section at a constant velocity. The number of propeller revolutions is a measure of the flow rate. Both pumped wells and wells at rest can be investigated. In pumped wells, flow meters allow one to quantify the relative contribution of screen sections to the overall pumping rate. An example can be found in Sec. 11.3.2.

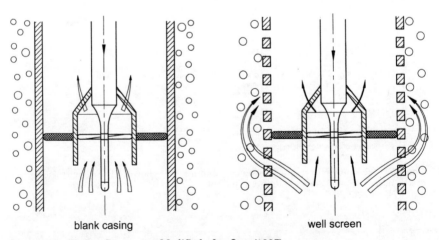

Figure 5.10 Packer flow meter. Modified after Lux (1997).

Identification of Ageing Processes and Performance 171

In many cases, a packer flow meter is used (Fig. 5-10). In this case the propeller gains its momentum from the movement of the system. It is especially useful to delineate zones of impeded water flow. When the packer passes through such a zone, the displaced water cannot evade it and must flow through the propeller. The diagram then shows a high number of revolutions (Fig. 5-11). In other sections, most of the displaced water flows through the permeable gravel pack, which leads to slow revolution of the propeller.

In many cases, a variety of geophysical techniques have to be used to solve questions regarding functionality and integrity of the well (Figs. 5-8, 5-11).

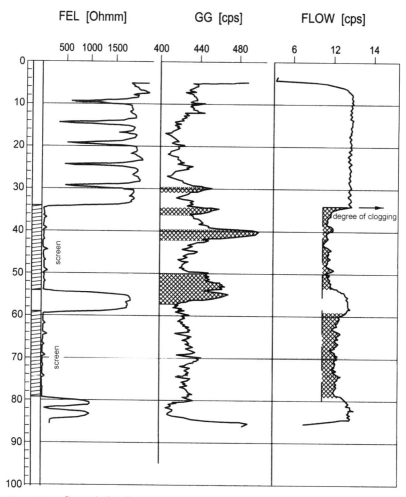

Figure 5.11 Control of well properties using several geophysical logging techniques: packer flow meter, FEL, and GG log. *Modified after Lux (1997).*

5.4 Measurements of Incrustation Deposition Rates

Various techniques can be used to investigate the type and amount of incrustations that have formed in a well. One method is based on small glass slides which are fixed to a cord and lowered into the well—or a piezometer installed in the annulus. If slides are fixed at designated depths, a depth-specific investigation is possible. The slides have to be recovered at regular intervals. The type and amount of incrustation can then be determined by appropriate mineralogical, chemical, and biological methods ($\rightarrow$ 5.5). A measurement of the corrosion potential of groundwater can be performed by a similar technique. Small pieces of different metals and varying metal quality can be lowered into the well (McLaughlan 2002; McLaughlan et al. 1993). Their material loss due to corrosion can be assessed after recovery.

A method for sample collection by Howsam (1991) uses a small acrylic glass flow box attached to the well head. This box is filled with gravel, sand, or glass beads. A small portion of the water pumped from the well flows through it. Incrustation buildup can be assessed visually or by opening and sampling the box from time to time. Of course, this method will not be able to emulate the conditions inside a well completely.

The formation of many types of incrustation is related to the presence of specialized bacteria ($\rightarrow$ 3.2.6, 3.6). These are present in natural aquifers and colonize the well from there. More detailed information on the types and numbers of such organisms in wells can be obtained from microbiological investigations (Cullimore 1993, 1999). Since full-fledged microbiological assays are costly and time-consuming, some simplified procedures are often used. Microscopic investigations often allow the identification of typical iron bacteria according to their distinctive morphology ($\rightarrow$ 3.2.6). Recent years have seen the advent of semiquantitative bacteria tests which can be performed by nonmicrobiologists. They are available for a variety of bacterial groups, including iron bacteria and sulphate-reducing bacteria.

5.5 Sampling and Investigation of Incrustation Samples

Knowing one's enemy makes it easier to select the necessary weapons to defeat him. If we know the type of incrustation we are dealing with, we can, e.g., select an appropriate technique or chemical to remove it. For the latter, lab experiments comparing the efficiency are useful (Houben et al. 2000a, Houben 2003b).

There are no specific regulations for the sampling of incrustations. In many cases the material has to be collected as obtained from pumps and riser pipes. In the authors experience, this is the same material as present in the gravel pack. Admittedly, the samples from the pump may be younger than the incrustation in the gravel pack, because the pump is commonly removed more often. More representative samples can be collected during mechanical rehabilitations.

Core drilling into the annulus of clogged wells is a useful but very expensive way to obtain excellent samples from "behind the screen" (Timmer et al. 2003; Houben 2006). Howsam et al. (1989) and Timmer et al. (2003) describe a type of apparatus which allows the collection of samples by shooting or hydraulic insertion of sample cups sideways (horizontally) into the borehole wall. This technique can only be applied to uncased boreholes.

Incrustation samples may contain substances that can be affected by contact with atmospheric oxygen, e.g., sulfides such as pyrite (FeS_2) [Eq. (5.7)]. The reaction products may differ significantly from the original, feigning a different type of incrustation.

$$2\ FeS_2 + 7\tfrac{1}{2}\ O_2 + 7\ H_2O \Leftrightarrow 2\ Fe(OH)_3 + 4\ SO_4^{2-} + 8\ H^+ \quad (5.7)$$

Sulfides, for instance, can completely disappear in favor of iron oxides. The acid produced during this process may dissolve calcium carbonate and eventually form gypsum. Therefore, samples should be transferred quickly into airtight glasses or plastic bags. They should be stored in a cool place in the dark. A refrigerator will usually serve the purpose. For ochre samples, drying is a tricky business. Young iron and manganese oxides recrystallize even at low temperatures (Cornell & Schwertmann 2003). Maximum permissible drying temperatures are 60°C (140°F) for iron oxides and 35°C (95°F) for manganese oxides. About 20 g (0.7 oz) of dry mass are required for a complete chemical and mineralogical investigation of the sample. Since the percentage of water in fresh samples can be very high, a much higher amount of wet sample is recommended. Prior to any further chemical analyses, the samples have to be ground to about 20 μm in size using a mill or hand mortar. Both mill inserts and mortar should consist of agate—not steel—in order to avoid contamination by metal abrasion.

A first quick test is to pour a few drops of dilute acid onto the sample. Carbonate minerals will make their presence clear by fizzing strongly. If iron sulfides are present, a strong smell of "rotten eggs" (hydrogen sulfide, H_2S) will develop quickly; even minor amounts are sufficient to produce an obnoxious odor.

A chemical analysis of incrustations should include at least the following parameters: iron, manganese, calcium, magnesium, silicon, sulfur, and organic and inorganic carbon. Analyses may be performed from solution after total dissolution of the sample, e.g., in aqua regia.

"Dry" analyses using melt assays and X-ray fluorescence (XRF) are also often used. A detailed analysis of the mineral phases gives good insight into the degree of crystallization of ochre minerals and thus their relative age. Useful analytical techniques include X-ray diffraction (XRD) and infrared spectroscopy (IR). The amount of reactive surface area is also a measure of crystallinity and a very important parameter to describe the reactivity toward chemicals. It can be measured by tracking the adsorption of nitrogen gas on the sample (BET technique).

Samples for microbiological investigation should not be dried and should be kept cool and in the dark. For microscopic studies, the moist material can be applied directly to a glass slide. Sample containers and handling equipment (including fingers or gloves!) must be sterile. A much more detailed description of the procedures for microbiological sampling and investigation can be found in Cullimore (1999). Samples intended to be used in scanning electron microscope (SEM) investigations should be only air-dried, and not mortared or ground. Otherwise the delicate structures of the iron bacteria ($\rightarrow$ 3.2.6) will be damaged.

In many practical applications we need to distinguish between ochre and corrosion products ("rust"). Both contain iron oxides, often as goethite (α-FeOOH), and they have similar appearance due to the ochreous colors. Rust often contains some typical minerals that are uncommon in ochre, including lepidocrocite (γ-FeOOH). The most important difference is the occurrence of magnetic minerals, which are absent in ochre. Magnetic minerals include magnetite (Fe_3O_4) and maghemite (γ-Fe_2O_3). Their presence can be checked easily on dry samples using a small magnet. If a substantial amount of magnetic minerals is present in the sample, it will adhere strongly to the magnet. Sulfate-reducing bacteria are also often involved in corrosion processes and produce black iron sulfides ($\rightarrow$ 3.1.3).

5.6 Particle Counting

The most convenient way to obtain information about the content of suspended solids in water is to use a particle counter. Surrogate parameters are turbidity and the membrane filtration index (MFI, $\rightarrow$ 4.7). However, turbidity is applicable only at fairly high concentrations. MFI measurements may take a long time and thus cannot detect sudden changes in concentration. These limitations can be circumvented by using a particle counter. These devices count the number of particles and present the result as the average number of particles of a chosen diameter interval per unit volume during a chosen time interval. As an example, Fig. 5-12 presents, for intervals of 1 min, the number of particles per milliliter that are larger than the diameter indicated. Increasing the pumping rate leads to an increase in particle entrainment. Disturbances

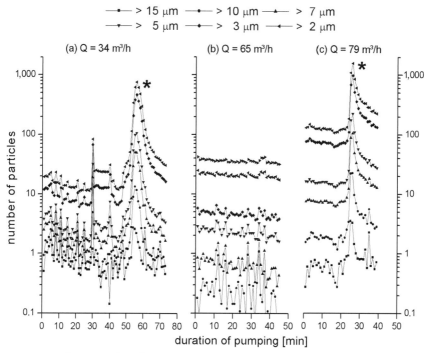

Figure 5.12 Concentration of suspended particles in the abstracted groundwater of Well 202 of Well Field Helmond (The Netherlands) at three different pumping rates. Note the logarithmic scale of the y axis. The peaks in concentration at 34 and 79 m^3/h (150 and 348 gal/min) were caused by switching the submersible pump on and off five times. *After data courtesy of Kees van Beek, KIWA, The Netherlands.*

during the measurement—switching the pump on and off—are clearly visible as peaks in particle concentration.

Particle counters are based on several principles: light blocking, light scattering, and laser-induced breakdown. The data in Fig. 5-12 were obtained by the light-blocking method. In this method a laser beam is directed through a cell, through which water flows, and onto a light-sensitive sensor. Each passing particle projects a shadow on the light-sensitive sensor, causing an electrical signal. The larger the particle, the larger the shadow, and the larger the electrical signal will be. The particle diameter interval and the time interval may be chosen freely. The light-blocking method is applicable for particles with a diameter larger than 1–2 μm. Both other laser methods are able to count smaller particles.

Particle counters are very sensitive to disturbances. For example, when the abstracted groundwater contains so much methane that bubbles are formed, these bubbles are counted as (large) particles. However, this method is also not free from complications: position of the particle in the laser beam, estimation of the diameter from the magnitude of the shadow, differences in transparency of particles at high concentrations coincidence of more particles which are counted as one particle, and so on.

5.7 Mass Balancing of Removed Material

The amount of sand or incrustation removed from a well during rehabilitation is a useful value with which to assess the success of the rehabilitation. For depth-specific rehabilitations, the amounts will also show where the bulk of the clogging material is hidden. Such sections can be treated more intensely, while less affected areas may be neglected. Only if, e.g., the turbidity of the pumped water falls below certain cutoff criteria will the rehabilitation equipment be transferred to the next section ($\rightarrow$ 10.2.3).

The mass of removed material can be calculated from repeated measurements of relevant parameters in the water pumped after the rehabilitation step:

- Dissolved ionic concentrations (measured on-site or in the laboratory)
- Turbidity
- Suspended-particle content (usually measured as volume, e.g., with an Imhoff cone)

The first measurement, of course, applies only to chemical rehabilitations. The latter two techniques have the disadvantage that the values obtained have to be converted to mass for a proper balance. Correlation curves of turbidity/mass and volume/mass, respectively, are therefore needed.

Figure 5-13 shows a typical curve of concentrations measured after pumping a rehabilitated section. Multiplying the concentration by the volume of pumped water gives the total mass of material removed, so we also need to measure the discharge rate of the pump. Corrections for any naturally occurring concentrations, e.g., of iron and manganese, may be necessary. Figure 5-13 shows that the dissolved concentrations and the turbidity are initially quite high. It also shows, however, that it takes a long time to reach the natural background levels again. Figure 5-14 shows a setup for such measurements. In most cases, measurements of suspended or dissolved concentrations will be discontinuous. The individual concentration values are then assumed to be representative for the short time span until the next measurement. They are then multiplied by the water volume pumped in this period of time. An example will clarify the concept:

A concentration of 250.2 mg/liter iron is measured during pumping after a rehabilitation. The natural iron concentration of the water is 0.2 mg/liter. The sampling interval is 5 min, the pumping rate is 10 m^3/h (= 167 liter/min, 44 gal/min). During this interval a mass of 250.0 mg/liter ×

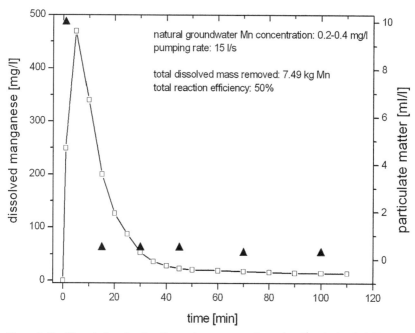

Figure 5.13 Mass balancing by discontinuous sampling of a chemical rehabilitation using a reductant on a well near Cologne (Germany). Rectangles = dissolved manganese; triangles = suspended particulate matter.

Figure 5.14 Sampling equipment for mass balancing of well rehabilitation with sampling port, water meter, and Imhoff cone. *Photograph: Christoph Treskatis.*

TABLE 5.3 Conversion Factors for Calculating Mineral Masses

Measured	Sought	Conversion factor
Fe	FeOOH	× 1.59
Fe	Fe(OH)$_3$	× 1.91
Mn	MnO$_2$	× 1.58
Mn	MnOOH	× 1.60
Ca	CaCO$_3$	× 2.50

835 liters = 208,750 mg (= 208.75 g, 0.46 lb) is removed. This procedure has to be repeated for each data point shown in Fig. 5-13. Summing all the values yields the total mass removed. During most chemical rehabilitations, not only is dissolved matter removed, but also some additional suspended matter. For a full balance, both have to be considered.

Masses calculated from measured concentrations of dissolved ions can be converted into mineral masses using the conversion factors in Table 5-3.

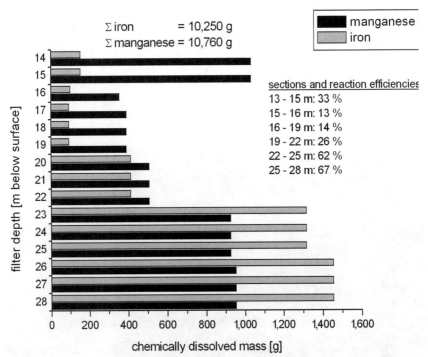

Figure 5.15 Depth-specific mass balancing (iron and manganese) of a chemical rehabilitation of a bank filtrate well near Cologne (Germany). The rehabilitation was subdivided into six sections. The percentages given to the right of the section depths show the calculated reaction efficiencies. The total removed mass (dissolved matter only) of Fe and Mn is 41 kg (90 lb), which translates into about 65 kg (143 lb) Fe and Mn oxides according to Table 5-3.

If we know the reaction stoichiometry and how much of a chemical has been injected into that particular screen section, we can even calculate the reaction efficiency. This is the ratio of the measured mass removed and the potentially removable mass according to the stoichiometry. As an example, a certain amount of chemical was injected, which according to the stoichiometry should have dissolved 100 g (0.22 lb) of incrustation mass. The measured amount of mass removed is 80 g. The reaction efficiency is thus 80%. The other 20% of the chemical reacted with other species or was lost into the aquifer. Efficiencies of >100% can occur in deeper sections when the rehabilitation is performed "top to bottom." Some of the chemical solution may flow downward owing to its higher density as opposed to "pure" groundwater, and reacts in a deeper part of the screen. Figure 5-15 shows the depth-specific distribution of removed incrustation mass of a bank filtrate well. The percentages given to the right of the bar graph show the calculated reaction efficiencies. The highest efficiencies are notably achieved in strongly incrusted sections.

5.8 Calculation of Saturation Indices

The natural hydrochemistry of groundwater is a good indicator of incrustation potential. Water that requires iron removal before being distributed to the consumer is probably from a well which is prone to ochre buildup. If the well is located in an area with limestone aquifers and waters with elevated hardness, carbonate scales are probably the main cause of well ageing. Sulfide incrustations are found only in strongly reducing groundwaters, which often smell like "rotten eggs" (hydrogen sulfide, H_2S). Aluminum scales are typical of groundwater of low pH and limited acid-buffering capacity.

A calculation of the saturation state of water with respect to one or more minerals is a good way to predict the precipitation or dissolution behavior in a quantitative way (e.g., Walter 1997).

For a hypothetical dissolution reaction of some mineral A, the dissolved products C and D, and their stoichiometric coefficients a, c, and d, we can write:

$$aA \Leftrightarrow cC + dD \quad (5.8)$$

For this reaction we can define a law of mass action with the equilibrium constant K:

$$K = \frac{[C]^c[D]^d}{[A]^a} \quad (5.9)$$

The constant K describes the activities of the ions involved at equilibrium. By definition, the activity of a solid mineral phase is 1. The ion activity product (IAP) describes the actual (measured) activities of the ions in the solution considered [Eq. (5.10)].

$$IAP = [C][D] \qquad (5.10)$$

The saturation state Ω of a system with respect to a mineral phase is simply the ratio of the measured ion concentrations (IAP) and the equilibrium constant:

$$\Omega = IAP/K \qquad (5.11)$$

To make the numbers easier to handle, we usually employ the saturation index SI, which is the logarithm of Ω:

$$SI = \log(\Omega) \qquad (5.12)$$

The interpretation of SI is straightforward:

SI = 0: equilibrium, mineral can neither be dissolved nor precipitated

SI < 0: undersaturation, mineral can be dissolved

SI > 0: oversaturation, mineral may precipitate

The SI is only an indication of the direction in which the reaction will proceed, not of the actual reaction progress. Oversaturated solutions can remain stable without precipitation of the oversaturated mineral when the precipitation process is kinetically inhibited. Taking into account all uncertainties that influence the calculation of saturation indices, e.g., analytical imprecisions, one should not overinterpret SI values. Values up to +0.2 and as low as −0.2 can be considered to be practically equilibrated.

Older literature is full of calculation schemes to assess the position of waters in the calcium carbonate equilibrium, which is a very important for its use in technical systems. Common methods include calculations of saturation indices after Langelier and stability indices after Ryznar. However, because they are all based on semiquantitative approaches, they should no longer be used. Today, we can use thermodynamical equilibrium models which yield much more accurate results and include many other mineral phases as well. One of the most popular models is PHREEQC (Parkhurst & Appelo 1999). It can be downloaded free of charge from various sites, including the U.S. Geological Survey homepage. Several interfaces are available to facilitate data input.

www.brr.cr.usgs.gov/projects/GWC_coupled/phreeqc

www.geo.vu.nl/users/posv/phreeqc.html

www.xs4all.nl/~appt

PHREEQC simply requires the input of a charge-balanced water analysis (of course, in its own format) and will then calculate the speciation of the dissolved constituents and saturation indices for several possible mineral phases. Figure 5-16 shows such calculations of the saturation indices for the carbonate minerals calcite and dolomite of 186 groundwater samples from the Kabul basin (Afghanistan). Since practically all samples show a $SI > 0$, wells in this environment are expected to suffer from scaling in the long run. The groundwater there tends to possess high hardness due to the abundance of carbonate detritus in the aquifer matrix.

The lower part of Fig. 5-17 shows the results of calculations of the saturation index for calcite in groundwater samples from a well row in Germany. The saturation indices fall below and above the equilibrium value of 0. PHREEQC also allows one to implement user-defined saturation indices of minerals for any given analysis. PHREEQC then calculates how much of this mineral phase has to precipitate or dissolve in order to attain this new equilibrium. In this case we have set $SI_{calcite} = 0$ and calculated the related amount of $CaCO_3$. A precipitation of several tens of milligrams of calcite from 1 liter of water—as would be the case for samples 1–3, 7, and 8—would of course involve a major incrustation potential.

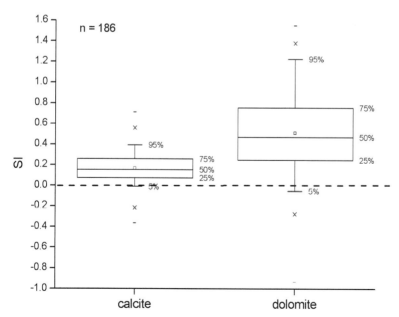

Figure 5.16 Box-whisker plots of saturation indices (SI) for the carbonate minerals calcite and dolomite in groundwater of the Kabul basin, Afghanistan. *From Houben et al. (2005).*

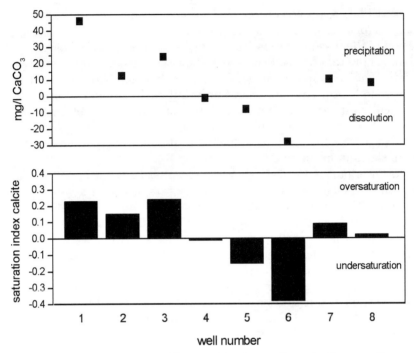

Figure 5.17 Saturation indices (SI) for calcite in a well row in Germany (lower graph) and the amount of calcite that would have to precipitate or dissolve to attain equilibrium ($SI = 0$).

PHREEQC also allows calculation of the effects on mineral saturation of mixing of two (or more) waters. This can be useful for predicting the incrustation potential of infiltration wells in which the infiltrated water interacts with naturally occurring groundwater. In many cases the infiltrated water has been exposed to atmospheric oxygen and may thus cause ochre precipitation when it encounters reduced groundwater. The top section of Table 5-4 shows some saturation indices for different mixing rations of a slightly oxic (2.0 mg/liter O_2) infiltration water and a reduced groundwater containing ferrous iron. A precipitation of calcium and magnesium carbonate minerals is unlikely due to the negative SI, whereas ochre buildup is highly likely. The amount of possible precipitates presented in the lower section of Table 5-4 is of course only a tentative approximation, because of the uncertainties discussed above. Nevertheless, the calculations have proven to be quite realistic: the well for which these calculations were performed showed signs of ochre buildup shortly after commissioning.

TABLE 5.4 **Example Results of Mixing Modeling of an Infiltration Water and a Natural Groundwater with PHREEQC**

Mineral phase	Saturation index			
	Mixing 20:80	Mixing 50:50	Mixing 80:20	
Calcite	−0.29	−0.33	−0.37	Undersaturated
Dolomite	−1.30	−1.34	−1.37	Undersaturated
$Fe(OH)_{3\ (am)}$	+2.74	+2.53	+2.19	Oversaturated
Goethite	+8.50	+8.30	+7.97	Oversaturated
Manganite	+4.88	+4.61	+4.15	Oversaturated
Diss./precip.*	Mixing 20:80	Mixing 50:50	Mixing 80:20	
Mineral phase	(g) mineral per m^3 of mixed water			
Calcite	−37.6	−41.0	−44.3	Dissolution possible
Goethite	+0.43	+0.29	+0.14	Precipitation possible

*Diss./precip. − dissolution/precipitation.

5.9 When to Rehabilitate?

The question when to initiate rehabilitation measures is not easily answered. Customers who have just paid a substantial amount of money for a new well are often quite reluctant to invest more money to rehabilitate it, especially when these costs are recurring on a regular basis.

The authors know of wells that have been operating continuously for about 25 years at the same yield with no ageing visible during camera inspections. Although the water is quite rich in iron, it is screened in a deep aquifer, protected from the inflow of oxygen. The fatal admixture of reactive components therefore does not occur. These wells have never been rehabilitated and have never had to be. Their life span is probably limited only by the ageing rate of the casing and screen material. On the other hand, we know of wells which had to be decommissioned after only 2 years of operation due to massive ochre clogging. In Chaps. 3 and 4 we discussed some of the reasons for the appearance of clogging. Naturally, combinations of adverse boundary conditions will considerably accelerate ageing processes. Negative factors may include:

- Elevated concentrations of incrustation-building chemical constituents
- Mixing of water from different hydrochemical zones, leading to strong mineral oversaturation
- High pH values (for ochre precipitation)
- High hardness (for scale precipitation)

- Inflow of oxygen through the well interior or screen (when drawdown is too deep)
- High nutrient supply promoting bacterial growth
- High flow velocities or even turbulent flow
- Low porosities in the aquifer and gravel pack
- Uneven grain-size distribution

Knowing how many of these adverse boundary conditions occur in the well and aquifer and to what extent they are present gives us a first indication of the ageing effect we may expect. A more quantitative approach is based on continuous monitoring of the well yield ($\rightarrow$ 5.2). Some requirements must be fulfilled to properly do so:

- A step-discharge pumping test immediately after construction, and the determination of well-yield coefficient (Q/s) (reference yield = 100%)
- Short pumping tests on a regular basis (once or twice a year), the determination of well-yield coefficient (in percent of original yield)
- Plot of well yield over time (... and, yes, somebody should have a look at it, too)

Such a graph is invaluable for our purposes. Right from the beginning it will show trends in well yield, especially deviations from the original performance. Sudden and unprecedented well-yield losses should suggest more detailed investigations, e.g., by camera inspection or borehole geophysical logging ($\rightarrow$ 5.1, 5.3). But how do we define the magic point of yield decline that makes a rehabilitation necessary? Experience from Germany has shown that deviations of about 20% from the original yield can usually be reverted with reasonable technical and financial expense. Driscoll (1989) recommends 25% as the permissible maximum yield loss, while McLaughlan (2002) sets the threshold at 10%. Higher yield declines may necessitate more complex and more costly rehabilitations, e.g., combined mechanical and chemical methods, to regain the yield. Customers who start thinking about rehabilitations when the well yield has decreased to less than 50% of the initial yield often face nasty surprises. If, e.g., the incrustations have recrystallized to insoluble phases over time ($\rightarrow$ 3.2.7), it may be practically impossible to regain more than a few percent in yield. Our recommendations are

Track your well's yield continuously.

Act early ("kill it before it grows").

The timing of the first rehabilitation and the temporal interval between the following ones therefore has to be determined individually for each well. Intervals may be months or years, depending on the boundary conditions described above.

5.10 Rehabilitation or Reconstruction?

Reconstructions cannot be avoided when:

- Structural damage to casing and/or screen that endangers the safe operation of the well is discovered prior to or after a rehabilitation.
- Well yield cannot be increased significantly even by repeated rehabilitations.

Again, well-yield curves, as described in Secs. 5.2 and 5.9, are a key element in assessing the yield gain obtained by rehabilitations. If repeated rehabilitation cannot improve the yield to more than 50% of the original value, we have to assume that the incrustations have reached a level of crystallinity that effectively prevents their complete removal. Rehabilitations will then only provide short-lived gains. We must presume that the yield will slowly but steadily decrease. The cost of the successive rehabilitations may soon exceed the cost of construction of a new well. The additional operation time that we have "bought" by this rehabilitation should be used to properly assess the alternatives, including construction of a new well or reconstruction of the given well. The financial calculations involved in the decision between rehabilitation and reconstruction are presented in more detail in Chap. 6.

Chapter 6

Economics of Well Rehabilitation and Reconstruction

6.1 Economic Principles and Evaluation Standards for Wells

The construction of a new well requires large expenditures and is of considerable financial importance to the operator. Zilch et al. (2002) recommend economic efficiency calculations and cost–benefit analyses prior to making such an investment. The targets of these procedures are

- Evaluation of an investment as to its profitability
- Examination of alternatives, e.g., rehabilitation, dismantling, and construction of new wells
- Determination of the optimal service life of the investment
- Examination of the possible forms of purchase and financing (buy, rent, lease)
- Comparison of internal and external contributions, e.g., for carrying out rehabilitation or redevelopment measures

The investigation period starts with the planning and ends with the planning horizon, constituting the calculatory end of the operating phase. The period between planning and planning horizon is designated as the calculation period or calculatory service life (Maniak 2001). The fixing of a near planning horizon (Fig. 6-1) to some extent prevents forthcoming uncertainties, as prospective maintenance measures are rated lower than benefits and costs. If, for example, a new well has to be built urgently to meet water demand, this will lead to a near planning horizon.

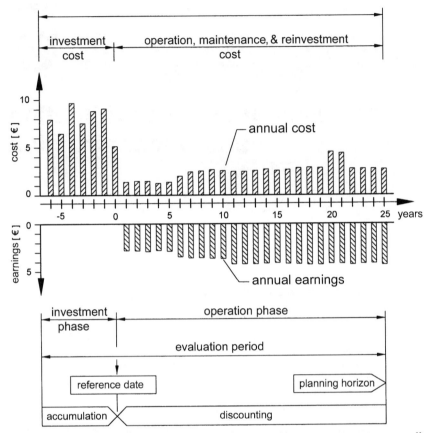

Figure 6.1 Economic phases in the life cycle of a technical investment, e.g., a water well. Modified after Maniak (2001).

The economic service life of a well, which ends when the expenditures for maintenance and operation start to exceed the remaining benefit, e.g., proceeds from water sales, confines the investigation period. Considering that the operating time of a water well may be prolonged nearly indefinitely by means of professional operation and gradual maintenance, e.g., regular rehabilitation measures, investigation periods of more than 30 years are common practice. Despite regular well rehabilitation measures, hydrogeological and hydrochemical as well as constructional conditions may lead to considerably shorter periods. In this case, operation and maintenance costs exceed the benefits.

From an economic point of view, wells are technical investments for water production, consisting of components with varying lifetimes. Submersible pumps have shorter service lives than well casings. Screen, casing, and annular seal have to be regarded as a functional unit which is subject mainly to material- and location-specific influences. The life of this functional unit depends above all on well dimensioning, construction, operation, and maintenance. Location-specific influences include, e.g., the hydrochemical zonation of the aquifer as well as its hydraulic features. All expenditures for well components, the lives of which end before the investigation period has been reached, are regarded as reinvestments. For water wells, a calculation period of 20–40 years is commonly considered as the service life range. When making comparative investigations of wells, one has to distinguish between the calculatory and the technical life. The actual average service life and the operating phase (amortization period) should roughly conform. In practice, one often sets a shorter amortization period, in order to be more flexible as to technical innovations, modification of prices, as well as changes in the water demand.

For an efficiency assessment of a well, both the monetary yield (gain, benefit) and the expenditure (cost) need to be considered. Economic analyses are based on the time value of money, which is not constant over time because of interest-rate and price changes. To allow viable comparisons between costs and benefits, both have to be converted to the same point in time (= reference point) (Helweg 1982; Maniak 2001). The numerous methods of economic investment assessment, according to Zilch et al. (2002), may be classified into three categories:

- *Monovariable efficiency calculations*—the monetary profitability of investments may be assessed by statistical and dynamic calculating methods.
- *Multivariable cost–benefit analysis,* allowing the inclusion of not-monetarily- rateable economic factors.
- *Programmed procedures,* being applied in case of complex optimization calculations and simulations of benefits, financing, investment, and operator models.

In terms of water production and distribution, the direct monetary benefit resulting from selling the product "water" is measurable in comparison to the direct costs (Helweg 1982; Maniak 2001). Among the direct costs are the investment costs and the operation costs, which may be further subdivided into personnel, material, and energy costs.

6.2 Financial Mathematical Principles

Benefits and costs of a well occur at different times during the economic investigation period. Capital costs (credit financing of investment) and expenses for operation and maintenance have to be made comparable for a monetary assessment in terms of their occurrence. They have to be related to a time or a period of time—e.g., to the overall operating phase. *An immediate one-off investment* (in monetary units, MU) as one-off payment with the present value P earning an interest of i monetary units within a specified period of time, e.g., 1 year (Maniak 2001). After the first period of time, the profit G amounts to (decursive interest) [Eq. (6.1)]

$$G = i \cdot P \qquad (6.1)$$

where G = profit [MU]
p = interest rate [%]
i = calulatory interest rate with $i = p/100$
P = investment payment [MU]

The investment has thus grown to the future amount, or the sum of the initial investment after the first year F:

$$F_1 = (1 + i) \cdot P \qquad (6.2)$$

where F = future amount of money [MU]

After the second period of time, the accretion accounts for Z if, after the first period of time, the total amount is reinvested.

$$Z = i \cdot (1 + i) \cdot P \qquad (6.3)$$

where Z = annual increase in money [MU]

At the end of the second year, the original investment P has reached the capital value F_2:

$$F_2 = P \cdot (1 + i) + i \cdot P(1 + i) = P \cdot (1 + i)^2 \qquad (6.4)$$

The sum F_n of the initial investment P after n periods of time has the accumulated value

$$F_n = (1 + i)^n \cdot P \qquad (6.5)$$

$$F_n/P = (1 + i)^n \qquad (6.6)$$

where F_n/P = accumulation factor of a one-off investment of 1 MU

Accumulation means to add interest or to convert investments to a reference point after payments have incurred. In contrast, the discounting of interest means the converting of investments to a reference point prior to the incurrence of payments. A one-off investment F_n, which is to be made in n future years, has an equivalent value P, which can be determined by means of Eq. (6.7):

$$P = (1 + i)^{-n} \cdot F_n \tag{6.7}$$

where $(1 + i)^{-n}$ = discounting factor of a one-off investment of 1 MU

The cash value of an investment is calculated by means of a discounting process, the accumulated value by means of adding on interest. The *capital value* is the difference between the benefit value and the cost value of an investment and constitutes an investment criterion.

For an unchanging, uniform deferred payment (annuity) R for the initial investment, the interest and the first rate—at the end of the first year—amount to

$$F_1 = (1 + i)^{-1} \cdot R \tag{6.8}$$

The summation value F_n after n periods of time accounts for

$$F_n = (1 + i)^{n-1} \cdot R + \cdots + (1 + i) \cdot R + R \tag{6.9}$$

The *annuity factor* $(F_n/R_{i,n})$ results from the division of F_n by R:

$$(F_n/R_{i,n}) = [(1 + i)^n - 1] \div i \tag{6.10}$$

The annuity factor is identical to the accumulation [Eq. (6.6)] in the case of a uniform yearly installment (annuity). The present value P of an investment is defined by the *capital recovery value* or the *amount of annuity A* as an equivalent uniform installment for n periods of time. *Amount of annuity* means the annual installment for the acquittance of a credit. It consists of an interest and a redemption share. Most common is the *constant annuity*, with unchanging installments: with redemption payments the amount owed decreases steadily, resulting in decreased interest charges.

The annuity (or capital recovery) A indicates the annual rate of an investment or loan made at time $t = 1$, comprising 1 MU, during a depreciation time of n years [Eq. (6.6)]. With $n \to \infty$ it corresponds to the interest rate.

$$A = \frac{i \cdot (1 + i)^n}{(1 + i)^n - 1} \cdot I \tag{6.11}$$

The annual installment of capital costs R_n (if credit has been raised for the investment) results from the correlation of Eq. (6.12):

$$R_n = \frac{i \cdot (1 + i)^n}{(1 + i)^n - 1} \cdot P \tag{6.12}$$

where P = investment costs [MU]

The *total annuity* A_n of an investment for an amortization period n is calculated by multiplying the capital recovery value by the invested capital P and adding the annual costs of operation, P_B.

$$A_n = P_I \cdot \left[\frac{i(1+i)^n}{(1+i)^n - 1}\right] + P_B \qquad (6.13)$$

Examples of the economic assessment of well operation and rehabilitation will be presented in the next two sections. Examples from the literature are, however, scarce (Helweg 1982).

Helweg (1982) performed calculations to determine the appropriate time for the exchange of a pump and also the appropriate time for rehabilitation of a well. In general, he concluded that a pump should be exchanged (or rehabilitation should be performed) when the maximum annualized benefit of the new pump (or the rehabilitated well) exceed the annual net benefit of the old pump (or the unrehabilitated well). While this concept works more or less straightforwardly for the calculation of the economic feasibility of the replacement of a pump, its application to well rehabilitation is much more uncertain. Since it is practically impossible to predict how costly a rehabilitation and how high the resulting yield increase from it will be, major calculation factors remain uncertain.

6.3 Economic Appraisal of Wells Applying the Annuity Method

In terms of financial mathematics, the efficiency of a well can be determined by its cost–benefit flow during a defined period of time (Fig. 6-1). In doing so, it is advisable to treat continuous payments, e.g., capital costs (in case of credit financing), operating costs, etc., as discretely apportioned payments in the form of a series of payments. The sum of all payments is calculated at the end of each year. In terms of temporal weighting of the cost–benefit flow, both the annuity method and the discounting method can be applied. Under the *annuity method,* all benefits and costs are converted into equivalent annual installments and are then compared. Under the *discounting method* (cash value or capital value method), all benefits and costs are converted into their present values—the values of future or previous payments with added or reduced interest up to the present time (cash values). In obvious cases, such as for calculation of yearly capital costs or for comparison of clearly structured project alternatives, the annuity method is more appropriate than the discounting method.

The economic viability of a well is calculated by means of the annuity and the yearly capital costs, taking the predicted life span into account. Capital costs, energy costs, and expenditures for recurrent maintenance are calculated considering different service times and amortization periods. The precondition is that the capital for investment and maintenance measures is financed through credit amortized during the well's life span, until further maintenance becomes necessary. The results of the annual capital costs acquired accordingly for two different example wells with service times (or amortization periods) of 15–40 years, are listed in Table 6-1. Equations (6.11) and (6.12) are the basis for the calculation of the annuity A and the annual capital costs R_n. Assuming a service life (amortization period) of 15 years and an interest rate of 5.5%, the annual capital costs for the first well (unconsolidated aquifer) amount to about 12,450 €/a ($15,700/a), for the second well (consolidated aquifer) to about 19,930 €/a ($25,170/a). With a service life extension to 30 years, the due redemption of capital costs would be about 8,600 €/a ($10,860/a) for the first well, and about 13,760 €/a ($17,380/a) for the second well.

The annuity method may also be used for a cost–benefit analysis of project alternatives. The annual costs for different types of wells involving different scopes of investment are calculated using Eqs. (6.12) and (6.13) and are then compared. It is common practice to search for possibilities to lower the costs for the drilling and completion of a new well. In the case presented in Table 6-2, a well with a drilling diameter of 400 mm (15 in) with continuous-slot screens and casing made of stainless steel (case 1) is compared to a well with identical drilling diameter but with casings and screens made of PVC (case 2). Case 1 is more capital-intensive than case 2. The differences in rehabilitation ability and the entrance resistance of both screen types are taken into account by assuming different operating costs. The head loss for operating the two wells and the rehabilitation frequency differ because in case 1 the head loss and thus the energy costs are assumed to be lower than in case 2. There is no need for a yearly rehabilitation, and the costs for both cases have been divided uniformly over the wells' service lives. The stipulated rehabilitation cost for case 2 is likely to be higher than for case 1.

The annuity method, furthermore, allows one to consider the running operating costs, which in turn depend on varying energy and rehabilitation costs. The case study of Table 6-2 clearly shows that a higher initial investment may result in lower annual annuities.

6.4 Economic Appraisal of Wells Applying the Discounting Method

Instead of determining the annual installments during the overall investigation period using the annuity method, one may equally consider the capital values for the calculation of the well's economic viability. The *capital value* is the cash or present value of the sum of all future return flow

TABLE 6.1 Calculation of Annual Capital Costs Resulting from Construction of Two Different Wells (Interest Rate 5.5%)

Amortization (service life) [a]	Investment Well in unconsolidated rock formation 125,000 € ($158,000)		Amortization (service life) [a]	Investment Well in consolidated rock formation 200,000 € ($253,000)	
	Annuity [dimensionless]	Annual rate of capital costs [€]		Annuity [dimensionless]	Annual rate of capital costs [€]
15	0.09963	12,453	15	0.09963	19,926
20	0.08368	10,460	20	0.08368	16,736
25	0.07455	9,317	25	0.07455	14,910
30	0.06881	8,601	30	0.06881	13,762
35	0.06497	8,122	35	0.06497	12,994
40	0.06232	7,790	40	0.06232	12,464

TABLE 6.2 Comparison of Two Project Alternatives with Determination of Annual Overall Costs by the Annuity Method

Case 1: Construction of a well with continuous-slot screen, stainless steel (400 mm)	Case 2: construction of a well with PVC screens (400 mm)
Investment costs: 110,000 € ($138,900)	Investment costs: 90,000 € ($113,700)
Assumed annual operating costs, including energy and rehabilitation costs: 25,000 € ($31,600)	Assumed annual operating costs: 34,000 € ($42,900) (assumption: lower yield and higher entrance resistance, increased expenditure for rehabilitation)
Amortization period $n = 25$ years	
Reflux of capital 15% of the initial investment ($= i$)	
Capital recovery factor $= 0.155$ according to Eq. (6.9)	
Annual overall costs: 42,050 € ($53,100)	Annual overall costs: 47,950 € ($60,600)
Difference between case 1 and case 2: 5,900 € ($7,500) *Conclusion:* despite lower capital costs, case 2 causes higher annual costs than case 1, due to increased operating costs.	

of funds while considering the acquisition costs. It is calculated by discounting with the adequate target interest rate. Contrary to this is the *earnings value,* which is calculated as value in monetary units by capitalizing (discounting of future yields) the expected net yield of an investment.

The capital value, being the difference between benefit value and cost value and the annuity, being the average annual benefit or cost rate are connected by the capital recovery factor. This consequently corresponds to the capitalized net benefit. Using the discounting method, all benefits and costs are converted into their present values. Using discounting, all yields and running costs of the project period are converted to the value they would have at the end of the first year (reference date). Thus later costs are of lower value than earlier costs. This implies that when evaluating the efficiency of a measure, one has to take into account the varying service lives, investments, and operation costs of a well. In terms of well operation, rehabilitation measures, energy, and reconstruction costs occur at different times during the life span of the well.

The current value of a well is calculated as cash value by multiplying the net yield for each year within the amortization period by the discounting factor [Eq. (6.7)]. The discounting factor considers the date of discounting and the interest rate. The net yields are subtracted from the gross yields, considering the investment on the cost side. The capital value results from the addition of the current values at the end of the investigation period (Table 6.3).

TABLE 6.3 Example for Discounting of a Newly Constructed Well with a Rehabilitation after $n = 5$ Years as an Additional Investment ($i = 8\%$), for an Investigation Period of $n = 10$ Years

Year	Investment [€]	Operating costs [€]	Yield [€]	Net yield	Discounting factor $(1 + i)^{-n}$	Current value [€]
1	80,000 (construction)			−80,000	0.926	−74,080
2		20,000	5,000	−15,000	0.857	−12,855
3		22,000	15,000	−7,000	0.794	−5,558
4		24,500	30,000	5,500	0.735	4,042.50
5	15,000 (rehabilitation)	21,500	40,000	3,500	0.681	2,383.50
6		27,000	65,000	38,000	0.630	23,940
7		28,500	66,000	37,500	0.573	21,487.50
8		29,700	69,000	39,300	0.540	21,222
9		30,900	69,500	38,600	0.500	19,300
10		31,500	70,000	38,500	0.463	17,825.50
Book value [€]:				+98,900	Capital value [€]:	+17,708

After 3 years of operation, the well presented in Table 6-3 will have reached a positive value, assuming an initial investment of 80,000 €. A rehabilitation carried out in the fifth year will lower the current value again. Starting from the sixth year of the investigation period, the operations costs will increase while the monetary yields will still grow. The rehabilitation measure will have increased the well's capital value at the end of the chosen 10-year period because, with the increased monetary yield, a higher profit will be reached.

The case study demonstrates that by an extension of the investigation period the capital value of the investment may be maximized up to the accumulated value, unless the operation costs or additional investments during this period have not compensated for the benefit. Rising operating costs, e.g., due to increased energy demand with decreasing hydraulic yields, as well as large additional investments for rehabilitation and redevelopment, mean lower gross yields and consequently decreasing net yields. The residual value of a well—with an assumed linear decrease—for the calculation period n results from the relation:

$$\text{Residual value} = [1 - (n/\text{economic service life})] \times \text{investment value} \times \text{discounting factor} \quad (6.14)$$

The residual value may lower the capital value of an object, as shown in Table 6-4.

In the case of wells, unfounded conclusions may arise from insufficiently long calculation periods. The calculated service life should

TABLE 6.4 Determination of the Capital Value of a Well Using Cost and Benefit Values

Input parameter for cost value:
Investment amount: 150,000 € ($189,400)
Operating costs: 38,000 €/a ($48,000)
Period of time (n): 10 years
Service live (economic life): 35 years
Interest rate: 6%

Input parameter for the benefit value:
Net yield: 45,000 €/a ($56,800)

Investment: 150,000 € ($189,400)
Discount factor $(1 + i)^{-n}$: $1.06^{-10} = 0.558$
*Running costs for a 10-year period
[from Eq. (6.11)]:* 279,555 € ($353,000)
Residual value: 59,786 € ($75,500)
Cost value: 489,341 € ($618,000)

Benefit value: 652,372 € ($823,900)

Capital value: benefit value − cost value = 163,031 € ($205,900)
Capital value without residual value: 222,817 € ($281,400)

---

therefore be equated with the economic service life and should be designated as average or actual service life:

Discounting period = calculation period

When applying the discounting method, the cost accounting for the net debt service does not need to be included. Instead, the problem of varying service lives has to be considered. The annuity method considers solely the annual net debt service defined for a certain period of time, e.g., for the amortization period of an object. Still, the service life of a well varies with location-specific boundary conditions, which are not considered when applying equal average annual costs for the investigation period.

6.5 Economic Considerations for Well Rehabilitation and Reconstruction

The operating costs of a well include:

- Annual capital costs for the well
- Energy costs for water production
- Costs for well rehabilitation
- Costs for reconstruction

Personnel, materials, maintenance, and management of the facilities are additional cost factors in well operation. These are incurred regardless of the efficiency of rehabilitation and/or redevelopment measures. Figure 6-2 shows a scheme for the temporal development of the cost factors schematically. With increasing operating time of the well:

- Annual net debt service for the capital costs decreases.
- Energy costs rise due to decreasing hydraulic yield of the well.

- Annual overall costs decrease during the amortization period, despite rising energy and maintenance costs.
- Annual capital requirements for rehabilitation and redevelopment rise gradually.

The increase in annual expenditure for rehabilitation is often due to shortening intervals or to the application of more complex technologies. At the end of the amortization period, the expenditure for a well consists solely of the operating costs. With a continuing rise of these costs, the annual overall expenditure may reach the cost level of the amortization period for a new well. The "termination criterion" for further rehabilitation and redevelopment measures will then be met (Fig. 6-2).

By applying the annuity or the discounting method, this termination criterion may be forecast taking into account the amortization of the well. The amortization period is the period of time during which the earned profit of the well equals the costs accumulated to this time. At this time the net benefit has been returned from the investment. In that case, rehabilitation measures may be separated from operating costs being considered as additional investment. Construction costs include not only the costs for the replacement of the well, but also the costs for the decommissioning of the old well, installation of the new pipeline, planning, and construction supervision, as well as fees, e.g., for

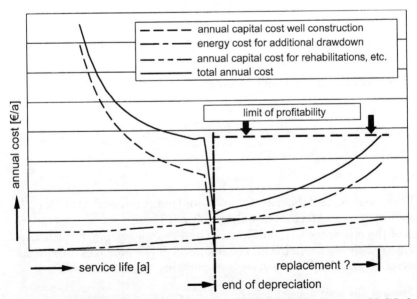

Figure 6.2 Development of the annual cost of a well as a function of time. *Modified after Walter (2001)*.

real estate purchase. If a new well pumps water of varying quality, the costs for the adaptation of a treatment system have to be considered.

In practice, the well owner's decision on whether to construct a new well or to rehabilitate/reconstruct an existing one depends mainly on the following factors, some of which are monetarily unpredictable:

- Legal conditions (is it right to approve a new construction in terms of watert?)
- Real estate (is there enough space for a new construction?)
- Unknown well-yield development
- Development of previous rehabilitation costs and the yield improvements achieved by this, compared to the energy costs and the profit (degree of utilization)
- Possible influence of construction flaws and/or groundwater quality on the yield and the operation of the well
- Which of these influential factors might be affected and/or eliminated by either a rehabilitation measure or a new construction
- What kind of rehabilitation interval can be expected in case of a proper new construction, considering the given hydrogeological and hydrochemical setting?

Before making a decision as to rehabilitation, reconstruction, or replacement of a well, one has to consider the economic, individual, location-specific hydrogeological, and hydrochemical as well as legal aspects. The influencing variables during the service life of a well, such as the energy costs, the rehabilitation intervals, or other necessary redevelopment measures may only be forecast by applying the discounting method. The annuity method assumes annual average costs, whereas the variable costs—averaged over the investigation period—have to be included. Regular recording of these variables during operation allows later comparison with the forecast so that a decision in terms of future operational proceedings may be made over yearly intervals. The construction of a new well is only economically and technically necessary when the amortization has been reached and the net yield, or the net benefit of the well, no longer covers the operating costs.

6.6 The Market for Rehabilitation and Reconstruction—Prices and Volumes

The cost for a rehabilitation or reconstruction can be substantial. Figure 6-3 shows cost ranges in Germany for rehabilitations, reconstructions, and decommissioning from the authors' practical experience.

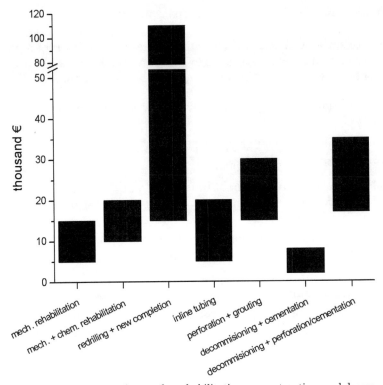

Figure 6.3 Cost range estimates for rehabilitation, reconstruction, and decommissioning of wells in Germany.

The values presented in Fig. 6-3 refer to "standard wells." For rehabilitation of radial collector wells, costs range from 50,000 € ($63,000) up to more than 85,000 € ($107,000) and may vary greatly, depending on the contractor and the technical setting. For the U.S. market, Williams (2002) estimated costs of about $55,000 for a mechanical and $120,000 for a combined mechanical-chemical rehabilitation (Fig. 6-4). The large discrepancy between the German and the U.S. data remains unexplained.

In most cases, rehabilitation work is offered on the basis of an hourly rate, which may vary with the experience of the contractor. Reliable suppliers display hours and expenses in their offer based on the information given by the customer in the transaction. It is therefore urgently advised to provide suppliers with ample information, in order to receive comparable and economical offers.

Considering that in Germany alone at least a minimum of 2,400 rehabilitations (Niehues 1999) and an unknown—but significantly smaller—number of reconstructions are performed annually, the total financial

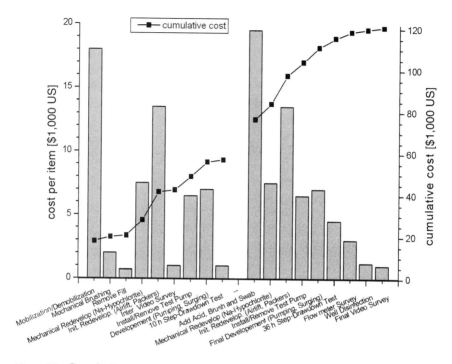

Figure 6.4 Cumulative cost range estimates for the rehabilitation of wells in the United States. *Modified after data by Williams (2002).*

volume cannot be neglected. Of the 2,400 rehabilitations in Germany, about one-quarter to one-third (600 to 800) involve the application of chemicals, although a previous mechanical step is always included. They are of course more expensive rehabilitations than purely mechanical ones given the additional costs for labor, chemicals, and treatment/disposal of wastewater. If we assume an average price of 10,000 € ($12,620) for a mechanical rehabilitation and an additional 10,000 € ($12,620) for the chemical part, we obtain:

$$2{,}400 \times 10{,}000 \, € = 24{,}000{,}000 \, € + 600 \text{ (or 800)} \times 10{,}000 \, €$$

$$= 6{,}000{,}000 \, € \text{ (or } 8{,}000{,}000 \, €)$$

$$= 30{,}000{,}000 \, € \text{ (or } 32{,}000{,}000 \, €)$$

$$= 37{,}890{,}000 \, \$ \text{ (or } 40{,}410{,}000 \, \$)$$

This is probably the minimum annual market volume in Germany. About 150 contractors with several hundred employees offer rehabilitations and reconstructions and compete in the German market. Many of them are also active in related fields such as drilling and pipeline construction. Only a few restrict their activities to rehabilitations.

The Dutch Water Research Institute KIWA found that maintenance and rehabilitation of clogged wells cause additional average costs of about 9.500 € ($12,000) per year. For the whole of the Netherlands, this amounts to about 12,000,000 €/a ($15,200,000/a) (KIWA 2002).

Chapter 7

Mechanical Rehabilitation

7.1 Processes of Mechanical Rehabilitation

The aim of mechanical rehabilitation is to remove all deposits that decrease the well's yield. Parts to be cleaned from such deposits include

- Inside of well
- Screen slots
- Outside of well screen
- Gravel pack
- Former borehole wall (at least in wells with thin annulus)

We can distinguish among hydraulic, thermal, and impulse separation methods, which are discussed in more detail in the following sections (Fig. 7-1).

Separation of incrustations may be achieved by hydraulic drag of flowing water ($\rightarrow$ 7.3), thermal expansion/contraction ($\rightarrow$ 7.4), or impulses ($\rightarrow$ 7.5). *Removal* of incrustations by pumping can either be continuous (during separation) or discontinuous (after separation). The former applies to methods that use an internal pumping system and the latter to methods such as blasting and CO_2 injection, in which the machinery has to be recovered before installing the pump. Continuous removal has an advantage, of course, in that repeated removal and installation of the equipment is not necessary, so less time is required.

All methods are subject to certain technical and sometimes hydraulic constraints, which are listed in Tables 7-1 and 7-2.

Many rehabilitation companies use self-developed, sometimes patented machinery. For reasons of competitiveness, information on

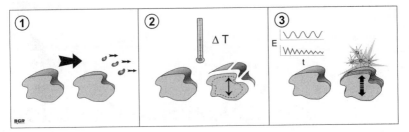

Figure 7.1 Separation methods for mechanical rehabilitation: (1) hydraulic erosion; (2) thermal expansion/contraction; (3) impulses.

modes of action, technical parameters, limitations, and effectiveness is often published restrictively. Nevertheless, the client is entitled to proper information about the advantages and disadvantages of the method offered for his wells. Undifferentiated slogans such as "is always working on all wells" are simply not enough. Serious documentation of past achievements and experiences is equally useful, as is information about limitations and situations in which this particular method should not be used. Good contractors will always provide such information and select or adapt the method to be used after taking a close look at the well specifications and the well's ageing history. One has to be aware that some method might work perfectly well at one well but may fail completely at another. The secret to success is selection of an appropriate method for the particular situation, and the experience and competence of the crew. Nevertheless, the authors present a qualitative view of the most common methods based on their experience (Tables 7-3 and 7-4).

Stepwise rehabilitations of individual screen sections have proven to be much more efficient than undifferentiated treatments of the whole screen. Methods which allow controlled implementation should be preferred to "black box" techniques. Protocols on removed incrustation or particle mass are one way to establish such control.

The efficiency of all mechanical rehabilitation schemes is higher with young, only partially solidified incrustations (→ 3.2.7). Certain adverse construction features may further limit the success possible. Small screen slots or fine, even preglued gravel packs will lead to a loss of mechanical energy, which significantly decreases the amount available for separation and removal of incrustations.

Some possible problems created by mechanical rehabilitations should be mentioned. One has to be aware that mechanical rehabilitations impose a certain amount of stress on the well's components. High pressure is often needed to agitate incrustations, which may in turn destroy weak(ened) parts of the screen. If the well was predamaged, e.g., due to corrosion, a collapse or sudden sand breakthrough is possible. The use

TABLE 7.1 Applicability of Mechanical Rehabilitation Methods for Wells of Various Types

Construction components	Brushing*	Over-pumping	Surge blocks	CO_2^- injection	Low-pressure jetting	High-pressure jetting	Jetting spears	Short-circuit pumping	Detonating gas, release of compressed air	Explosive	Ultra-sound
Tubing	++										
Wire-wound screen	++	++	+	++	++	++		++	++	++	++
Slot bridge screen	++	++	++	++	++	++		++	++	++	++
Slotted screen, PVC	++	++	++	++	+	+		+	+	–	++
Slotted screen, metal	++	++	++	++	+	++		++	++	++	++
Ceramic	++	++	+	+	++	+		+	+	–	++
Laminated plywood	++	++	+	–	+	+		+	+	–	++
Preglued gravel pack	++	+	–	–	–	+		–	+	+	++
Well sump	++			–	++	++				–	
Single gravel pack		++	++	++	++	++	+	++	++	++	++
Differentiated gravel pack		++	+	+	–	+	–	+	+	+	++
Depth-differentiated gravel pack		++	+	+	+	+		+	+	+	++
Piezometer	++	+		++		++			++	–	

Key: ++ = fully applicable and useful; + = applicable and partially useful, checking for potential damage required; – = not applicable, not recommended; no entry = application not useful or not possible.
*Plastic brushes, usually as pretreatment.
Modified after DVGW W 130 (2001).

TABLE 7.2 Applicability of Mechanical Rehabilitation Devices in Combined Mechanical and Chemical Well Rehabilitation for Various Well Types

Construction components	Surge blocks	Single-chamber devices	Multichamber devices without reversal of flow direction and flow control	Multichamber devices with reversal of flow direction and flow control
Tubing				
Wire-wound screen	+	++	++	++
Slot bridge screen	++	++	++	++
Slotted screen, PVC	+!	++	++	++
Slotted screen, metal	++	++	++	++
Ceramics	+	+	++	++
Laminated plywood	+	+	++	++
Preglued gravel pack	−	+	++	++
Well sump	−	+	−	−
Single gravel pack	++	++	++	++
Differentiated gravel pack	+	+	+	++
Depth-differentiated gravel pack	+	++	++	++
Piezometer	−			

Key: ++ = fully applicable and useful; + = applicable and partially useful, checking for potential damage required; − = not applicable, not recommended; no entry = application not useful or not possible.
Modified after DVGW W 130 (2001).

of compressed air in well rehabilitations is subject to debate, because air includes large amounts of oxygen, which oxidizes and precipitates iron and manganese oxides.

The authors know of a few cases in which the well yield actually decreased after a mechanical rehabilitation. Settling of loose gravel packs through agitation or pumping-induced colmation was probably the cause. Unfortunately, such failures are usually not published, although doing so would provide much insight. Contractors, of course, are inclined to sweep such things under the carpet.

7.2 Brushing

Partially solidified incrustations can be removed by brushing. This cheap and simple process is limited to the well interior. Brushing is often used as a preparative step before other rehabilitation techniques. Diameter and material of the brushes have to be selected in accordance with the size and material of the casing/screen. Care has to be taken

TABLE 7.3 Qualitative Assessment of Mechanical Rehabilitation Methods

Principle of rehabilitation	Brushing	Over-pumping	Surge blocks	CO_2^- injection	Low-pressure jetting	High-pressure jetting	Jetting spears	Short circuit pumping	Detonating gas, release compressed air	Explosive	Ultrasound
Operating principle	Mechanical scraping	Suffossion	Suffossion	Thermal expansion	Suffossion	Suffossion	Impulse	Impulse	Impulse	Impulse	Impulse
Removal	Discont.	Continuous in sections	Discont.	Discont.	Continuous in sections	Continuous in sections	Continuous in sections	Continuous in sections	Discont.	Discont.	Discont.
Process monitoring	Discont.	Continuous	Discont. (contin. if pump installed)	Discont.	Continuous	Continuous	Continuous	Continuous	Discont. (contin. if pump installed)	Discont.	Discont. (contin. if a pump is installed)

Discont. = Discontinuous; Contin. = Continuous.

TABLE 7.4 Qualitative Appraisal of Devices for Combined Mechanical and Chemical Rehabilitation

Principle of rehabilitation	Single-chamber devices	Multichamber devices without reversal of flow direction and flow control	Multichamber devices with reversal of flow direction and flow control
Working principle	Suffossion	Suffossion	Suffossion
Removal	Reversal of flow direction necessary; discontinuous removal possible without retrieval of device	Reversal of flow direction necessary; discontinuous removal possible without retrieval of device	Continuous removal possible through reversal of flow direction
Process monitoring	Discontinuous	Discontinuous; possible without retrieval of device	Continuous

when coated tubing, e.g., plastic-coated steel, has to be brushed. Metal brushes will damage the coating and are therefore not recommended. Plastic brushes are generally the best choice.

Incrustations can be removed by vertical and rotational movement of the brush (Fig. 7-2). Screens with vertically arranged slots are usually treated using vertical strokes, while rotational movement is used to remove deposits from horizontally arranged slots. A general problem

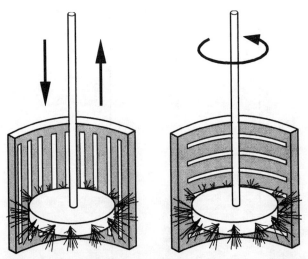

Figure 7.2 Brushing of wells with different screen slot arrangements. *Drawing: Schröder.*

with brushing is that soft deposits may be smeared into the screen slots, which can further decrease their permeability.

Loosened incrustations can be removed during or after the brushing. The former is recommended because it allows a simple form of process control: when the turbidity decreases, the brushes can be lowered into the next section. Simultaneous removal also decreases the risk of fines being displaced into the gravel pack. Because some parts of the loosened incrustations will always be missed by pumping and will settle, the sump should be cleared afterwards.

7.3 Hydraulic Methods

7.3.1 Background: hydraulic suffossion

The process of removing particles from subsurface strata by water flowing through its pore spaces is called suffossion. This process is applied for desanding and the removal of incrustations. Obviously, the forces exerted by the flowing water must exceed the forces that hold the particles. Flowing water exerts drag-and-lift forces on particles it passes by ($\rightarrow$ 4.1). Particles, on the other hand, are subject to gravitational force according to their mass and are interconnected with other grains. Finer grains, e.g., clay and silt, additionally interact by electrostatic attractive forces, rendering their cohesion stronger. For incrustations, the chemical bonding force of the crystal lattice or between the incrustation minerals and their substrate has to be taken into consideration, too. The theory behind this is discussed in more detail in Chap. 4.

7.3.2 Surge blocks

Surge blocks are simple and inexpensive tools that have found widespread use in well development but are also used for rehabilitation. The setup can be quite simple: a packer disk that fits loosely within the inner well diameter is moved up and down intermittingly. The downward movement displaces water into the annulus (piston effect). A relief hole or valve is usually provided in the packer to limit pressure surges and deposition of loosened material on the packer. The following upward movement displaces the water column above and thus sucks water and loose(ned) solids from the annulus. The whole cycle thus creates a wash and backwash action (Fig. 7-3). Loosened material settles and collects in the sump and has to be removed subsequently. This is usually performed using robust pumps without movable parts (injector pumps). Surging by slugs is not recommended because of the danger of mechanical damage to the well. An alternative to packer surging blocks is pistons which are expanded—and retracted—by compressed air or other means. The diameter of these pistons has to be slightly smaller than the well diameter.

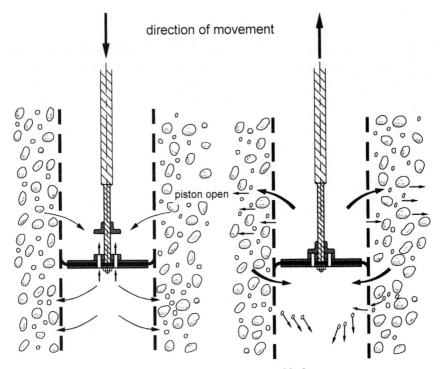

Figure 7.3 Mechanical well rehabilitation using a surge block.

Surging is usually performed from top to bottom of the screen. Slow movement is recommended to limit the stress on well casing and screen. The penetration depth into the annulus is limited due to the limited amount of energy applied. Inappropriate use may lead to unwanted compaction and modification of the gravel pack texture. This can result in decreasing yield or sand intake.

A general disadvantage of surge blocks is that process control is nearly impossible, because the amount of removed material can only be measured while clearing the sump. If the amounts removed are very high, a camera inspection is recommended to clarify the causes (e.g., broken screen slot bridges).

7.3.3 Intense pumping

Intense pumping is a method that was originally developed for well development and desanding. A pump—or at least its inlet—is installed between two packer disks (rubber disks or inflatable sleeves) (Fig. 7-4). Water, sand, and loosened incrustations enter within this small section. Usually an entrance velocity five times higher than during regular

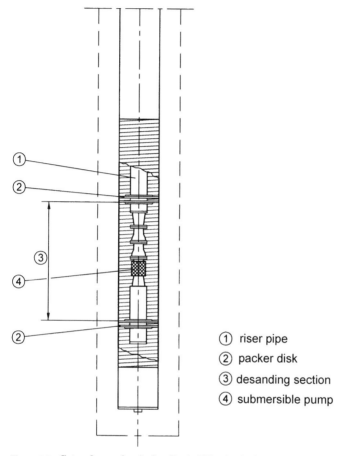

Figure 7.4 Setup for mechanical well rehabilitation by intense pumping.

operation is recommended. Packer spacing and pump discharge have to be adjusted to the screen length and nominal pumping rate of the well.

Using the same pumping rate, the packer spacing should be one-fifth of the screen length. Example calculations on packer spacing and pumping rate will make the concept clear [Eq. (7.1)].

Calculation of packer spacing

$$L = \frac{L_s}{5} \cdot \frac{Q_I}{Q_n} \tag{7.1}$$

where L = packer spacing, L
L_s = screen length, L

Q_i = intense pumping rate, L^3/T
Q_n = nominal pumping rate (during regular operation), L^3/T

Example
Given: Q_i = 80 m³/h (47 ft³/min)
Q_n = 125 m³/h (74 ft³/min)
L_s = 30 m (98 ft)
To be calculated: L

Using Eq. (7.1), we calculate a packer spacing of 3.84 m (12.6 ft). For practical purposes we would employ a spacing of 4.0 m.

Calculation of intense pumping rate

$$Q_i = \frac{5 \cdot L \cdot Q_n}{L_s} \qquad (7.2)$$

Example
Given: Q_n = 50 m³/h (29.4 ft³/min)
L_s = 15 m (49 ft)
L = 2.5 m (8.2 ft)
To be calculated: Q_i

From Eq. (7.2), we obtain an intense pumping rate of about 42 m³/h (25 ft³/min).

The packer disks or sleeves should not be too tight, to avoid wedging of the apparatus or damage to the screen. The overlap of individual treatment sections is usually 0.5–1.0 m (1.5–3 ft).

Intense pumping is useful to remove loose sand and soft or slightly solidified incrustations. The effective range is restricted to the screen slots and the neighboring gravel pack. The penetration depth is a function of the flux and the flow velocity of the groundwater (Williams 1981). Based on the continuity equation and the work of Huisman (1972, cited in Williams 1981), the radial distance from which fine material can be mobilized and removed can be calculated. This assumes that the gravel pack is adjusted to the surrounding formation and functions as planned ($\rightarrow$ 2.3). The pumping rate necessary for mobilization of fines from a defined radial distance can be calculated following Eqs. (7.3a) and (7.3b):

$$\frac{Q_i}{Q_n} = \frac{v_e \cdot A_2}{v_e \cdot A_1} = \frac{d_2}{d_1} \qquad (7.3a)$$

$$Q_I = \frac{Q_B \cdot d_2}{d_1} \qquad (7.3b)$$

where v_e = entrance velocity, L/T
A_1 = outer mantle surface, L^2
A_2 = inner mantle surface, L^2
d_1 = well diameter (outer screen diameter), L
d_2 = penetration depth of intense pumping, L
Q_B = nominal pumping rate (during normal operation), L^3/T
Q_I = intense pumping rate, L^3/T

Example
Given: Q_n = 100 m³/h (59 ft³/min)
d_1 = 440 mm = 0.44 m (outer screen diameter) (1.44 ft)
d_2 = 850 mm = 0.85 m [borehole diameter (= screen + annulus) + desired radial penetration depth] (2.79 ft)
To be calculated: Q_i

After substitution into Eq. (7.3b), we obtain an intense pumping rate of about 193 m³/h (114 ft³/min).

Intense pumping allows direct progress control, because the turbidity or particle content can be measured easily from above ground. Turbidity meters or *Imhoff cones* (sedimentation cones) can be used. Measurement and treatment of a section have to be repeated until a stable and tolerable amount of particles or turbidity has been obtained. Spontaneous increases of sand intake are an indication of mechanical failures in the screen.

Intense pumping is useful for desanding wells, but erroneous calculations of gravel pack granulometry and screen slot size may result in a subsequent colmation or sand intake from the formation. Preglued gravel packs should not be treated using this method, because the mobilized sand can easily plug the limited pore space, or the pack might break. It has to be noted that the amount of water used during such rehabilitations may be substantial and has to be discarded properly.

7.3.4 Closed-circuit pumping (chambered systems)

Closed-circuit pumping systems consist of one or more chambers with separate injection and extraction (Fig. 7-5). Many setups have been available on the market since the early 1990s (Paul 1990). Differences involve the packer type, the injection and extraction system, and the water supply. In many cases, differences are quite minor. Such systems are often referred to as "hydromechanical rehabilitation techniques" or simply "gravel washers."

The chambers are separated from each other by packers, which may be inflatable or simply consist of rubber disks. Water is injected via a freely rotating pump impeller or nozzles. The water flux hits the screen slots and proceeds—with diminished energy—into the gravel pack and

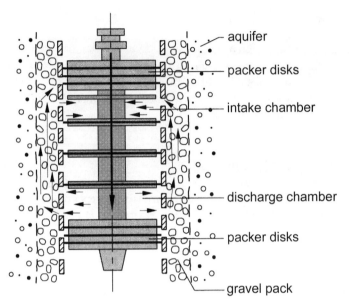

Figure 7.5 Multichamber pumping system for well rehabilitation. *From Detay (1997).*

the formation. The same amount of water is extracted by a separate pump with its inlet installed in a different chamber. Therefore, a local flow cell develops through the gravel pack (short circuit). The extraction pump must be robust enough to handle substantial amounts of suspended particulate matter. In some cases the injected water volume is taken directly from the well by a separate pump installed above the system. In other cases clean water from other wells must be used. It is essential not to use the water from the extraction chamber, because of its turbidity and potential for replugging.

Systems with one chamber contain both the injection and the extraction in one chamber. Packers are installed above and below this chamber. In multichamber systems the injection and extraction chambers are separated by an additional packer system (Figs. 7-6 and 7-7).

The water from the extraction chamber is pumped to the surface, which easily allows semiquantitative inspection of the amounts of mobilized incrustation or particles. *Imhoff sedimentation cones* or turbidity measurements are particularly useful for this purpose. If the pumped water regains clarity, the system may be lowered to the next section. The general advantage of these systems is that separation, removal, and process control can be achieved simultaneously and for individual sections.

As with other methods, sudden increases in particle content during an application could be an indicator of screen failure. In this case the

Figure 7.6 Chambered pumping system for well rehabilitation with attached brushes. *Photograph: Brunnen- und Pumpenservice, Pulheim, Germany.*

application must be interrupted and the causes investigated. Otherwise the sand may bury the whole system and require costly measures for retrieval and well reconstruction.

Penetration depth and suffossional energy are functions of the pumping injection rate and the permeability of the screen slots and the gravel pack. According to studies on a 1:1 scale model, multichambered systems were the only hydraulic systems able to penetrate to the former borehole wall (DGFZ 2003). Frequency-regulated pumps are often employed to adapt to a larger suite of well conditions. The packered setup allows treatment of individual screen sections. This feature also makes such

Figure 7.7 Chambered pumping system for well rehabilitation. Note the inflatable packers and the open impeller in the middle. *Photograph: Aquaplus, Kronach, Germany.*

systems very useful for chemical rehabilitations in which a loss of chemicals into the formation must be avoided (→ 8).

Application is restricted to screen material which can handle the flow magnitudes involved. Screens with corroded slot bridges and such made of aged PVC or laminated wood may not be able to stand the stresses. Preglued, resin-bound gravel packs are also difficult to treat with this method.

7.3.5 Jetting

Jetting relies on the erosional effects of thin but fast-flowing water jets. Water, either from inside or from a source outside the well, is pumped into it and exits through nozzles. The user must be aware that the pressures quoted by the manufacturers all refer to the source pressure. The pressures that reach the outside of the screen or even the former borehole wall will be substantially lower due to scattering and attenuation of energy. Fountain and Howsam (1990) therefore correctly postulate that the distance between nozzle and the surface to be cleaned must be as small as possible. Their experiments showed that the pressures decrease rapidly with increasing distance from the nozzle, e.g., to 5% of the nozzle exit pressure at a distance of only about 40 nozzle diameters. With nozzle diameters usually in the range of a few millimeters (1–3 mm), this translates into an effective range of a few centimeters (40–120 mm). Fountain and Howsam (1990) obtained an equation that describes the three-dimensional diffusion of pressure emanating from a jet nozzle [Eq. (7.4)]. The first term on the right-hand side describes the pressure diffusion in the axial direction, and the second term the diffusion in the radial direction.

$$\frac{P_x}{P_0} = 332 \cdot \left(\frac{D_0}{x}\right)^{2.64} \cdot \left[\exp^{-y^2/2 \cdot (0.05 \cdot x^2)}\right]^2 \qquad (7.4)$$

where P_x = pressure at distance x, L
P_0 = pressure at nozzle exit, F/L^2
D_0 = nozzle diameter, L
x = axial distance from nozzle (in direction of jet flow), L
y = radial distance from centerline of jet (assuming rotational symmetry, y = z), L

We can distinguish between low-pressure and high-pressure jetting. The former relies on low pressures [<5 bar (73 psi)] and high flow rates (several 10s or 100s of m^3/h), while the latter applies high pressures [>>10 bar (145 psi)] and low flow rates [<20 m^3/h (<12 ft^3/min)].

Low-pressure systems contain up to 30 fixed nozzles with an opening width of about 10 mm (0.4 in) (Fig. 7-8). Up to 100 m^3/h (440 gal/min) are injected during operation at pressures of about 1.5 bar. The water then exits at velocities of 5–20 m/s (16–65 ft/s).

High pressure systems involve much higher source pressures, adjustable in a wide range from 50 to 500 bar (725 to 7,250 psi) (Schultes & Moses 2002). Typical injection rates are in the range of <10 m^3/h (44 gal/min). Two or three nozzles are the most common setup (Figs. 7-9 and 7-10). The very high-flow velocities of the water jets produce very high stresses on small areas of the screen (blowpipe effect). This may lead to serious damage, especially when the water jet acts on weak spots—such as screen slot bridges and coatings—for a longer time. Fountain and Howsam (1990) found that jet pressures exceeding 300 bar (4,350 psi) cut through plastic tubing at 25-mm (1-in) nozzle-to-surface distance, with 1.5- to 2.0-mm nozzle diameters. In general practice, the

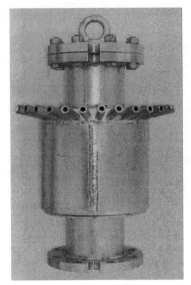 

Figure 7.8 Low-pressure jetting system. *From DGFZ (2003).*

recommended maximum pressures for plastic pipes should not exceed 200 bar (2,900 psi). Pressures of up to 550 bar (8,000 psi) have been observed not to cause damage to steel well tubing (Fountain & Howsam 1990).

As can be easily imagined, the design and geometry of the screen slots have a profound influence on the efficiency of jet rehabilitation (Fountain & Howsam 1990). Screens with high open area and continuous slots, e.g., wire-wound screens, impose the lowest resistance to jet flow. Screen slot bridges, on the other hand, are a serious impediment due to the 90° deflection (and dampening) of the jet.

The potential damage caused by high-pressure jets hitting the screen surface precludes the use of fixed nozzles. Rotation of nozzles may be achieved either by the recoil of the exiting water jet or by electrical drives. The former is often employed with systems of very high pressure (several hundred bar of pressure at the nozzle head). The rotation can be fast enough to produce cavitation and thus secondary impulses (500 Hz) which act on incrustations (Etschel & Schmidt 2001; → 7.5). The rotational speed of electrically driven nozzle heads is between 20 and 100 rev/min. Secondary impulses are much less pronounced at these speeds with frequencies in the range of 500–700 Hz (DGFZ 2003). Electrical drives require a more sophisticated control system but have the advantage that unwanted stops of the nozzle rotation can easily be monitored. Sometimes the blowpipe effect of high-pressure jetting is used deliberately to cut apart well tubing as a preparation for retrieval and subsequent reconstruction (→ 9.2).

Figure 7.9 High-pressure jetting system for well rehabilitation with electrically driven nozzle head with two nozzles. *Photograph: Aquaplus, Kronach, Germany.*

The water jet is assumed to break up incrustations and erode particles from the gravel pack. Pulsation may occur due to the rotation of the nozzle head and the subsequent rhythmic increase and decrease of water pressure. Gravel grains might be excited to start rotating and peel off incrustations.

Most jetting systems include an independent pump to remove the loosened material during application (Fig. 7-10). The amount of removed material can then be assessed aboveground. Sometimes the jetting system and the pump are installed between packers, which effectively resembles the systems described in the preceding section. The pumping rate must be higher than the infiltration rate, often by a factor of 2.

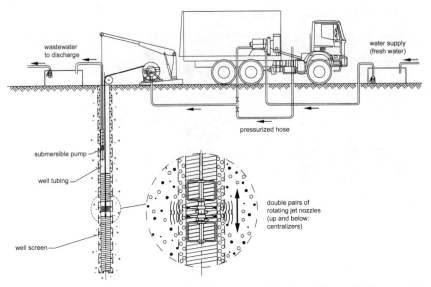

Figure 7.10 High-pressure jetting system for well rehabilitation with recoil-driven nozzle head (two nozzles) *Drawing courtesy of E & M Bohrgesellschaft mbH, Hof, Germany.*

Wells equipped with casing or screen material of low strength, e.g., thin steel, stoneware, laminated wood, PVC, and preglued gravel packs, should not be treated with high-pressure jetting or at least at reduced pressures. Localized elevated pressure acting on the inside of the well casing and screen may damage coatings (Fig. 7-11). Contractors usually have experience in the applicable pressures and limitations.

Figure 7.11 Damaged coating of ceramic well screen after high-pressure jetting. *From Munding (2005).*

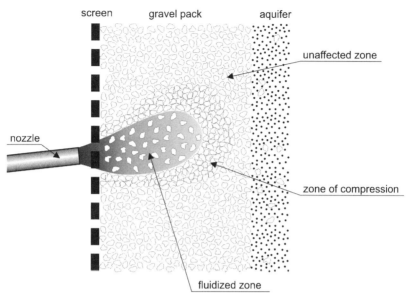

Figure 7.12 Effects of water jets on the texture of gravel packs. *Redrawn after DGFZ (2003).*

The injection of substantial volumes of water at high flow velocities may lead to a redistribution of grains, disruption of textures, and compaction of the gravel pack (Fig. 7-12).

7.3.6 Jetting spears

The use of jetting spears for mechanical rehabilitation is mostly of historical interest but still should be mentioned. The method employs jetting spears (drive points) which are jetted into the annulus from the ground surface or the floor of the well shaft. Once the drive-point screen is in place, the gravel pack can be flushed directly. Since there is no interference by the well screen—which dampens the applications of all methods acting from inside the well—the efficiency is quite high. Although this implies a high potential for breaking up incrustations, the method has some serious disadvantages:

- Damage to annular seals
- Disruption of gravel pack texture, especially in differentiated gravel packs
- Settling of gravel pack
- Damage to screen and casing

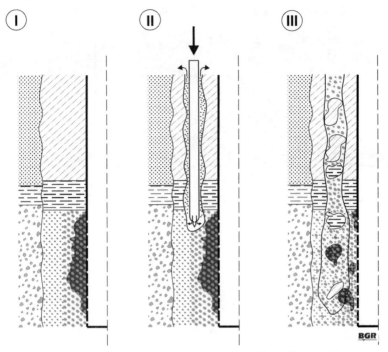

Figure 7.13 Rehabilitation by jetting spears and resulting effects on the texture of the gravel pack and the annular seal. I = well prior to application (dark = incrustation); II = jetting spear being inserted; III = broken-up incrustation and destroyed texture of annulus after application.

These features may lead to an uncontrolled inflow of surface water through the annulus, breakthrough of sand, and loss of yield. Therefore the authors strongly advise against the use of jetting spears. Figure 7-13 shows the application and the possible disruption of the annulus.

7.4 Thermal Methods

7.4.1 Carbon dioxide (CO_2) freezing

The phase diagram presented in Fig. 7-14 shows the different phase transitions of carbon dioxide as a function of temperature and pressure. Low temperatures and high pressure are needed to liquefy it. The simplest way to apply CO_2 in well rehabilitation would be to dump blocks of solid CO_2 (dry ice) into the water column (Howsam 1990c). However, this would lead to its rapid conversion to gas, producing a violent reaction due to the massive volume expansion. The danger of a sudden outburst of a suffocative gas/water mixture precludes the use of this method. For this reason, more sophisticated concepts had to be developed.

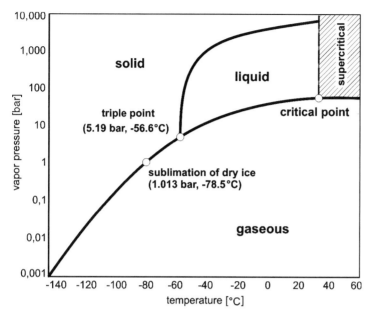

Figure 7.14 Phase diagram of carbon dioxide.

The currently used technique of deep-cold carbon dioxide application was originally developed in the United States around 1970 by Igor Jawrowsky and it can look back now on some 20 years of experience (Saunders 1996; Mansuy 1998). Since the end of the 1990s the technique has also been available outside the United States.

The effect of carbon dioxide on incrustations can be attributed mainly to the mechanically and thermally induced stresses. Therefore this method is correctly listed under mechanical methods even though it uses a chemical agent. The injection of a chemical, however, may still require formal authorization by the authorities in some countries (→ 8.1).

All materials react to changes in temperature by expanding or contracting. This causes mechanical stresses which can cause the cracking of solids or detachment of incrustations from surfaces. Most solids contract under low temperatures, yet as we all know, water actually expands. The thermal contraction of mineral grains under very cold conditions is also quite minor, but any water enclosed in pores will expand by about 9.1 vol% when turning from liquid to solid (ice).

Carbon dioxide is a weak acid which forms some H^+ ions when dissolved in water [Eq. (7.5)]. The minimum obtainable pH is in the range of pH 5 and is thus practically negligible for the proton-assisted chemical dissolution of incrustations (→ 8.2).

$$CO_2 + H_2O \Leftrightarrow HCO_3^- + H^+ \qquad (7.5)$$

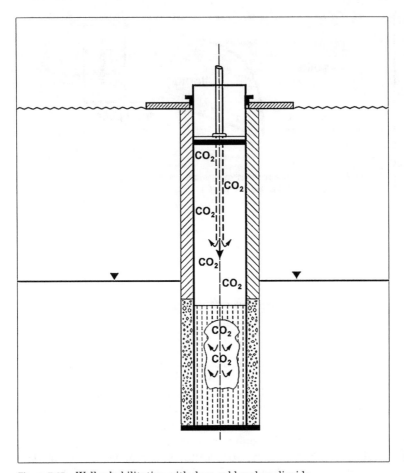

Figure 7.15 Well rehabilitation with deep-cold carbon dioxide.

The actual procedure can be divided into several steps (Fig. 7-15). It begins with the installation of a packer to seal off the well. In the following step, the water volume inside the well is expelled by the injection of gaseous carbon dioxide. This procedure is essential because freezing of the present water volume would require an immense amount of energy and would severely damage the well. After that, liquid carbon dioxide is injected at >12 bar (174 psi) and −40°C (−40°F), which is then transformed into gas by agitation. This phase transformation results in an up to 570-fold increase in volume, which induces mechanical stresses that can loosen incrustations. Penetration of the gas phase into the aquifer can be substantial [20 m (60 ft) or more]. The low temperatures also cause water remaining in pore spaces of the gravel pack to freeze and expand. The whole procedure may be repeated several times and may be applied in sections using packers. The amount of CO_2 used can be quite substantial, depending on the well diameter, screen length, and number of repetitions. Usually a tank car of CO_2 is needed.

Gaseous CO_2 may spread quite widely within the aquifer and may even reach neighboring wells. There it will rise from the screen into the well interior and possibly the shaft. Therefore these wells must be checked before anyone enters the housing.

Because parts of the injected CO_2 will dissolve in the groundwater, the water quality may be affected. According to Eq. (7.5), the pH will be lowered while at the same time the dissolution of carbonates will be promoted. During an application in Germany, Jüttner and Ries (2001) noticed marked alterations of water chemistry (Fig. 7-16). They found a decrease in pH from 7.52 to 6.80, an increase of calcium from 69 to 113 mg/liter, an increase of total hardness from 14.7 to 21.7dGH and of carbonate hardness from 12.0 to 19.2 dGH (dGH = German Hardness, 1 dH = 10 mg/liter CaO, 1 American hardness = 1 mg/liter $CaCO_3$). Specific electrical conductivity rose from 556 to 722 µS/cm, due to the enhanced carbonate dissolution. Turbidity increased from 0.1 to 0.3 FNU (FNU = Formazin Nephelo metric unit), probably due to incomplete removal of loosened ochre. A slight increase in iron and manganese concentrations was also noticed. Other parameters were not affected. Such changes may have an effect on the water treatment scheme prior to distribution into the

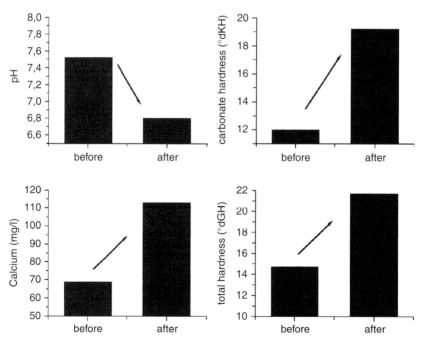

Figure 7.16 Changes in water chemistry during a well rehabilitation with deep-cold carbon dioxide. *After data by Jüttner & Ries (2002).*

network. The duration of these changes was not reported. Nevertheless, carbon dioxide has the positive side effect of being microbicidal.

Occupational safety is certainly an issue of concern in using this method. To prevent frostbite, personnel must not come into contact with the deep-cold CO_2. Suffocating CO_2 gas may collect in shafts and morphological depressions. Natural and anthropogenic CO_2 occurrences were responsible for many fatalities in the past. Specialized equipment, enforced aeration of well shafts, hand-held gas indicators, protective clothing, and special training are inevitable.

The authors performed some laboratory-scale experiments on the application of deep-cold liquids on iron oxide incrustations. We used liquid nitrogen which allows for much lower temperatures [$\leq -195°C$ ($-319°F$)] than liquid carbon dioxide. Soft ochre incrustations submersed in liquid nitrogen froze immediately. Upon evaporation of the nitrogen, the broken fragments disintegrated into a pumpable viscous suspension. The usefulness of this scheme for soft incrustations is therefore easily shown. On the other hand, hardened older ochres cannot be treated successfully, according to our experiments. Some wet goethite-rich incrustations were completely immersed in liquid nitrogen for about 45 min. We were able to retrieve them almost unaltered after the specified time. The effect on such incrustations is therefore strongly limited.

7.4.2 Steam injection ("geyser")

The boiling of wells is one of the oldest methods of rehabilitation. The first patent on well rehabilitation in Germany was based on the injection of hot steam [Böttcher 1905, Patent no. 181 578, *Reinigen von Rohrfilterbrunnen mittels Dampf* (= "cleaning of tube wells with steam")]. When the uplift of the injected steam exceeds the gravitational force of the water column, the well will "blow" and erupt like a geyser. The effect of the boiling of the water column inside the well is thus twofold:

1. Thermal expansion
2. Strong erosion during geyser-like eruption of water

Hot water and steam also have a microbicidal effect ("pasteurization") and can thus kill off ageing-related bacteria such as iron-related bacteria (IRB).

General disadvantages of heating methods include:

- Extremely high energy demand
- Safety issues (eruption of hot water possible)
- Heat may negatively affect casing and screen material as well as annular seals
- Biomass may coagulate (like boiled eggs) and cause further plugging

The most important disadvantage in terms of cost is of course the energy demand. Water has a heat capacity of 4.2 kJ/kg · K. This means that we have to invest 4.2 kJ of energy to increase the temperature of 1 liter (or 1 kg) of water by 1°C. Compared to other substances, this value is quite high (sand, 1.2; solid rock, 0.7–1.4 kJ/kg · K).

The method is not applied in Europe anymore but is sometimes still used in the United States (Alford & Cullimore 1998).

7.5 Impulse Methods

7.5.1 Background: impulse generation and attenuation

Impulses can trigger various processes which may affect the adhesion of incrustations to casing, screen, and gravel pack:

- Differential acceleration of components (casing/screen, gravel pack, and sediment particles, etc.) due to different elastic properties causing relative movement against each other and thus abrasion
- Resonance effects
- Liquefaction of thixotropic fluids, e.g., iron/manganese oxide gels, bacterial slimes

Resonance effects are known to anyone who drives a car. The motor obviously emits an exciting or source frequency f_s depending on speed. Sometimes some small part suddenly begins to vibrate vigorously, e.g., the lid of the glove compartment or a CD cover. Accelerating will usually stop the vibration, but maybe some other component will then act up. The physics behind this phenomenon is that bodies will start to vibrate when excited at a certain frequency, the so-called eigenfrequency f_0. At this frequency the amplitude of the resonance vibration suddenly reaches a maximum ("resonance catastrophe," Fig. 7-17). It is mainly a function of the mass of the body: lighter parts will vibrate much more easily. Therefore, the glove compartment will often rattle while the heavier motor will hopefully not do so.

Thixotropy is the capacity of a suspension to pass from a solid state (gel) to a liquid state while being agitated and to return to the initial state when the agitation ceases. A common household example is ketchup, which will not flow from the bottle unless we shake it well. Whether young iron and manganese oxide gels or bacterial slimes are actually thixotropic and can be liquefied by impulses has yet to be investigated.

Several techniques are available to produce impulses for well rehabilitation. Water is practically incompressible and therefore transmits

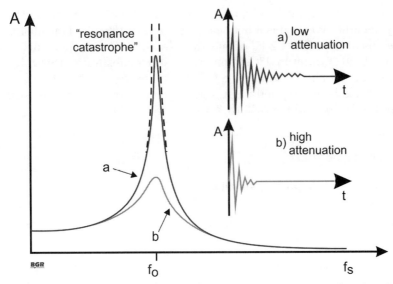

Figure 7.17 Amplitude A of the resonance oscillation of a body as a function of the excitation by a source of frequency f_s.

impulses very well. Short impulses can be achieved by explosions, while ultrasound produces periodic impulses. The frequencies involved can be around 20–40 Hz for the former and 20,000–25,000 Hz for the latter.

When energy, e.g., as sound, travels through a medium, its intensity diminishes with distance. In idealized materials, the sound pressure (amplitude) is reduced only by the spatial spreading of the wave. Natural materials, however, all produce an effect which further weakens the sound. This is due to two basic factors, scattering and absorption. The combined effect of the two is called attenuation. The energy we put into an impulse will therefore not completely reach its target, i.e., the incrustation. Attenuating processes are

- Reflection at interfaces, e.g., water/screen
- Refraction at interfaces, e.g., water/gravel
- Transformation into other energy types (sound, temperature, movement, other wave types), e.g., absorption
- Destructive interference of waves

Let us first consider spatial spreading. The energy released in impulses usually comes from a very small source (point source) and dissipates spherically over its surroundings. The input power P is therefore distributed over the surface area of the sphere A (Fig. 7-18a). This surface increases with

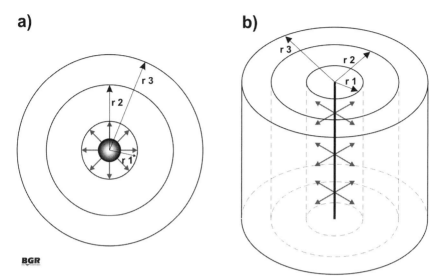

Figure 7.18 Dissipation of energy during the propagation of a wave front from (a) a point source (radial/spherical spreading) and (b) a line source (cylindrical spreading).

increasing distance from the source. The intensity I—i.e., the ratio between power and surface area of the sphere—thus decreases [Eq. (7.6)]. Since the surface area of a sphere is $A = 4 \cdot \pi \cdot r^2$, the intensity decreases with the square of the radius, which involves quite dramatic intensity decreases.

$$I = \frac{\text{power}}{\text{area}} = \frac{P}{A} = \frac{P}{4\pi \cdot r^2} \qquad (7.6)$$

For a line source with a cylindrical dissipation of energy, the area equals $A = 2 \cdot \pi \cdot r \cdot h$ (Fig. 7-18b). An example of energy dissipation of ultrasonic sound is discussed in Sec. 7.5.5.

Reflection of sound waves at interfaces is yet another limit to the penetration beyond the screen and into the gravel pack (Fig. 7-19). The effects of reflection are strongly influenced by the geometry of the system and the density of the interface components. Interfaces can be the transition between water, screen, or gravel. Here we consider only density contrasts at interfaces, because the complex geometries of screen slots and the gravel pack are basically impossible to assess. The relative amount of reflection R at an interface of components with differing densities can be calculated according to Eq. (7.7):

$$R = \left(\frac{\rho_{\text{water}} - \rho_{\text{solid}}}{\rho_{\text{water}} + \rho_{\text{solid}}}\right)^2 \qquad (7.7)$$

where ρ = density of material, M/L^3

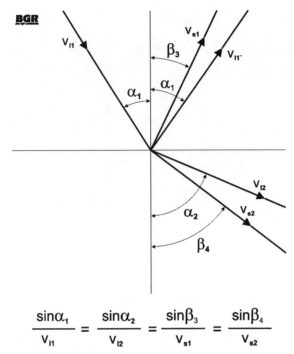

Figure 7.19 Reflection and refraction of longitudinal waves and generation of shear waves at a material interface. v_{l1} = longitudinal wave velocity in material 1. v_{l2} = longitudinal wave velocity in material 2. v_{s1} = shear wave velocity in material 1. v_{s2} = shear wave velocity in material 2.

$$\frac{\sin\alpha_1}{V_{l1}} = \frac{\sin\alpha_2}{V_{l2}} = \frac{\sin\beta_3}{V_{s1}} = \frac{\sin\beta_4}{V_{s2}}$$

An example on the effects of reflection is shown in more detail for ultrasonic sound in Sec. 7.5.5.

When a sound wave passes an interface between two materials with different indices of refraction at an oblique angle, both reflected and refracted waves occur (Fig. 7.19). This is a well-known phenomenon for light, which makes objects you see across an interface, e.g., a water surface, appear to be shifted relative to their real position. Refraction at interfaces occurs as a result of the different velocities of the acoustic waves within the two materials. The velocity of sound in materials is determined by the physical properties of the material (elastic modulus; density). Refraction will cause dissipation of the energy of the waves.

When sound travels in a solid material, one form of wave energy can be transformed into another form (mode conversion). Longitudinal waves hitting an interface between materials of different acoustic impedance at an angle not normal to the interface can cause particle movement in the transverse direction. This causes a shear wave to form. Finally, all the reflected, refracted, and converted waves present at one time can interact with each other. Both constructive and destructive interactions are possible. Interacting waves superimpose their amplitudes onto each other. The amplitude at any point of interaction is the sum of the amplitudes of the individual waves. When two waves are *in phase*—peaks and valleys of one are exactly aligned with those of the other—they combine to add the amplitudes of either wave. When they are *out of phase*—peaks of one wave are exactly aligned with the valleys of the other wave—they combine to cancel each other out. When the two waves are not completely *in* or *out of phase*, the resulting wave is the sum of the wave amplitudes for all points along the wave.

As mentioned above, some portion of the wave energy of a wave traveling through solid material, e.g., rocks in the subsurface, will be converted to heat, sound, or particle movement within the material. This conversion of wave energy represents a net loss of energy for the propagating wave. Attenuation is generally proportional to the square of sound frequency. Because the absorption of sound increases with increasing frequency, ultrasonic sound suffers higher attenuation than audible sound.

Impulse methods have some noteworthy technical limitations which should be considered before using them indiscriminately on wells:

- Damaging of brittle or predamaged casing or screens is possible, especially when they are exposed to very high accelerations (explosions).
- Settling/compaction of the gravel pack may result in a decrease of hydraulic conductivity and dislocation of the annular seal.

From the authors' experience, several wells have experienced a decrease in yield, were damaged, or even collapsed as a result of impulse rehabilitation. Caution in selection and application of impulse methods is therefore strongly advised. Camera inspections and geophysical logging are useful to existing present damage beforehand—e.g., corrosion holes or broken screen slot bridges—which might lead to a complete collapse of the well with vigorous agitation.

Before applying impulse techniques, the well interior has to be cleaned mechanically, e.g., by brushing. After the impulse application, section-wise pumping is usually required to remove loosened incrustations.

7.5.2 Explosive charges

The use of explosives to perforate casings and stimulate aquifers is a well-known practice in the oil industry. Explosive charges are sometimes used to stimulate open-hole water wells in consolidated rocks. The charges commonly range between 10 and 100 kg of explosives, but results are somewhat erratic.

Explosive charges for the rehabilitation of water wells were first tested in the United States at the beginning of the 1950s. In Europe, the

Berlin water works took the leading role in establishing this method ("blasting") in the 1990s (Rübesame 1996; Steußloff & Wicklein 1999; Steußloff & Steinbrecher 2001). Charges are usually in the range of a few to about 100 g of TNT equivalent. This is sufficient to stimulate the well but low enough to preclude damage to surrounding buildings, etc. Both inorganic, e.g., ammonium nitrate (NH_4NO_3), and (nitro)organic explosives, e.g., trinitrotoluene (TNT, $C_7H_5N_3O_6$), are used. The main reaction products are gases, which cause the high pressures. After the explosion they escape into the atmosphere as large gas bubbles. The reaction for ammonium nitrate proceeds as follows [Eq. (7.8)]:

$$2\ NH_4NO_3 \Leftrightarrow 2\ N_2 + O_2 + 4\ H_2O \qquad (7.8)$$

Reaction products of nitroorganic explosives are also chiefly gases, including carbon dioxide (CO_2), carbon monoxide (CO), nitrogen (N_2), and minor amounts of nitric oxides (NO), hydrogen (H_2), and oxygen (O_2). Pumping the well after treatment ensures that dissolved remnants are removed. Chemical analyses by the Berlin water works did not show negative alterations of water quality after blasting (Rübesame 1996).

The high accelerations induced by explosions are the main process that loosens incrustations. Additionally, the highly compressed gas bubble flushes the gravel pack during its passage. Repeated compression and decompression causes vibration. The gas bubble rising in the well interior has a slight piston effect, removing loosened material from the gravel pack. The expanding gas bubble can be utilized to introduce chemical rehabilitation agents.

In many cases, rehabilitation by blasting consists of several treatments, each time interrupted by intermittent pumping to remove loosened incrustations and chemical residues. The explosives usually come as line charges (detonation cords) fixed between centralizers in order to ensure radial action and to avoid damage to the screen (Fig. 7-20). Line charges allow treatment of the whole screen or at least longer sections

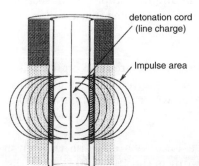

Figure 7.20 Explosive line charges (detonation cord) for well rehabilitation. *Drawing courtesy of Pigadi, Berlin, Germany.*

at once. They contain between 5 and 25 g of explosive per meter. Both vertical and horizontal wells can be treated.

Explosive charges potentially endanger brittle or predamaged well casings and screens. The application is therefore limited or even impossible for wells made of PVC, ceramics, stoneware, and laminated wood. Resin-bound gravel packs are also difficult to handle. Sometimes such wells are deliberately destroyed by blasting after the previous insertion of a wire-wound screen (→ 9.2).

Handling and using of explosives is subject to strict legal regulations—at least in most countries. Personnel need to have special training and applicable certifications. In some countries it may be required to notify the local authorities prior to performing such rehabilitations. Because the gases formed through the explosion are suffocative and toxic, the well shafts need to be ventilated before anyone enters.

7.5.3 Explosion of gas mixtures

Explosions can also be produced by ignition of appropriate gas mixtures in the well. Such methods were originally developed in Russia but have found some application in central Europe as well. The current method is based on the electrolytic decomposition of water into hydrogen gas (H_2) and oxygen gas (O_2) by a direct current (DC) [Eq. (7.9)]:

$$2\ H_2O \xrightarrow{e-} 2\ H_2 + O_2 \qquad (7.9)$$

The time needed for electrolysis is approximately 20–60 s, depending on the desired magnitude of the explosion. Gases collect under a bell and are ignited by a spark plug (Fig. 7-21). This causes a violent back-reaction to water [reversal of Eq. (7.9)]. This explosion and the subsequent implosion of the steam bubble produce short-lived (milliseconds) pressure impulses ranging from 13 to −0.6 bar (190 to −8.7 psi) which loosen incrustations. The implosive character sucks water—and loosened incrustations—back into the well. Pumping is nevertheless still necessary.

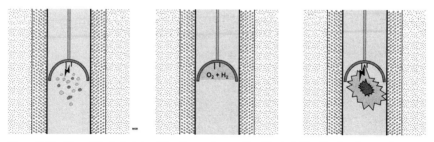

Figure 7.21 Simplified sketch of the three stages of well rehabilitation by detonating gas: (left) hydrogen and oxygen gas are electrolytically generated by a direct current; (middle) gas bubbles collect under a bell; (right) gas mixture is ignited.

Two main impulse frequencies were detected during test rehabilitations in physical scale models (DGFZ 2003). The main frequency of about 100 Hz produces an acceleration of 68 m/s^2 (223 ft/s^2) in the gravel pack. The secondary frequency of 3,000 Hz causes accelerations of 49 m/s^2 (161 ft/s^2). The impact of the explosion is much more localized than with explosive line charges. Applications can be repeated several times without recovering the apparatus from the well.

7.5.4 Release of compressed fluids

The sudden release of compressed fluids can also be used to create pressure pulses. The equipment ("air guns") was originally developed for offshore petroleum exploration but is now in use for well rehabilitation (Moore 1998). Pumps are used to build up the pressure in small chambers of 200–300 ml (0.05–0.08 gal) volume, which is then released abruptly (within milliseconds) by opening a valve. Pressures ranges can be selected between 10 and 130 bar (145 and 1,885 psi). An impulse generator of this type is shown in Fig. 7-22. Some systems include a packer system to create a more focused effect. Centralizers are employed to keep the impulse source away from the tubing, in order to prevent damage (Fig. 7-22).

Compressed media are usually gases such as air or nitrogen. Nitrogen is preferred because of its chemically inert nature. The impulse strength can be regulated by adjusting the pressure. It is generally somewhat smaller than with explosive charges. Although the cleaning effect is

Figure 7.22 Impulse generator chamber with centralizers *Photograph: Pigadi, Berlin, Germany.*

hence smaller, this procedure has the advantage that applications in wells with more brittle casing and screen materials are possible. Similar to the gas explosion technique described in the previous section, the effect is more localized than with detonation chords but may be repeated several times without recovering the tool. Impulse frequencies can be 1–3 pulses/s. The expulsive effect of the expanding fluids may be used to introduce rehabilitation chemicals into the gravel pack. Some systems contain a built-in pump to remove loosened material, which allows more or less continuous operation.

During test rehabilitations in physical scale models, again two main frequencies were detected (DGFZ 2003). For the two types of equipment used, the main frequencies were 100 and 200 Hz, respectively. The corresponding accelerations in the gravel pack were 39 and 971 m/s^2 (128 and 3,186 ft/s^2). The secondary frequencies of 1,000 and 10,000 Hz caused accelerations of 17 and 777 m/s^2 (3.3 and 2,549 ft/s^2), respectively. As one can easily see from this, individual setups of the same idea can result in quite different outcomes.

7.5.5 Ultrasound

Sound at frequencies above the human range of audibility (>16–20 to 10^6 kHz) is called ultrasound. The intensity of ultrasonic sound is in the range of 5–10 W/cm^2, compared to 10^{-3} W/cm^2 for a close-by cannon shot or 10^{-9} W/cm^2 for a speaker at normal volume. Ultrasonic sound therefore has a mechanical effect that is much more pronounced than that of audible sound.

Ultrasound is generated by electrical and magneto-restrictive oscillators. They are based on the piezoelectricity of certain crystals, e.g., quartz. When exposed to an alternating current (AC), such crystals start to expand and shrink slightly at a rate related to the imposed AC frequency (Fig. 7-23). This rhythmic movement (oscillation) excites the

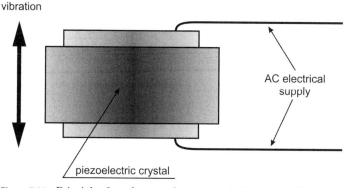

Figure 7.23 Principle of an ultrasound generator. *Redrawn after Serway & Faughn (1992).*

surroundings of the source. The crystals vibrate with a wavelength twice their thickness. Piezoelectric crystals therefore have a thickness of one-half of the desired radiated wavelength.

The usefulness of ultrasonic sound in material cleaning has long been known. The effect is based on the elastic stimulation of the exposed materials and the formation of tiny cavitation bubbles. The latter are formed by the local development of a vacuum, especially at interfaces. At their subsequent implosion, local temperatures and pressures of up to 10,000 K and 10,000 bar (145,000 psi) are possible. The associated mechanical stresses cause detachment of adhering material.

The idea of ultrasonic sound for well rehabilitation has been known since the early 1960s (Halfmeier 1960) but was developed into a practical application during the 1990s. Typical sound parameters of such a device are presented in Table 7-5.

Frequencies applied a range from 20 to 25 kHz. Some devices comprise several sound sources which emit more or less radially (Fig. 7-24) with individual intensities of about 6.8 W/cm^2 (Bott & Wilken 2002; Wilken & Bott 2002). The sources cover a length of about 0.3–1.2 m (1– 4 ft), which allows a stepwise treatment of screen sections with subsequent removal of loosened incrustations by pumping. The sound exposure per section is in the range of 10–15 min. The well is usually brushed and pumped prior to application (→ 7.2).

Experiments on test installations showed that the effect of ultrasound in wells is strongly dependent on water pressure: with increasing pressure (= depth of water), the amount of mobilized turbidity increases strongly. Between 0 and 16 bar (232 psi), a 10-fold increase was measured (Wilken & Bott 2002). Below 2 bar (29 psi)—i.e., a water column of 20 m—the amount of mobilized turbidity was negligible. Ultrasound should therefore be much more useful for the rehabilitation of deep wells than for shallow ones. Other reports state a decrease in efficiency

TABLE 7.5 Characteristic Parameters of an Ultrasound Device Used for Well Rehabilitation

Sound frequency	f	20 kHz = 20,000 s^{-1}
Circle frequency	ω	125,664 s^{-1}
Wavelength	λ	7.2 cm
Source power		2,000 W
Net source power (efficiency 70%)		1,400 W
Sound intensity	J	68 kW/m^2
Energy density	w	46 J/m^3
Amplitude	A	4.5 ×·10^5 Pa = 4.5 bar

From Wilken & Bott (2002).

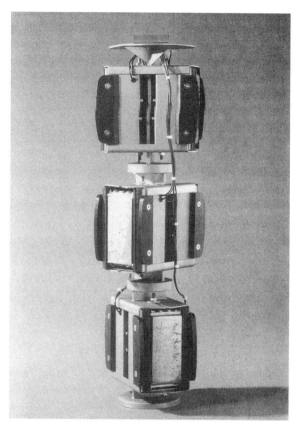

Figure 7.24 Ultrasound unit with multiple sound sources for well rehabilitation. *Photograph: Sonic Umwelttechnik, Bad Mergentheim, Germany.*

at a pressure above 5 bar (72.5 psi), probably due to a decrease in cavitation bubble formation (Kiwa 2004a). Because increasing pressure creates a decreased affinity for cavitation, other processes might also be important:

- Abrasion by oscillating grains
- Induced material stresses
- Liquefaction of thixotropic materials (ochre gels, bacterial slimes)

The strong forces acting during ultrasonic irradiation can tear apart the cell walls of microorganisms. Therefore ultrasound always has a microbicidal effect.

As we have seen in Sec. 7.5.1, the intensity of a point source decreases with the square of the radius from the source. Calculations for two hypothetical

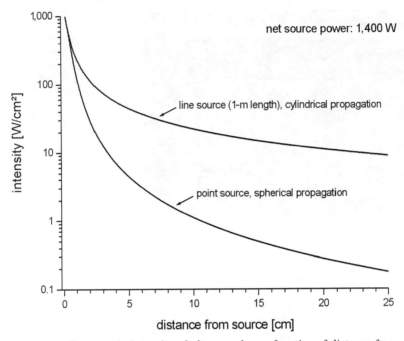

Figure 7.25 Decrease in intensity of ultrasound as a function of distance from source. Note the log scale of the y axis.

ultrasound sources, one a point source, the other a line source 1 m (3.28 ft) in length, both with a net power output of 1,400 W, show the drastic effect on the intensity as a function of the distance from the source (Fig. 7-25).

Experiments on physical scale models by Wilken and Bott (2002) are in good agreement with these theoretical considerations. Wilken and Bott found that the maximum effective penetration depth of ultrasound at water pressures of 5 bar (72.5 psi) is about 25 cm (0.82 ft). Their experiments showed that the amount of mobilized turbidity decreases strongly with increasing distance from the source.

The density-dependent reflection of ultrasound waves at interfaces also limits the penetration beyond the screen and into the gravel pack—and thus the effectiveness. As can be expected from Eq. (7.7), materials of high density, e.g., steel and copper, strongly reflect and limit the penetration beyond the screen (Fig. 7-26). Of course, this applies only to solid materials and does not apply to the openings (screen slots). High percentages of open screen area should therefore be favorable.

Again, experiments by Wilken and Bott (2002) support the theoretical considerations on reflection (Fig. 7-27). These authors measured the cavitation efficiency behind different screens using three different measuring methods, among them the counting of holes caused by cavitation

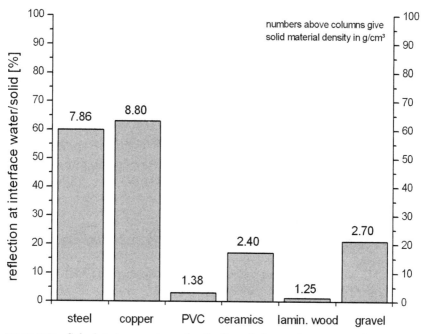

Figure 7.26 Calculated percentage of reflection of ultrasonic waves at the interface between water and various solid materials.

in aluminum foil fixed behind the screen. The efficiency behind steel screens is low, probably due to the strong reflection encountered by the ultrasonic waves.

Another limitation is well known from geophysical exploration. The absorption of waves traveling through solid material depends on the wave frequency. The higher the wave frequency, the higher is the absorption. Ultrasonic waves hence suffer much more from attenuation than impulse waves of lower frequencies.

Taking into account all limitations, ultrasound well rehabilitation should show the best results under the following circumstances:

- Moderately deep wells [20- to 50-m (65- to 164-ft) water column]
- Small diameters of well and borehole
- Low-density screen material, e.g., PVC
- High percentage of open area of screen, e.g., wire-wound screen

Practical applications of ultrasonic sound for well rehabilitation in wells in Germany yielded ambiguous results. Some wells were rehabilitated successfully; the yield of others could not be improved significantly. In at least one case the yield decreased, probably due to

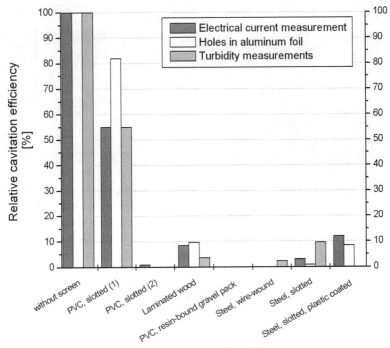

Figure 7.27 Energy losses of ultrasound measured behind screens of various materials. *Redrawn after data by Wilken & Bott (2002).*

settlement of the gravel pack (Berger 1997). Test rehabilitations in the Netherlands (one well, 2003) and in Belgium (two wells, 1997) showed only weak performance (KIWA 2004a).

7.5.6 Enforced jet cavitation

The method of enforced jet cavitation was originally developed in the Netherlands to stimulate and rehabilitate oil and gas wells. The principle is based on the development of cavitation bubbles (→ 7.5.1). In this case they are not formed by ultrasonic sound but by the fast injection of water through nozzles in a rotating head (Fig. 7-28). The rotor head is designed to spread the cavitation bubbles to the outside. Flow rates of several hundred liters per minute are usually employed. The impulses caused by the implosion of the cavitation bubbles and the vibration of the tool act on sand bridges and incrustations and suck out material. Loosened particles can be pumped from the well during operation using an additional pump. Tools of different diameters are available (3.2, 4.5, and 6 in). Treatment starts at the bottom of the screen, and the tool is then slowly lifted at a rate of about 1 m/min. Repeated treatments are possible.

The method has been applied in six water wells in the Netherlands (KIWA 2004b). It is still under development, because the first experiments

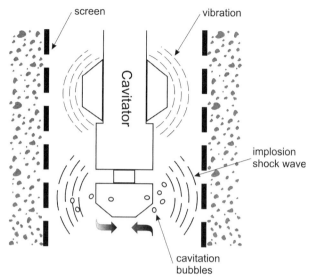

Figure 7.28 Schematic drawing of a rotational tool for enforced jet cavitation. The pump is not shown. *Redrawn after a brochure from Whirlwindi, The Netherlands.*

were not fully successful. The yield gains obtained were small, and in one case the well was damaged.

7.6 Comparison of Methods

On behalf of the DVGW, the German Groundwater Research Centre (DGFZ e.V.) in Dresden performed quantitative tests on two different model wells at 1:1 scale (DGFZ 2003). The models were equipped with regular well screens, gravel pack, and unconsolidated aquifer sediment. Several sensors were installed both in the gravel pack and the aquifer material to monitor penetration depth and intensity of the different methods. Measured parameters included electrical resistivity as well as acceleration, amplitude, and frequency of impulses. From basic physics we can derive some simple concepts to assess the energy contained in impulse waves. The maximum amplitude of a wave can be related to the maximum acceleration and the wave frequency by Eq. (7.10):

$$x_{max} = \frac{a_{max}}{\omega^2} = \frac{a_{max}}{(2 \cdot \pi \cdot f)^2} \qquad (7.10)$$

where x_{max} = maximum amplitude, L
a_{max} = maximum acceleration, L/T^2
ω = circular frequency, T^{-1}
f = wave frequency, T^{-1}

The maximum oscillation velocity v_{max} can then be derived from Eq. (7.11):

$$v_{max} = \omega \cdot x_{max} = 2 \cdot \pi \cdot f \cdot x_{max} \qquad (7.11)$$

As can be easily seen, waves of higher frequencies, e.g., ultrasound, produce much lower amplitudes due to the square dependency in the denominator of Eq. (7.10). The mechanical energy E_{mech} of an oscillation can be defined as kinetic energy, with m being the mass of the accelerated body [Eq. (7.12)]:

$$E_{mech} = \frac{1}{2} \cdot m \cdot v_{max}^2 = 2 \cdot m \cdot (\pi \cdot f \cdot x_{max})^2 \qquad (7.12)$$

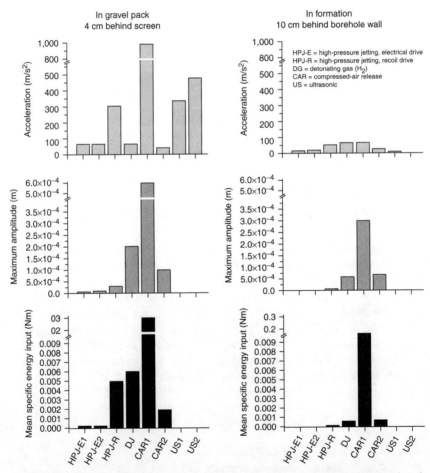

Figure 7.29 Comparison of acceleration, amplitude, and energy input in the gravel pack and in aquifer material obtained from various mechanical rehabilitation techniques applied on 1:1 physical scale models. *After data by DGFZ (2003).*

These simple calculations allow a comparison of the energy input of different impulse methods by measuring the maximum amplitude and the frequency. Often energy is expressed as specific energy E/m, because the mass of the accelerated body is difficult to assess.

Since the participation in the tests was voluntary and dependent on budgets and time schedules of the participating companies, not all currently available methods could be tested. Two different high-pressure jetting techniques (electrically-driven nozzle head), one high-pressure rotational jetting method (recoil-driven nozzle head), two schemes based on the release of compressed air ($\rightarrow$ 7.5.4), and two different ultrasonic probes ($\rightarrow$ 7.5.5) were tested (Fig. 7-29). The main results—measured or calculated using the equations presented above—are summarized in Fig. 7-29. As expected, for almost all methods, much of the energy only reaches to the gravel pack and does not reach past the former borehole wall. Only methods with high accelerations and low frequencies can successfully pass this limitation.

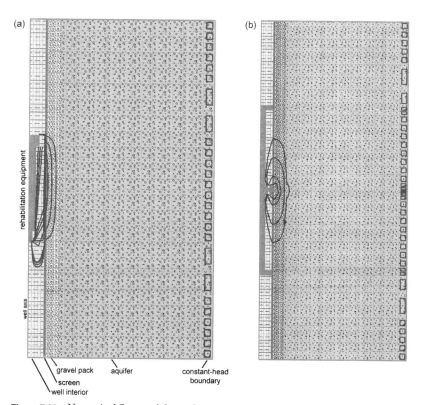

Figure 7.30 Numerical flow models used to assess the penetration depth of different types of rehabilitation equipment into annulus and aquifer (DGFZ 2003). Calculated with MODFLOW. (a) Low-pressure jetting system. (b) Multichamber system.

Numerical flow models were calculated to assess the penetration depth of the different techniques in more detail (DGFZ 2003, see, e.g., Fig. 7-30). They confirmed the findings of the physical test models that most methods cannot penetrate to the former borehole wall. Their effectiveness is thus limited to the gravel pack. Multichambered systems are a noteworthy exception and can penetrate farther than other methods.

The main results of the DGFZ study can be summarized as follows.

- All methods suffer from strong relative energy losses with distance from the source.
- Penetration depth of most methods is limited to the gravel pack; clogging at the borehole wall or even in the aquifer can hardly be treated.
- Rehabilitation of sections of limited spatial dimensions is preferable to treatment of the whole screen at once.
- Methods that separate and remove incrustations in one step are more efficient.

Chapter 8

Chemical Rehabilitation Techniques

8.1 Legal Constraints

In many countries the injection of chemicals into a well and thus into the groundwater may be restricted by legal regulations. A chemical injection may therefore require a formal application to the proper authorities well in advance. Including the following specifications in the application that will facilitate getting permission:

- Location and number of wells
- Reasons for rehabilitation (why a mechanical rehabilitation may not suffice)
- Contractor (and his experience with chemical rehabilitations)
- Type and quantity of chemical(s) to be used
- Time schedule
- Safety precautions
- Solid and liquid waste disposal concept
- Monitoring concept for both rehabilitation and waste disposal

8.2 Reaction Mechanisms

8.2.1 Introduction

In principle, all chemical rehabilitations are based on a phase transfer from solid incrustation minerals to dissolved constituents. This can usually be achieved by marked changes in the physicochemical environment, especially in pH and redox conditions. It does not require complete removal of the incrustation mass. If the chemicals destroy enough intercrystalline bonds, smaller incrustation particles can be flushed out by mechanical agitation. Strong turbidity is an indication of this process.

We can distinguish two basic dissolution reaction types: (1) solution-controlled and (2) surface-controlled reactions. In the former, saturation of the solution close to the mineral surface is quickly achieved and the reaction proceeds only when this solution is removed and fresh solution is replenished. Reaction progress can thus be enhanced by circulating the solution. We all know reactions like this from our daily lives. If we put sugar into our coffee, we can hasten its dissolution by stirring. Most reactions in chemical well rehabilitation—such as the dissolution of iron and manganese ochre—are surface-controlled. The detachment of an ion dissolved from the mineral surface is much slower than its transport in the solution. The flow velocity of the surrounding solution therefore does not influence the reaction progress. "Stirring" will not accelerate the reaction, but it may help to remove loose particles.

The influence of the reactive surface area of mineral grains on the reaction progress cannot be overestimated. From a chemical point of view, more surface area means more sites where the solution can interact with the mineral. As can be seen from Fig. 8-1, surface area is a function of grain size. If we cut a mineral cube into increasingly smaller cubes, the surface area increases dramatically. Again, we all know this from our daily lives: fine-grained sugar will dissolve in coffee much

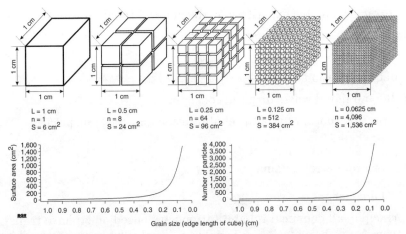

Figure 8.1 Total surface area S as a function of grain size L and the number of particles n in a constant volume.

more quickly than clotted chunks. This provides some insight into the boundary conditions of well rehabilitations: older, highly crystalline incrustations of low-surface area are much more difficult to dissolve than younger incrustations of low-surface area. On the other hand, mechanical pretreatment which breaks down the minerals in size can significantly enhance the success of a chemical rehabilitation.

Surface-controlled reactions can be inhibited by certain ions. They form stable complexes at the mineral surface and prevent the access of dissolving agents. Small and highly charged ions are usually efficient inhibitors. Accordingly, the presence of anions such as phosphate, borate, or arsenate decreases the dissolution rate of iron oxides significantly (Biber et al. 1994).

8.2.2 Dissolution of iron and manganese oxides

Ochre incrustations are the most common form of well ageing in many countries (→ 3.2). They consist predominantly of ferric iron [Fe(III)] and manganese [Mn(III, IV)] phases. All of them are practically insoluble at neutral pH. Three main processes are available to transfer them into more soluble species, i.e., to dissolve them:

- Proton-assisted dissolution
- Ligand-assisted dissolution
- Reduction

Proton-assisted dissolution is based on a strong decrease of pH by the addition of protons (H^+ ions) through acids. The protons sorb to the mineral surface and interact with surface hydroxyl groups, which causes the release of ferric iron and manganese [Eqs. (8.1)–(8.3), Fig. 8-2].

$$FeOOH + 3\,H^+ \leftrightarrow Fe^{3+} + 2\,H_2O \tag{8.1}$$

$$MnOOH + 3\,H^+ \leftrightarrow Mn^{3+} + 2\,H_2O \tag{8.2}$$

$$MnO_2 + 4\,H^+ \leftrightarrow Mn^{4+} + 2\,H_2O \tag{8.3}$$

The application of acids is probably the oldest chemical rehabilitation scheme. The oldest documented reports from Germany date back to the 1920s (Geißler & Wiegand 1924; Wiegand 1929). Since the dissolution of ochre consumes acid [Eqs. (8.1)–(8.3)], measurements of pH are a good way to control the progress and to allot acid during rehabilitation.

The dissolution of iron oxides in acids is strongly pH-dependent (Figs. 8-3, 8-4). In simple words, the more acid we add, the more iron oxide will dissolve. Nevertheless, some practical constraints keep us

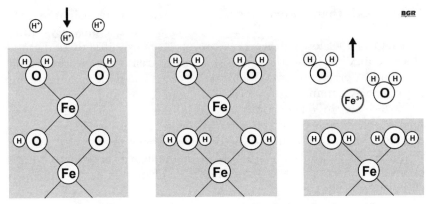

Figure 8.2 Reaction steps of proton-assisted (acid) dissolution of an iron oxide.

from using acids of arbitrary low pH value. Acids, especially hydrochloric acid, have a high potential to corrode metals and damage other materials, too. For all practical purposes, the pH value should not be lower than about pH 1, at least not for long periods of time. The cheapest grades of hydrochloric acid ("technically pure")—often used in the past—

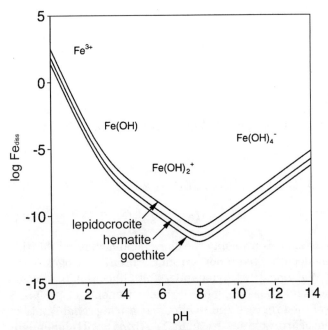

Figure 8.3 Solubility of various iron oxide minerals as a function of pH. Note the log scales. *Modified after Cornell & Schwertmann (2003).*

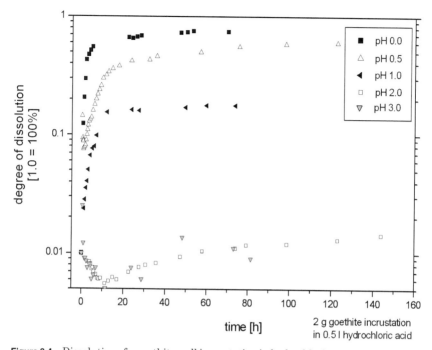

Figure 8.4 Dissolution of a goethite well incrustation in hydrochloric acid at varying pH.

may contain traces of arsenic. They must not be used in wells because of the possible formation of toxic arsenic hydride (AsH_3) (Naumann 1936). The hydrogen required for AsH_3 formation stems from the contact of the acid with metal casings (→ 3.1). The minimum requirement for acids used in rehabilitations is "chemically pure."

The strength of an acid is defined by its potential to release protons (= deprotonation). The logarithm of the equilibrium constant for this reaction, called pK_a, is used to quantify and compare the strength. At a low pK_a, protons are expelled more easily, rendering the acid stronger. The amount of protons of course defines the resulting pH of the solution. Some acids can set free more than one proton (usually two or three), e.g., sulfuric acid (H_2SO_4) and phosphoric acid (H_3PO_4). For those acids, a pK_a must be defined for each deprotonation step. Usually the first deprotonation step is the easiest. Depending on the pK_a and the chemical setting, the second or third proton might not be liberated at all. Phosphoric acid, for example, often deprotonates only two H^+ ions ($H_3PO_4 \leftrightarrow H^+ + H^+ + HPO_4^{2-}$). The secondary and tertiary steps are thus often not important for the resulting pH of the solution. Table 8-1 shows the pK_a of several acids commonly used in chemical well rehabilitations and the resulting pH of a 0.1 M solution.

TABLE 8.1 Comparison of Acids Used in Well Rehabilitations

Name of acid	Chemical formula	pK_a (1)	pK_a (2)	pK_a (3)	pH of 0.1 M solution
Hydrochloric	HCl	~−6	—	—	1.00
Sulfuric	H_2SO_4	~−3	1.89	—	0.70
Ascorbic	$C_6H_8O_6$	4.10	11.79	—	2.55
Citric	$C_6H_8O_7$	3.14	4.77	6.39	2.07
Sulfamic	NH_2SO_3H	0.99	—	—	1.00
Glycolic	$C_2H_4O_3$	3.83	—	—	2.42
Malonic	$C_3H_4O_4$	2.83	5.69	—	1.92

The solubility of iron oxides at a pH above 1.5 is rather small (Figs. 8-3, 8-4). Figure 8-4 shows the dissolution of a crystalline (goethite-dominated) iron oxide incrustation in hydrochloric acid at different values of pH. As expected, above pH 2.0, the solubility is basically negligible. At a pH of 1.0—which is often employed in real-life rehabilitations—dissolution is rather limited, too. Only a pH significantly below 1.0 results in fast and efficient mobilization of iron. In the older German literature, suggestions of an "optimal" or "ideal" pH value of about 0.9 can sometimes be found. This is misleading and simply not correct. The optimum pH is the lowest pH we can use without causing corrosion of the well material.

Hydrochloric acid is the most common acid for many industrial processes, including chemical well rehabilitation. This is due to its efficiency, chemical stability, and low cost. The downside is the high corrosion potential due to the acidity and the chloride content (→ 3.1). Sulfuric acid (H_2SO_4) is a feasible alternative with a slightly lower corrosion potential. Phosphoric and nitric acids are usually not used—except sometimes as additives—because of their higher cost and the potential of the phosphate and nitrate ions to foster microbial growth. The described acids are available as liquid concentrates which can be diluted with water to a desired pH. During dilution, it is essential that the acid be poured into water and not vice versa. Otherwise the water will most likely boil over (the reaction is strongly exothermal). The boiling water and some of the acid will spatter around—with dreadful consequences, of course, for anyone unfortunate enough to be standing close by.

Some acids are available as crystalline solid powders. For these acids, transport is somewhat easier because the material is less likely to spill. On the other hand, the acid has to be prepared on-site by adding water, which involves some handling in the open. A "solid" acid that is popular for well rehabilitations in the United States is sulfamic acid (amidosulfuric acid, NH_2SO_3H) (Spon 1999). Somewhat less popular is glycolic acid (hydroxyacetic acid, $C_2H_4O_3$), which has a limited microbicidal effect.

Acids can dissolve not only iron and manganese oxides but also carbonates, sulfides, and aluminum hydroxides ($\rightarrow$ 8.2.3, 8.2.4, 8.2.5). What may at first sound like an advantage can in fact be a substantial disadvantage. If the aquifer and the gravel pack material of a well infested by ochre contain some carbonates (which is very common), a large part of the injected acid reacts with those instead of attacking the incrustation. The removal of carbonates can seriously diminish the natural acid-buffering capacity of the aquifer. Berger (1997) analyzed the amounts of iron, manganese, calcium, and magnesium removed during chemical rehabilitations with hydrochloric acid (pH $\approx$ 1.0) at several wells in Wiesbaden (Germany). Considering the reaction stoichiometry ($\rightarrow$ 8.2.2, 8.2.3), between 40% and 54% of the acid reacted with calcium or magnesium carbonates from the aquifer (Fig. 8-5). Since the present incrustations contained four to 10 times more iron and manganese than calcium and magnesium, the bulk of the dissolved carbonates was likely to stem from the aquifer. As we can easily see from this, a substantial amount of acid was wasted destroying several kilograms of natural aquifer carbonates instead of battling the ochre incrustations.

Acids possess some additional impediments: they promote corrosion of metals (especially hydrochloric acid) and can cause shrinkage of clays, e.g., in annular seals. On the other hand, higher concentrations of acids

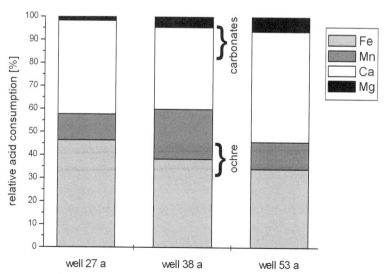

Figure 8.5 Mass balance for relative acid consumption in the dissolution of iron and manganese oxides, calcium, and magnesium carbonates for three chemical rehabilitations with hydrochloric acid. *After data from Berger (1997)*.

can have a microbicidal effect. According to Barbic and Savic (1990), a 0.12 M solution of hydrochloric acid kills off iron bacteria such as *Thiobacillus ferrooxidans*.

Both iron and manganese oxides are amphoteric substances. Their solubility curve shows maxima under both acid and basic pH (Fig. 8-3). Therefore, dissolution of ochre in strong bases should be possible. On the other hand, Fig. 8-3 shows clearly that the solubility in bases is several orders of magnitude lower than under acidic pH (note the log scale of the y axis). Bases are thus not a recommendable choice as rehabilitation agents.

Ligand-based dissolution (complexation) is based on ions or molecules which have one or more free electron pairs. They employ this charge to attach themselves to a metal ion and release it as complex from the crystal lattice (Fig. 8-6; Furrer & Stumm 1986). Effective ligands have two electron pairs (bidentate). Many are organic molecules, e.g., carboxylic acids such as oxalate, citrate, salicylate, and tartrate. The most famous ligand is probably ethylenediaminetetraacetic acid (EDTA), a very efficient complexing agent used in many industrial applications. Because of its very low biodegradability and its potential to scavenge and transport metal ions, it should not be used in natural environments and wells. The remaining ligands are usually too low in efficiency to be used in well rehabilitations but are often employed as additives with acids or reductants. There they help to transport ions away from the mineral surface and prevent reprecipitation.

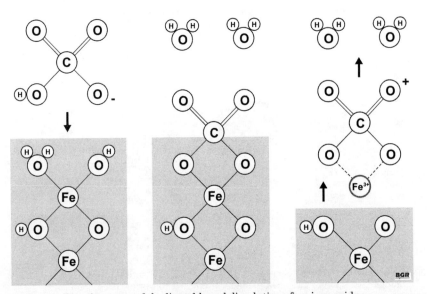

Figure 8.6 Reaction steps of the ligand-based dissolution of an iron oxide.

As the name implies, *reductive dissolution* is based on chemicals acting as reductants. They donate electrons to their reaction partners and can thereby transform insoluble oxic species into more soluble reduced species (Fig. 8-7). In the case of iron and manganese oxides, ferric iron and manganic manganese are transformed to ferrous iron and manganous manganese, respectively (e.g., $Fe^{3+} + e^- \to Fe^{2+}$; $Mn^{4+} + 2\,e^- \to Mn^{2+}$).

Reductants can be both organic and inorganic. Inorganic reductants include sulfides (e.g., H_2S), reduced metal species (e.g., Fe^{2+}, Ti^{3+}, V^{2+}), and sodium dithionite ($Na_2S_2O_4$) [Eqs. (8.4) and (8.5)].

$$2\,FeOOH + Na_2S_2O_4 + 4\,H^+$$
$$\leftrightarrow 2\,Fe^{2+} + 2\,Na^+ + 2\,HSO_3^- + 2\,H_2O \tag{8.4}$$

$$MnO_2 + Na_2S_2O_4 + 2\,H^+ \leftrightarrow Mn^{2+} + 2\,Na^+ + 2\,HSO_3^- \tag{8.5}$$

Organic reductants include: phenols, fructose, lactate, pyruvate, formiate, glycolate, fulvic acids, tannic acids, and cysteine (Hering & Stumm, cited in Hochella & White 1990). A strong and often-used organic reductant is ascorbic acid, better known as vitamin C (Fig. 8-7).

The mineralogy of an incrustation has a profound influence on its solubility. This can be seen easily by comparing the dissolution curves

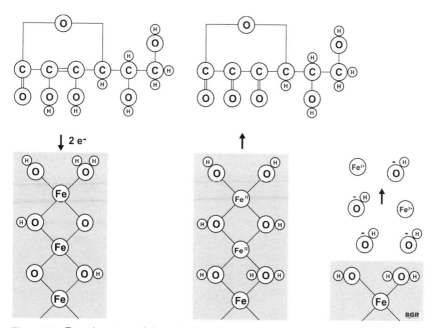

Figure 8.7 Reaction steps of the reductive dissolution of an iron oxide by ascorbic acid.

of several ochre incrustations with different mineral contents (Fig. 3-24). Young incrustations containing minerals of low crystallinity and high surface area are much more soluble than recrystallized older incrustations containing chiefly goethite. The surface area is a good measure of this (Fig. 8-8; → 3.2.5, 3.2.7).

Some organic acids may use several of the dissolution mechanism discussed above. Ascorbic acid is a good example. As a regular acid it can release a H^+ ion for proton-assisted dissolution. In addition, the remaining ascorbate anion can act as ligand and as reductant.

8.2.3 Dissolution of carbonates

Carbonate incrustations (scales) can be dissolved using inorganic acids, e.g., hydrochloric and sulfuric acid [Eq.(8.6)].

$$CaCO_3 + 2\ HCl \rightarrow Ca^{2+}_{(aq)} + CO_{2(g)}\uparrow + H_2O + 2\ Cl^-_{(aq)} \tag{8.6}$$

A pH probe can be used to measure the pH inside the well and control the reaction. Carbonate buffers the acid and leads to rising pH. Again, section-wise treatment of the screen is recommended. When additional amounts of acid are not buffered (decreasing pH), one can assume that no carbonates are left and proceed to the next section.

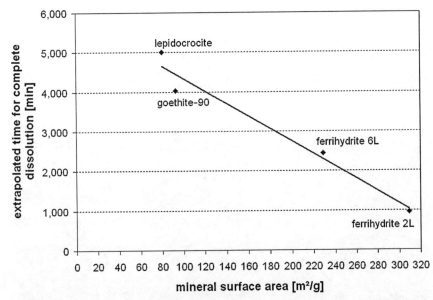

Figure 8.8 Specific surface areas of synthetic iron oxides and time required for complete dissolution in 0.1 M hydrochloric acid. *Extrapolated from rate laws by Houben (2003b).*

Acids can also be used to enhance the yield (development) of newly constructed wells in carbonate aquifers (Harker 1990; Banks et al. 1993).

As we can see from Eq. (8.6), a large amount of carbon dioxide gas is formed. In the worst case this may lead to an effervescence and expulsion of acid to the surface, endangering personnel and equipment. About 22.7 liters (6 gal) of gas are formed per 100 g (0.22 lb) of dissolved calcium carbonate. Therefore the acid should be filled in slowly and under constant surveillance.

Carbon dioxide gas is suffocating and may collect in shafts, etc., because of its higher density compared to air. The reader is urgently requested to refer to Sec. 10.2.2 for details on the potential problems of carbon dioxide in wells.

8.2.4 Dissolution of aluminum hydroxides

Aluminum hydroxide incrustations (rarely occur) can also be removed by dissolution in inorganic acids [Eq. (8.7)]. Solubility increases strongly with decreasing pH below pH 3.5 (Fig. 8-9). Since the dissolution consumes H^+ ions, the resulting pH rises slightly.

$$Al(OH)_3 + 3\,H^+ \rightarrow Al^{3+}_{(aq)} + 3\,H_2O \qquad (8.7)$$

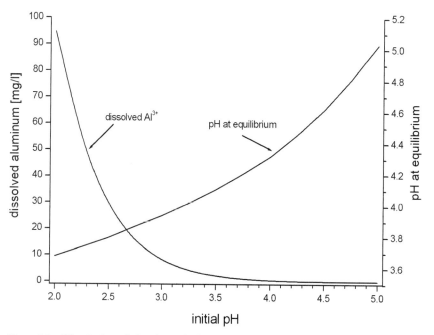

Figure 8.9 Dissolution of aluminum hydroxide (gibbsite) as a function of pH. Modeled using PHREEQC-2. *From Houben (2000).*

The wastewater to be pumped from the well after such a treatment is very acidic and contains much dissolved aluminum. Because of the resulting toxicity for aquatic life such as fish, it must not be disposed off in aquatic environments such as lakes and rivers. Neutralization is necessary ($\rightarrow$ 8.7).

As an alternative to acids, strong bases can also be used to dissolve aluminum hydroxides (McLaughlan 2002).

8.2.5 Dissolution of metal sulfides

Monosulfides such as FeS and greigite (Fe_3S_4) are well soluble in inorganic acids, e.g., hydrochloric acid. The reaction results in the generation of hydrogen sulfide gas [Eq. (8.8)].

$$FeS + 2\ HCl \rightarrow Fe^{2+}_{(aq)} + H_2S \uparrow + 2\ Cl^- \tag{8.8}$$

Hydrogen sulfide gas (H_2S) is poisonous, and the reader is urgently requested to refer to Sec. 10.2.2 for more details. Concentrations greater than 0.1 vol% are toxic. It can be recognized even at very low concentrations by its distinctive smell of rotten eggs.

Sodium hypochlorite (NaClO) is also a useful chemical to remove sulfides (van Beek & Kooper 1980), as shown in Eq. (8.9a). The ferrous iron is usually oxidized to ferric iron by hypochlorite [Eq. (8.9b)] and has to be removed as iron oxide fluffs.

$$FeS + 4\ NaClO \rightarrow Fe^{2+}_{(aq)} + SO_4^{2-} + 4\ Cl^- + 4\ Na^+ \tag{8.9a}$$

$$2\ Fe^{2+}_{(aq)} + NaClO + 5\ H_2O \rightarrow 2\ "Fe(OH)_3" \\ + Cl^- + Na^+ + 4\ H^+ \tag{8.9b}$$

The overall reaction shows the generation of acid (H^+), which assists in dissolving the ferric iron or may be buffered by other reactions, e.g., when carbonates are present.

Higher sulfides, e.g., pyrite and marcasite (both FeS_2), which may form from monosulfides are much less soluble in acid. They have to be oxidized by hydrogen peroxide or sodium hypochlorite [Eq. (8.10)], before being dissolved further in acids.

$$FeS_2 + 7\ NaClO + H_2O \rightarrow Fe^{2+}_{(aq)} + 2\ SO_4^{2-} \\ + 7\ Cl^- + 7\ Na^+ + 2\ H^+ \tag{8.10}$$

8.2.6 Oxidation/dissolution of biomass

Microbial biomass ("slimes") is usually battled using strong oxidants (Brown 1942; Klink 1994). Their principle of operation is simply based on the mineralization of the biomass (→ 11.1.2). Common oxidants include hydrogen peroxide (H_2O_2) [Eq. (8.11a)] and chlorine-releasing agents, e.g., sodium hypochlorite (NaClO) [Eq. (8.11b)].

$$\text{"}CH_2O\text{"} + 2\ H_2O_2 \rightarrow CO_2 + 3\ H_2O \qquad (8.11a)$$

$$\text{"}CH_2O\text{"} + 2\ NaClO \rightarrow HCO_3^-{}_{(aq)} + 2\ Cl^- + 2\ Na^+ + H^+ \qquad (8.11b)$$

In both cases organic matter is transferred into inorganic carbon. Carbon dioxide gas (CO_2) may collect in shafts, etc., and require safety precautions as described in Sec. 10.2.2.

At least in theory, potassium permanganate ($KMnO_4$) or ozone (O_3) are also possible oxidizing agents. A reaction product of the former, however, is manganese oxide, which in turn can cause clogging. Therefore it is easy to see why we do not want to replace one evil by another. Ozone, on the other hand, has to be produced on-site at high expense and requires elaborate safety precautions.

Oxidants have the general disadvantage of causing the oxidation of ferrous iron and manganese (if present) and their subsequent precipitation causing clogging (→ 11.1.2).

The application of chlorine-containing chemicals can lead to the formation of chlorinated hydrocarbons, which is undesirable from a hygienic point of view. If the mineralization of the biomass is incomplete, fragments of the organic substances may cause microbial growth. One has to keep in mind that all disinfections in wells are short-lived. New bacteria are continuously supplied from the aquifer that can thrive on any remaining organic substrate. The extremely high (exponential) growth rates of bacteria are well known: a bacterial cell dividing every 30 min can have—at least in theory—2^{48} descendants after 24 h. Repeated stress—including disinfections—cause bacteria to emit a slimy protective cover (EPS, → 3.6). Rehabilitations following disinfections need a lot of chemicals to battle the slime before being able to penetrate to the bacteria themselves.

Practical experience has shown that hydrogen peroxide has noticeable effects on the dissolution of manganese oxide incrustations. Since they do not contain more biomass than iron oxide—at least according to our investigations—another explanation has to be found. We suspect that part of the manganese is present as Mn(III), e.g., as MnOOH. This can be converted into Mn(IV) by oxidation, causing phase transitions [Eq. (8.12)].

$$2\ MnOOH + H_2O_2 \leftrightarrow 2\ MnO_2 + 2\ H_2O \qquad (8.12)$$

The mechanical stress during this recrystallization probably leads to the disintegration of manganese oxide crusts.

8.2.7 Removal of drilling fluid remnants

As a result of insufficient removal of the drilling fluid during well completion or well development, sometimes bentonite and/or CMC (carboxymethylcellulose) remnants can be found, especially at the former borehole wall and its immediate surroundings. They plug pore spaces and therefore diminish the well yield. Their removal is not a chemical rehabilitation in the strict sense, but should be mentioned due to the commonness of the problem.

The most common drilling fluid additive is bentonite, a naturally occurring mixture of expanding clay minerals. Bentonite in contact with water takes up some water molecules into the crystal interlayers, which causes the mineral to expand (Fig. 8-10). The expansion helps to seal off the borehole wall and to diminish circulation losses. At approximately neutral pH, the particles have negative charges on the mineral surfaces and positive charges at the edges (Fig. 8-11). The electrostatic attraction between these opposite charges leads to the development of a voluminous "house of cards" structure (Fig. 8-11, left) and a cohesive force between the particles which is responsible for the high viscosity.

The addition of high concentrations of salts causes flocculation of the bentonite. Although this decreases the volume, the emerging bentonite clots are too big to be removed and still clog the pores. Therefore the aim is to decrease the viscosity without flocculation. To annihilate the electrostatic interaction between the particles, highly negatively charged molecules, usually (poly)phosphates, are added. They strongly attach themselves to the positive edge sites and thereby saturate those (Fig. 8-11, right). This counteracts the attractive forces between individual crystals. The "house-of-cards" structure thus collapses and the shear forces necessary to overcome the attraction between the particles (= viscosity) decrease to a degree that the suspension can be pumped.

The following phosphate salts are commonly used:

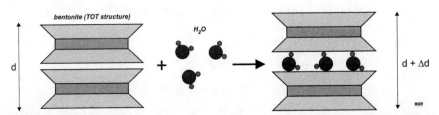

Figure 8.10 Uptake of water molecules into the interlayer between two bentonite crystals, leading to volume expansion.

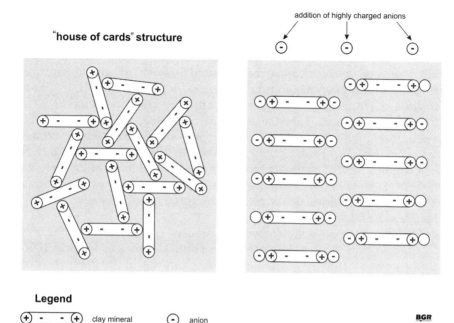

Figure 8.11 "House-of-cards" structure in a bentonite–water system due to electrostatic interaction of surface and edge charges (left), and collapse of this structure through addition of highly charged anions and resulting saturation of the edge charges (right).

- Sodiumhydrogenpyrophosphate (SAPP)
- Tetrasodiumpyrophosphate (TSPP, $Na_4P_2O_7 \cdot 10H_2O$)
- Trisodiumpolyphosphate (TSPP)
- Sodiumhexametaphosphate ("Graham's salt," SHMP, $(NaPO_3)_n$, $n \approx 20$).

Driscoll (1989) advises dissolving the polyphosphates in water before injecting them into the well. He recommends using about 6.8 kg (15 lb) in 400 liters (100 gal) of water (17 g/liter). Experiments by the authors of this book led to a similar conclusion: about 20 g/liter.

Because phosphates are a perfect nutrient for many bacteria and algae (eutrophication), a secondary microbial pollution or algal bloom is possible. Therefore, in many cases a microbicidal agent is injected together with the phosphate. Wastewater from such a treatment scheme should—for the same reason—not be discharged in lakes or rivers. Phosphate is removed quite effectively during soil passage. When this is not an option, the wastewater should be filtered through granulated iron hydroxide pellets, which have a very high sorption capacity for phosphates (→ 3.2.8).

8.3 Combinations of Chemicals

In many cases the efficiency of chemicals can be enhanced by combining them with other chemicals. One of the most common combinations in the lab is aqua regia, a mixture of concentrated hydrochloric and nitric acids. Its superior dissolution potential is due partially to its low pH, but more to the generation of very aggressive chlorine radicals. Another well-known combination of chemicals, used since 1883, is Fenton's reagent, a mixture of hydrogen peroxide and iron(II) salts. Its extremely powerful oxidizing effect is also attributable to the effects of free radicals.

A combination of hydrogen peroxide (H_2O_2) and hydrochloric or sulfuric acid is a well-established practice in chemical well rehabilitations. This allows simultaneous attack on iron and manganese oxides, carbonates, sulfides, and biomass.

The efficiencies of acids in wells as reducing agents can be enhanced by the addition of complexing agents (ligands) (Mehra & Jackson 1960; Banwart et al. 1989). They prevent reprecipitation of the dissolved iron or manganese and aid in the removal of dissolved species from the mineral surface. Ligands are usually organic molecules, e.g., oxalate, EDTA, citrate, or picolinate (Hering & Stumm, in Hochella & White 1990). Their potential as a food source for bacteria should not be ignored (→ 10.4.2). The usefulness of combinations of reducing agents with ligands and pH buffers in chemical well rehabilitations has been demonstrated by Houben et al. (2000b).

8.4 Additives

Additives are chemicals which do not participate in the actual reactions that dissolve incrustations. They may fulfill two purposes:

- Support the reactive chemicals by providing adequate physicochemical conditions
- Block negative effects of the reactive chemicals

The corrosive effects of acids on metal surfaces are well known (→ 3.1). For this reason, some acids contain corrosion inhibitors. Generally, acidic phosphates are employed, which form a thin protective layer of iron phosphate on the metal surface. This is a common procedure in the metal industry and is known as "bondering." An alternative chemical is diethylthiourea ($C_5H_{12}N_2S$).

Emulsifiers (tensides, surfactants) are used by everybody in day-to-day life, for example, in dishwashers to help clean plates. They decrease the surface tension ("softening" the water") at the interface between the fluid and the solid stains to be removed. In well-rehabilitation applications, emulsifiers decrease the tension at the interface between the fluid rehabilitation chemical and the solid incrustation mineral to be dissolved. Decreased surface tension facilitates wetting of the mineral surface and the intrusion of the fluid into microcracks and fine pores. The chemicals employed are mostly organic, so their potential as a food source for bacteria should therefore again be considered ($\rightarrow$ 10.4.2).

Chemicals used for rehabilitations are usually ionic, e.g., acids, which means simply that they contain charged molecules. If the mineral surface has the same charge as a dissolved species, a small repulsive force develops and diminishes the "docking" onto the mineral surface. Ochre particles show this behavior. Simple salts are mostly sufficient to subdue the surface charge and promote the dissolution reaction. In the case of hydrochloric acid, table salt (NaCl) is a useful additive (Sidhu et al. 1990).

Reducing agents as discussed in Secs. 8.2.1 and 8.5 work best under neutral pH conditions. Because the dissolution reaction alters the pH—which would lead to a decrease in reaction progress—a pH buffer substance is often added (Mehra & Jackson 1960).

Organic chemicals used in well rehabilitation might be utilized by bacteria as a food source and cause microbial pollution ($\rightarrow$ 10.4.2). Therefore some of those agents contain microbicidal additives, e.g., soluble silver salts such as silver nitrate ($AgNO_3$).

The use of drilling fluid additives (bentonite or polymers) during rehabilitations has not been shown to be advantageous. The benefits of the additional particle-carrying capacity are more than outweighed by the sealing of porosity caused by the additives (Blackwell et al. 1995).

Chemical reaction progress and efficiency are functions of temperature. We all know this phenomenon from our daily lives: hot water is much more useful for dishwashing or removing stains than cold water. Therefore chemicals to be used in rehabilitations and even the well water are sometimes heated to up to 70°C (158°F) (ADITCL 1996; Alford & Cullimore 1998). Injection of hot water or steam is mostly employed. Although the energy cost involved can be substantial, hot water has an additional benefit in being microbicidal ("pasteurization"). If the well is host to a substantial amount of biomass, this material may coagulate in hot water—similar to an egg being boiled—and clog the pores of the gravel pack (Alford & Cullimore 1998).

8.5 Comparison of Rehabilitation Chemicals

The efficiency of a chemical agent can be described by the chemical conversion it causes during a defined time span, e.g., the amount of mineral incrustation removed during a chemical rehabilitation. High reaction rates are favored because they allow fast operation and minimize residence time in the well and possible migration into the aquifer.

A series of laboratory experiments on the efficiencies of various rehabilitation chemicals on ochre incrustations was performed by Houben (2003b). In order to obtain reproducible experimental conditions, synthetic iron oxides were used. Only iron oxide types that had been identified previously in natural incrustations were synthesized ($\rightarrow$ 3.2). Lepidocrocite (γ-FeOOH) and two types of ferrihydrite ($\approx$ $Fe_5HO_8 \cdot 4H_2O$) and goethite (α-FeOOH) were produced. They were precipitated following the procedures described by Schwertmann and Cornell (2000). Ferrihydrite 2L represents a material with very low crystallinity and high surface area. It is similar to most natural iron oxide incrustations of low crystallinity. Ferrihydrite 6L is slightly more crystalline. It was observed in only one natural incrustation. Lepidocrocite sometimes appears as an accessory mineral phase with goethite. Goethite-90 represents a goethite similar to a species commonly found in natural soils and has a surface area of ~90 m^2/g (968 ft^2/g) (Schwertmann & Cornell 2000). Goethite-20 is somewhat more crystalline and has a lower surface area of ~20 m^2/g (215 ft^2/g). The methods applied for synthesizing are summarized in Table 8-2.

Dissolution experiments were carried out under standardized conditions: 0.5 mmol of iron oxide was continuously stirred at $T = 25°C$ (77°F) in 0.5 liter (0.132 gal) of an aqueous solution of the dissolving agent. Deionized water was used as the solvent. Before and during the experiments, nitrogen gas was bubbled through the solution to avoid the oxidation of reducing agents and dissolved ferrous iron. When using acids, the pH was monitored during the experiments. Because of the high solute-to-mineral ratio, the changes were always negligible. Samples of about 1 ml each were taken at regular intervals and analyzed for dissolved iron. Maximum experiment time was 7 h, although several dissolution curves had not reached equilibrium at that time. The time frame of 7 h was used to approximate a working day during a chemical well rehabilitation. A few experiments of longer duration were also performed with natural iron oxide incrustations.

TABLE 8.2 Synthetic Iron Oxides Used for Dissolution Experiments by Houben (2003b)

Iron oxide	Surface area [m^2/g]	Method of synthesization (Schwertmann & Cornell 2000)
Ferrihydrite 2L (Fh 2L)	309	Precipitation from dissolved $Fe(NO_3)_3 \cdot 9H_2O$ with 1 M KOH
Ferrihydrite 6L (Fh 6L)	229	Hydrolysis of dissolved $Fe(NO_3)_3 \cdot 9H_2O$ at $T = 75°C$
Goethite-20 (Gt-20)	22	Thermal ageing of ferrihydrite 2L at $T = 70°C$
Goethite-90 (Gt-90)	93	Oxidation of dissolved buffered $FeCl_2 \cdot 4H_2O$ with air
Lepidocrocite (Lc)	80	Oxidation of dissolved $FeCl_2 \cdot 4H_2O$ at constant pH (pH 6.7–6.9) by air

Chemical Rehabilitation Techniques 263

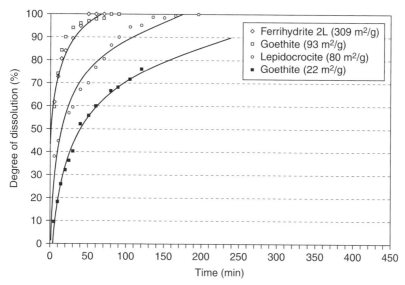

Figure 8.12 Dissolution of synthetic iron oxide minerals in 0.1 M sodium dithionite solution. From Houben (2003b).

Sodium dithionite ($Na_2S_2O_4$), a strong reductant, and oxalic acid turned out to be the most effective chemicals considering both efficiency and reaction rate (Figs. 8-12, 8-13). Both chemicals are able to quickly dissolve ferrihydrite as well as goethite. Oxalic acid has the disadvantage

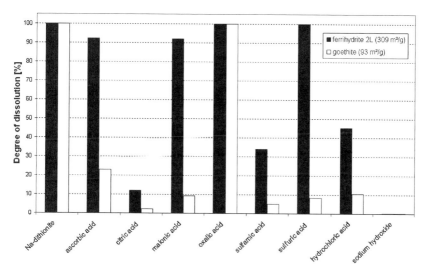

Figure 8.13 Degrees of dissolution obtained by different chemicals on synthetic iron oxide minerals of low (ferrihydrite) and high (goethite) crystallinity after 7 h. *From Houben (2003b).*

of being a potential microbial substrate. Dithionite has some additional advantages at surface areas of less than 70 m^2/g (753 ft^2/g). The high potential of sodium dithionite for iron oxide dissolution has been known for a long time. Because the highest reduction rates are achieved at neutral pH, buffering, e.g., with bicarbonate, is essential. The efficiency can be further enhanced by adding a complexing agent, e.g., citrate. This combination has evolved into a standard method for iron oxide removal in soil science and in several industrial processes (Mehra & Jackson 1960). It is the only one of the tested chemicals that dissolves significant amounts of highly crystalline goethite (Gt-20). Although it is often used in well rehabilitations, hydrochloric acid shows only mediocre results at pH > 0.0. Sulfuric acid is more efficient, although not for goethite. Ascorbic and malonic acid are useful in removing ferrihydrite but fail to remove substantial portions of goethite. Sulfamic and citric acid performed rather poorly. Bases such as sodium hydroxide proved to be generally useless, at least in the pH range studied. The essentials of the dissolution experiments are summarized in Fig. 8-13.

Crystallinity ($\approx$ age) of the iron oxides has a major influence on their solubility, regardless of the chemical. The specific surface area of a mineral is a good indicator of its reactivity (Fig. 8-8).

Efficiency and reaction rate are not the only criteria for the selection of rehabilitation chemicals. Additional factors to be considered are

- Corrosive effects on well material (acid/steel)
- Additionally required chemicals (additives)
- Expenditures for neutralization (for acids and bases)
- Possible microbial growth (organic chemicals, e.g., ascorbic and citric acid → 10.4.2)
- Toxicity, e.g., titanium(III) is a strong—but quite toxic—reductant
- Cost-efficiency

8.6 Chemical Rehabilitations in the Field

8.6.1 Introduction

According to the DVGW survey introduced in Sec. 1.3, of about 2,400 well rehabilitations performed each year in Germany, one-quarter to one-third involve the application of chemicals (Niehues 1999). This means an annual number of chemical rehabilitations between 600 and 800, practically all of which are combined mechanical-chemical rehabilitations.

The multiplicity of chemicals available for chemical rehabilitation and their differing properties require a high level of knowledge from both the planners and the executing personnel. Generally, a mechanical rehabilitation should always be performed as a first step. It is counterproductive to waste expensive chemicals to dissolve material that could have been removed much more easily by brushing or pumping. This also allows one to minimize the amount of chemicals to be injected. In addition, mechanical pretreatment steps break up incrustations and thereby create pathways and reactive surfaces for the chemicals to act on. This can significantly enhance the efficiency of the chemical ($\rightarrow$ 8.2).

In the early days of chemical well rehabilitation, a large volume of chemicals, usually several tons of acid, was poured into the well and left there to react for some time, often 24 h (Kruse 1953). The authors strongly discourage the use of this simple technique, for several reasons (Fig. 8-14):

- The rehabilitation solution will exit through the remaining open screen slots and open pore spaces into the gravel pack instead of reacting with the incrustations in the clogged parts.
- In permeable formations, the exiting solution will penetrate far into the surrounding aquifer. Uncontrolled reactions with the aquifer matrix are possible. Because of the dilution caused by dispersion and

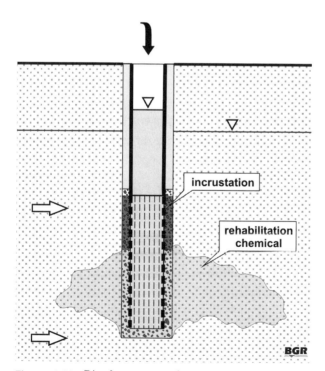

Figure 8.14 Disadvantages of uncontrolled chemical rehabilitations.

diffusion, the amount of water to be pumped after the rehabilitation to completely remove the chemicals becomes quite large.
- Operating wells in the vicinity may attract the rehabilitation solution, rendering the water quality inappropriate or even dangerous for consumption.
- Chemical solutions may collect in the well sump as a result of its higher density compared to water.

Nowadays, chemical well rehabilitations are performed after a mechanical pretreatment and in vertically staged sections. Because of the higher density of the chemical solution, the latter will try to flow downwards. Therefore the solution should be circulated in the section using a closed-circuit pump system and packers ($\rightarrow$ 7.3.4). High pumping rates are not necessarily important, because the reactions taking place are surface-controlled and do not depend on the flow velocity.

A small portion of the chemicals will almost always escape and seep downwards. Therefore it is recommended that the screen be treated "from top to bottom." Chemicals that have "sagged" can then be caught in subsequent steps. As a final step, the sump has to be cleared thoroughly. If a stepwise rehabilitation cannot be completed in one day, the well should be pumped overnight and the remaining chemicals should be stored safely.

8.6.2 Safety precautions during transport of chemicals and on-site

In most countries, the storage of chemicals is regulated by national laws and bylaws. Again, we cannot discuss all regulations of all countries or states here. Some general comments can be made, however. Storage rooms must fulfill some basic requirements:

- Secured against unauthorized access
- No connection to canalization
- Impervious floor
- Fireproof walls and ceilings
- Alarm plan and storage list available
- Escape route arranged and marked

The maximum storage height for fragile containers should not exceed 0.5 m (1.6 ft); for unbreakable containers, 1.5 m (5 ft). The combined storage of different chemicals can pose problems in case of fire. Strong oxidants and reductants must be kept separate to prevent violent reactions. Corrosive substances should be kept away from metal containers.

Radioactive substances, gas bottles, and explosives must always be kept in separate rooms and never together with other chemicals.

It is advisable to have chemicals for well rehabilitation delivered just-in-time to the rehabilitation site. This minimizes transport and storage times at the contractor's yard. There is good reason to assume that the distributor of the chemicals knows more about proper storage and transport than your regular rehabilitation company does. The distributor usually has trained and experienced personnel, adequate storage facilities, and is prepared for emergencies, while the rehabilitation companies often keep acid drums in shedlike buildings.

In practically all cases, chemicals will be brought to the rehabilitation site by road vehicles. The transport of hazardous materials is regulated by national and international laws. The U.S. Department of Transportation (DOT) has issued *Hazardous Materials Regulations* for the United States, while the *Transport of Dangerous Goods (TDG)* regulations are in use in Canada. The United Nations Economic Commission for Europe (UN/ECE) has issued the so-called *ADR* (*Accord européen sur le transport des marchandises dangereuses par route*) for the transport of dangerous goods within and between European states. As both the DOT regulations and the ADR are based on the *U.N. Recommendations on the Transport of Dangerous Goods*, they have many similarities. The ADR method of classifying materials is very similar to the DOT system. However, the ADR proceeds one step further and assigns specific item numbers for dangerous goods. The ADR regulations were elaborated by the Inland Transport Committee of UN/ECE and published as document *ECE/TRANS/175*. It has become obligatory for the EU member states as of July 1, 2005, and must be implemented into national laws. The ADR comprises two volumes totaling about 1,100 pages (much useful information can be found on the Internet, too). Every chemical can be classified into one of nine ADR hazard classes. Explosives belong to class 1 (explosive substances), carbon dioxide to class 2 (gases), and acids to class 8 (caustic substances). Depending on further properties (e.g., state of aggregation, acidity) a more detailed subdivision is possible. The ADR regulations have subgroups (a), (b), and (c) for dangerous goods that correspond to the packing groups (I, II, or III) that DOT assigns. The subgroups define how much of the individual chemical may be transported in one vehicle and which safety precautions have to be followed. For quantities which fall below a certain limit, exemptions from some of the ADR regulations are possible. These quantities are summarized in 29 LQ (*limited quantity*) groups. Combined packaging and transport of different chemicals is also regulated. Explosives, for instance, must not be transported together with any other chemical. Oxidants—including disinfectants—must not be transported together with reductants.

Producers or distributors of chemicals must supply safety data sheets for each chemical. The contents of these sheets are also regulated. In the European Union, EU directive *91/155/EWG* applies. The safety data sheet contains important information on transport, application, disposal, toxicity, environmental aspects, and occupational safety. People performing a chemical well rehabilitation must know the contents of the data sheet and keep a copy of it on-site at all times. In case of accidents or spills, it should be shown to the medical personnel or the firefighters who respond.

Protective clothing must be worn at all times while handling chemicals. This includes goggles, gloves, full cover of arms and legs (a coverall is best), and dust masks (the latter when applicable). Water, e.g., a washing bottle, must always be available to rinse splashes of chemicals from skin or eyes. Precautions for spills must be taken, and adsorbent material needs to be kept close by.

Many rehabilitation chemicals, especially acids, are supplied as liquids. They have the advantage of easy handling because they can be pumped directly from the container into the well. On the other hand, spills may occur during transport, storage, and handling, and may be difficult to control. Solid rehabilitation chemicals are easier to control but have to be dissolved on-site (Fig. 8-15). Fine-grained material may be blown away during this process and may cause respiratory problems. For reductants,

Figure 8.15 Dissolving a solid rehabilitation chemical in a mixing tank on-site. The smaller chamber contains a pump for water circulation during dissolution and for later injection into the well. The scraper is used to mobilize potential solid residue from the bottom of the tank. Note the protective clothing (gloves, goggles, dust respirators) worn by the personnel. *Photograph: Treskatis.*

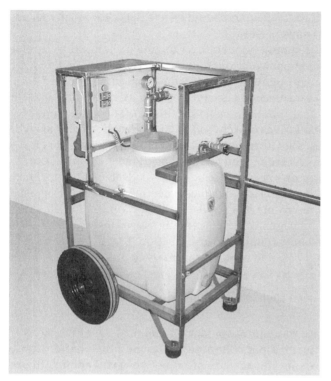

Figure 8.16 Example of a mixing and circulation system for chemical well rehabilitations. *Photograph: Aquaplus, Kronach, Germany.*

oxygen contact can diminish the efficiency of the chemical. In this case, closed mixing and circulation systems are useful (Fig. 8-16). They also decrease the chances of exposure of personnel to the chemicals.

Dumping of solid chemicals directly into the well is strongly discouraged. This will only lead to agglutination and accumulation in the sump, which will limit further dissolution and render subsequent removal difficult. Some chemicals can be transformed into gels by using certain additives. This "pudding" is then backfilled into the well and slowly releases its reactive agents. From time to time the gel mass is circulated.

Mixtures of reductants and oxidants—including disinfectants—in the well have to be avoided at all cost, because this might result in violent reactions and the formation of hazardous gases.

Chemicals that are not actually in use—for instance, during an interruption of the rehabilitation (night, weekend, technical problems)—must be stored safely, especially in well sites outside fenced-off terrains. Curious animals and/or vandalism always have to be considered. Lockable containers, trucks, or garages are the best choices to avoid this problem. Storage in the well shaft should be avoided. Storage in cool and dark locations is always preferable.

Neighboring wells should be turned off during a chemical well rehabilitation to prevent chemicals from flowing ino their direction and getting into the treatment plant or even the distribution network.

The occurrence of hazardous gases is possible at all well sites, especially during chemical rehabilitations (→ 10.2.2). The same applies to the application of explosives and carbon dioxide. Reactions of acids with carbonates, sulfides, and/or oxidants with biomass can result in massive gas generation. Effervescence may cause backflow of chemicals to the surface. Suffocating or explosive gases may collect in shafts. These therefore must be ventilated at all times. Gas indicators must be used, and recovery equipment must be present. The absence of such safety precautions and simple ignorance has already caused a significant death toll among drillers and rehabilitators. Section 10.2.2 discusses gas-related dangers and possible remedies in more detail.

8.6.3 Do we really need chemical rehabilitations?

Considering the environmental impact of chemical rehabilitations, it would be best to avoid them completely. On the other hand, purely hydromechanical rehabilitations are often not powerful enough to produce a sufficient regain in well yield. According to well yield tests performed on several wells

- Prior to rehabilitation
- After a mechanical rehabilitation step
- After a chemical rehabilitation step

We found that the application of chemicals contributes about 40–60% of the total yield gain obtained (Fig. 8-17). This clearly shows that chemical rehabilitations are still necessary, and this necessity will remain until the advent of more powerful mechanical techniques.

From our experience, the best results were obtained using a multistep approach:

Step 1: mechanical rehabilitation

Step 2: chemical rehabilitation

Step 3: mechanical rehabilitation

The third step sounds like an unnecessary idea at first because of the additional cost imposed. The authors have found, however, that the chemical rehabilitation step weakens bonds between incrustation grains.

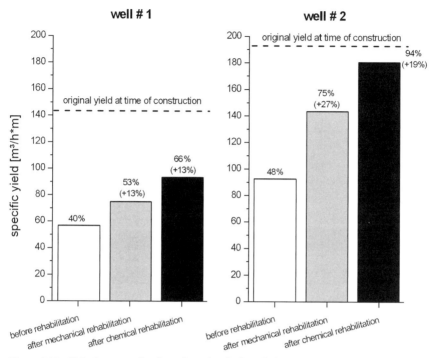

Figure 8.17 Relative contribution of mechanical and chemical rehabilitation steps for total yield gain of two wells near Krefeld, Germany. *Modified after Houben (2001).*

This material is still locked so tightly that it cannot be recovered through pumping. An additional mechanical or chemical step can crack open the weakened bonds and result in surprisingly high mobilization of material. Figure 8-18 shows the dramatic increase in well yield obtained after a third chemical rehabilitation step.

8.7 Disposal of Liquid and Solid Waste

8.7.1 Introduction

The proper disposal of liquid and solid waste is one the biggest concerns in chemical well rehabilitations. Similar to the injection of chemicals into a well, the disposal of wastewaters into surface waters or into the soil may be regulated by national laws. The classification of the environmental hazard of individual chemicals is often part of bylaws. A common partition comprises four classes (no hazard, low, medium, high). Since the regulations and classifications vary widely from country to country and often even between federal states or provinces, we cannot go into much more detail here. The reader is referred to the laws

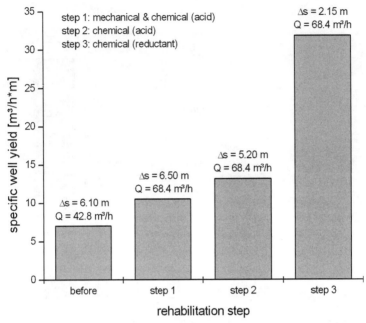

Figure 8.18 Relative contribution of mechanical and chemical rehabilitation steps for total yield gain of a well in the Lower Rhine area (Germany). The third chemical step (using a reductant instead of an acid as applied in the first step) finally constituted the breakthrough, more than doubling the well yield.

and regulations of his or her home country. The Internet is most often a good starting point to find more detailed information.

Disposal of liquid waste can be achieved by various means:

- Soil infiltration
- Infiltration into surface waters (rivers, lakes)
- Discharge into the canalization
- Intermediate storage in containers and subsequent disposal elsewhere

Disposal of the transport containers of the chemicals must not be neglected. Sometimes they can or even must be handed back to the distributor of the chemicals. Sometimes farmers and gardeners like to take over the containers for their sturdiness and closability. Just make sure that you have cleaned out all remains of the chemical properly before you give it to your brother-in-law to store his chicken feed. The remaining containers should also be cleaned thoroughly and recycled or disposed of in a landfill.

Concerning soil infiltrations, a sufficient safety distance is essential to prevent backflow of the wastewater into the treated well or other wells. The chosen infiltration area should be outside the cone of depression or at least downstream of the direction of groundwater flow. It should definitely not be inside the inner groundwater protection zone. Because the volumes pumped from the well to remove the chemicals can be substantial, it is a wise idea to check the terrain beforehand, especially when the soil is impermeable. A flooded cellar is unlikely to win you many new friends. It might be necessary to split up the infiltration over several locations.

The highest concentrations and turbidities will most likely occur during the first few minutes after the pump has been switched on ($\rightarrow$ 10.2.4). Sometimes it is a good idea to collect this first splash of highly contaminated water and take it by tank truck to a sewage treatment plant. The remaining water can then be discarded as described above.

8.7.2 Neutralization of inorganic acids

Acid injected for chemical well rehabilitation often may not react completely and has to be abstracted with a pH of less than neutral, often around pH 3. Prior to disposal, the pH has to be restored to a value between 6.5 and 8.5. This process is called neutralization and can be performed continuously (flow-through reactor) or batch-wise in containers. The pH before disposal has to be continuously measured and documented.

Neutralization requires the addition of defined amounts of bases corresponding to the pH of the pumped water. Highly and quickly soluble hydroxides such as sodium or calcium hydroxide [$NaOH$, $Ca(OH)_2$] are easy to handle and are available both in liquid and solid forms. Equation (8.13) shows that a sodium chloride solution will form when hydrochloric acid is neutralized. With an originally 0.1 M hydrochloric acid, the resulting NaCl content of the disposed water amounts to 5.85 g/liter!

$$H^+ + Cl^- + NaOH \rightarrow Na^+ + Cl^- + H_2O \tag{8.13}$$

Solid calcium carbonate (limestone, $CaCO_3$) or soda (Na_2CO_3) granules can be used instead of the highly reactive but strongly caustic hydroxides. The reaction progress is slower and determining the proper amount is therefore more complicated. A reaction product which has to be considered is carbon dioxide [CO_2; see Eq. (8.6)].

During neutralization, dissolved iron and manganese reprecipitate as solid oxides and release the same amount of acid as was necessary to dissolve them beforehand. This amount of acid must be neutralized as well. Iron and manganese oxides cause strong turbidity and give a reddish or blackish color of the water. They should be separated from the remaining solution through sedimentation and disposed off separately ($\rightarrow$ 8.7.6).

8.7.3 Neutralization and disposal of organic acids

Organic acids have to be neutralized just like inorganic acids. The procedure must comply with the ones described in Sec. 8.7.2. The organic component can—at least in theory—be destroyed (mineralized) by treatment with hydrogen peroxide. Because of the masses involved and corresponding safety constraints, however, this is not recommended. In most cases the organic acids are biodegradable—e.g., ascorbate, citrate—and are quickly and thoroughly mineralized in soil or wasterwater treatment plants.

8.7.4 Neutralization of bases

The comments in Sec. 8.7.2 apply equally to the neutralization of bases, except, of course, that acids must be added in this case to obtain a neutral pH, instead of bases.

8.7.5 Disposal of reductants

Commonly used reductants, such as sodium dithionite, work best at neutral pH conditions. Subsequent neutralization is therefore not necessary. This reduces the amounts of chemicals that need to be transported and applied quite significantly.

The water pumped from wells treated with dithionite contains sulfite and has a pronounced lack of oxygen (reducing character). Contact with oxygen (air) will solve this problem. Injecting air through the water via a compressor will accelerate this process. The authors are aware of one case in which the wastewater was treated by the addition of hydrogen peroxide. This procedure must be used only when the user is sure that no remaining dithionite is present in the wastewater (total conversion to sulfite or sulfate), otherwise violent reactions and gas formation are possible. Oxidation by air or peroxide may lead to reprecipitation of the dissolved iron and/or manganese ($\rightarrow$ 3.2.2). The solution remaining after oxidation is a more or less concentrated sodium sulfate solution.

8.7.6 Disposal of sludges and solid waste

The wastewater pumped from the well contains chemically dissolved and often particulate matter also. The latter may comprise loosened incrustations as well as sand, causing significant turbidity. Particle contents may be high enough to form sludges. Similar sludges may form during neutralization after an acid treatment ($\rightarrow$ 8.7.2, 8.7.3). Trace elements mobilized from incrustations during chemical dissolution will be immobilized again during sludge precipitation ($\rightarrow$ 3.2.8). Though the sludge contains harmless, naturally occurring material (ochre, carbonates, sand, silt), sedimentation prior to disposal seems a good idea. The German federal state of Bavaria has issued a decree that the water from such settling ponds must contain less than 50 mg/liter of filterable solids, have a pH value between 6.5 and 9.0, and a maximum arsenic concentration of 0.1 mg/liter (Bayer. Landesamt Wasserwirtschaft, Merkblatt 1.4/3).

Ochre and carbonate sludges are generally not considered dangerous to the environment, although the former may contain significant concentrations of trace elements such as arsenic. They can be disposed of in landfills after dewatering or cached with the sludges obtained during iron and/or manganese removal in the water works during regular water treatment. Ochre sludges with low trace element contents can be used in wastewater treatment plants for the removal of excess phosphate and sulfide.

Chapter 9

Repair, Reconstruction, and Decommissioning of Wells

9.1 Definition of Terms

We define *reconstruction of wells* as technical intervention into the current structural construction of a well (Table 9-1). Reconstruction is usually employed to handle

- Fundamental flaws in planning and construction, e.g., inappropriate gravel pack, defective annular seal
- Material deterioration (corrosion)
- Physical damage
- Clogging that cannot be removed even by repeated rehabilitations

Reconstruction measures may include the removal of casing, screen, and even the annular filling material and the subsequent insertion of new components (Table 9.1). Less advanced reconstruction measures are often referred to as "well repair." This term describes limited technical measures employed to handle construction flaws and material failures, e.g., corrosion and leaky seals, which may disturb or endanger well operation. Casing and annular filling material are not affected by repairs.

The decommissioning of wells comprises all measures after a well has fully been put out of operation. This may be necessary for any of several reasons (Table 9-1):

- Irreparable damage
- Decreasing water demand

TABLE 9.1 **Reconstruction and Decommissioning Techniques**

Method	Process	Measures and aim
Partial sealing of well screen	Reconstruction	Closure of undesired inflow
Inline tubing	Reconstruction	Run-in of inline tubing over full or partial length
Insertion of new screen	Reconstruction	Recovery of screen, removal of gravel pack, insertion of new screen
Redrilling	Reconstruction	Exchange of existing tubing and annular filling via drilling (clamshell or rotary)
Repair of well-head construction	Reconstruction	Elimination of structural damage, sealing, exchange of machinery and fittings
Sealing of well interior	Decommissioning	(Partial) backfilling of well tubing and shaft with granular or sealing material (bentonite, concrete)
Perforation	Decommissioning	Generation of perforations, e.g., via explosives, for subsequent insertion of annular seal
Cutting of tubing	Decommissioning	Separation of tubes to prepare for later recovery
Removal of well tubing	Decommissioning	Redrilling of well and recovery of tubing and annular filling
Injections	Decommissioning	All sealing techniques, e.g., in the annulus
Removal of well-head constructions	Decommissioning	Demolition of protective building, removal of machinery and fittings, backfilling of shaft

- Decreasing groundwater table, e.g., in mining areas
- Pollution of groundwater which prohibits distribution to consumers
- Relocation of water works

There is more to decommissioning than simply turning off the pump and locking the lid. Although the well may seem sturdy enough at the time of abandonment, corrosion, weathering, burrowing animals, root growth, and vandalism may soon lead to its collapse. Construction workers, animals, and playing children, unaware of its presence, are in danger.

Aquifer protection requires that abandoned wells do not act as a conduit of contaminated surface water or shallow groundwater into deeper formations. The function of aquitards must not be endangered. Even small disruptions in the aquitard, e.g., through a missing annular seal of a well, may cause serious influxes of pollutants. This may necessitate sealing the well interior, reinstatement of aquitards, or even total removal of the well head, casing, and annular filling, with subsequent hydraulic sealing of the borehole.

Camera inspections (→ 5.1) and geophysical well logging (→ 5.3) are very useful tools to assess the current state of a well and its components. Indirect hints at constructional flaws and material failure include:

- Unprecedented sudden changes in water chemistry and microbiology
- Sudden increase in turbidity and particle content
- Very fast decline in well yield
- Failure of a fairly new pump
- Ground settling around the well

The main causes of structural damages are:

- Material fatigue
- Corrosion (due to missing or deficient corrosion protection)
- Deficient material
- Processing deficits
- Inappropriate dimensioning
- Handling errors, e.g., during insertion and removal of the pump

Structural damage enhances the influx of water of undesirable quality, e.g., surface water, and lowers the stability and the well yield by promoting mixing, corrosion, and incrustation buildup (→ 3). Influx of polluted water can also occur from locations outside the well. This includes leaky wells or piezometers in the vicinity, preferential flow paths, e.g., from the settling pond of the drilling or the excavation pit of the well head, and intrusion of plants with deep roots.

Even a new well might require reconstruction when material defects, improper design, or poor construction severely interfere with its operation. The problem is usually monocausal and can be solved with limited financial expense if it is identified early enough. Older wells may be subject to multicausal and overlapping problems which often require substantial investments for reconstruction, especially when design flaws necessitate the exchange of the annular filling. The extent of the identified damage, of course, dictates the measures to be taken.

While the cause study must include identification and appraisal of the present damage, technical feasibility and economical constraints need to be considered in planning the actual reconstruction. Legal constraints should also be taken into account, e.g., protection zones. All in all, sound knowledge of technical feasibility, cost ranges (→ 6), and the legal background require the involvement of experienced planners and contractors.

Well construction companies offer a variety of reconstruction methods. An indispensable prerequisite is the study of all available technical and hydraulic data on the well. Operational characteristics should be documented continuously and evaluated on a regular basis. This includes water levels, well yield, results of previous rehabilitations, optical and geophysical inspections, water analyses, and pumping tests (→ 2.5). Evaluation of these data and special investigations should clarify the causes of the damage. Because we may have to deal with a variety of causes and technical constraints, the qualification of contractors becomes a major issue. Tables 9.2 and 9.3 summarize causes and triggering incidents as well as reconstruction methods and their inherent risks. In addition, they provide estimates of the financial risks involved and possible consequences for structural stability and well yield.

Criteria to be considered before the reconstruction of a well include:

- Drilling technique
- Well construction (partially or fully penetrating)
- Well material (PVC, steel, etc.)
- Depth and drilling diameter
- Type of aquifer (consolidated/unconsolidated)
- Use of well (drinking water, recharge well)
- Technical risk (feasibility)

Wells equipped with additional protective casings (lost drill casings) often require measures only below this casing. In this case, even deep wells can still be reconstructed with limited technical and financial expenditure. Shallow wells with a fully cemented annulus, on the other hand, pose a major difficulty for redrilling. This should show that the individuality of wells does not allow any generalizations.

The risks to be expected during execution are a function of the intensity of the reconstruction measures (Tables 9-2, 9-3). Reconstruction of wells with structurally stable casings of course presents a much lower risk than wells that show defective annular seals or even heavy corrosion. The latter require massive interventions, including an exchange of the annular seal, perforations/injections, redrilling, and casing/screen recovery.

The decision in favor of or against a reconstruction should always take into account the future use of the well. Even if the well is not needed any more, reconstructions may still be advisable or even compulsory if the well acts as a conduit for polluted water into deeper formations, e.g., via a leaky annular seal.

TABLE 9.2 Causes and Triggers for Well and Piezometer Reconstruction

Causes and triggers	Techniques	Risk assessment		Appraisal		
		Execution	Cost	Stability	Yield	Ageing
Sand intake	Mechanical, hydromechanical rehabilitation, ex-post desanding Installation of suction flow control device Recovery and exchange of screen tubing	Low Medium	Low Medium	No significant influence if applied correctly Improvement	Improvement (if no compaction or settlement occurs) Improvement	Decrease of sand intake (prerequisite: correct dimensioning of gravel pack and screen slots) Improvement
Mixing of waters of different origin	Perforation of tubing and injection of sealing material	High	High	Risky	No significant influence	Improvement
Water flow from adjacent aquifers, e.g., through the annulus	Removal and reinstallation of tubing	Medium to high	Medium to high	Improvement	Slight deterioration possible	Improvement
Structural damage to well tubing	Partial or complete run-in of inline tubing and annular filling	Low	Low	Improvement	Decline possible	Decrease of sand intake Deterioration through elevated intake velocities

(*continued*)

TABLE 9.2 Causes and Triggers for Well and Piezometer Reconstruction (*Continued*)

Causes and triggers	Techniques	Risk assessment			Appraisal		
		Execution	Cost	Stability	Yield	Ageing	
Decline in well yield over time	Mechanical, hydromechanical, and/or chemical rehabilitation	Low	Low	No significant influence if applied correctly (caution in corroded wells)	Improvement (when applied in time)	Decline in sand intake possible (ex-post desanding) Complete stoppage of ageing usually impossible as to dependency on naturally occurring processes	
Corrosion of well tubing	Recovery of Tubing, redrilling	High	High	Improvement	Mostly improvement (if ageing products are removed)	Improvement	

TABLE 9.3 Appraisal of Technical and Financial Constraints for Well Reconstruction Schemes

Measure	Risk assessment — Execution	Risk assessment — Cost	Probability of success	Methods for performance assessment
Recovery of pump and installations	Low	Low	High	Visual inspection
Closure of pipeline	Low	Low	High	Visual inspection; pressure test, if necessary
Removal of well head	Low	Low	High	Visual inspection
Removal of well tubing; removal of sediment	Low to medium	Low	High	Depth sounding; visual inspection of recovered material
Sealing of well interior using loose material (only if annular seal is still intact)	Low to medium	Low	High	Depth sounding; material control prior to filling
Sealing of well interior with cement plug (only if annular seal is still intact)	Low	Low	High	Visual inspection; mass balancing of filling material
Demolition of well shaft	Low	Low	High	Visual inspection
Removal of tubing and screen	Medium to high (depending on material and subsurface conditions)	Medium to high	Medium (collapse of formation)	Visual inspection of recovered material; borehole geophysical measurements (in stable formations)
Redrilling, perforation, and injection to seal off inflow through the annulus	High	High	Medium to low	Visual inspection and borehole geophysical measurements before and after completing the measure; mass balancing of injected goods
Injection sealing (partial or full)	Medium	Medium	High	Depth sounding; mass balancing of injected goods
Rehabilitation as precursor to injection	Low to medium	Low to medium	Medium to high (depending on incrustation type and residual permeability)	Depth sounding; visual inspection of removed material

9.2 Reconstruction

9.2.1 Partial reconstruction

We can distinguish between partial and complete reconstruction. Table 9-4 summarizes the currently available reconstruction techniques.

Partial reconstruction includes a multitude of possible measures. The simplest reconstruction measure is, of course, replacement of the pump and riser pipe. Installation of centralizers to prevent damage to the casing is another cheap and efficient remedy.

Insertion of a smaller inline casing and/or screen into the interior of an existing well is a common method for prolonging the operation life of an aged well that is on the brink of structural collapse (Fig. 9-1). This is a very ancient technique. The wooden-frame Neolithic dug well mentioned in Sec. 2.1.2 was reconstructed at least twice by insertion of smaller wooden frames (Weiner, in LVR 1998). The insertion was intended to stabilize damaged older framing and to deepen the well. Causes for the reconstruction were damage to the original frame by violent acts of war and a natural decline of the groundwater table.

The overall statics of a well to be reconstructed must be intact prior to insertion. An inline casing protects the pump from collapsing and may keep sand away. Both partial and full inline casings/screens are possible. Partial relining is often used to seal off certain sections which, for instance are structurally damaged, contain water of undesirable quality or cause sand intake due to broken screen slots. Full relining is usually employed to stabilize the whole well. The remaining space between the original casing and the inline casing/screen is sometimes backfilled with filling material (gravel, bentonite, or cement grout) to stabilize the setup. The technique is used in both vertical and horizontal wells. It is not applicable for wells with small diameters and defective annular seals.

Insertion of inline casings and screens reduces the net diameter of the well. A new, smaller pump must be selected. Well yield might be negatively affected by the increased head losses. Rehabilitation efficiency is strongly reduced because of the flow impediments caused by the additional installations. Recovery of inline casings and screens is often impossible. Redrilling such wells is difficult and requires highly sophisticated equipment.

Relining is possible for all casing material and for wells of almost any depth. Some features of the old casing which may hamper the insertion are rough surfaces, protrusions, open connections, deviations of the well axis from straightness, and plumbness. If a metal inline casing is to be inserted into a well that is already equipped with a metal casing, issues of electrochemical compatibility have to be considered (→ 3.1.2). Otherwise, contact corrosion may quickly destroy the well.

TABLE 9.4 Reconstruction Techniques and Their Applicability to Wells of Various Materials

x possible
o limited
− not possible

material		relining — tubing: annular grouting	relining — screen: gravel pack insertion	redrilling — tubing: ex-post sealing	redrilling — tubing: protective casing	redrilling — screen: crushing	under-reaming gravel pack
steel	uncoated	x	x	x	x	x	x
steel	coated	x	x	x	x	x	x
stainless steel		x	x	x	x	x	x
plastic		x	x	−	−	−	o
stoneware		x	x	o	o	o	o (in open borehole)
laminated wood		x	x	o	o	x	o

Redrawn from Rübesamen & Nolte (1999).

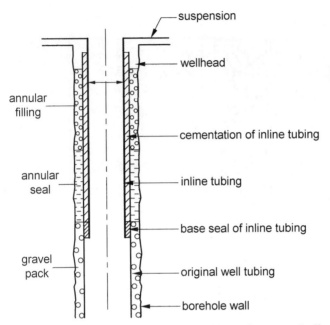

Figure 9.1 Partial relining tube cemented into place to seal off leaky tubing.

A specialized version of relining was developed in Berlin, Germany, for wells with ceramic and brittle PVC tubing. After the installation of a steel inline casing—usually with a wire-wound screen— a strong explosive line charge is inserted (→ 7.5.2) which causes the collapse of the old screen. The fragments of the old screen and the disrupted gravel pack remain in the well. The blasting cracks open flow paths and thus enhances the well yield. The slots of the new screen must of course be able to retain the grains of the gravel pack. This method should not be applied where annular seals are damaged and cause an influx of polluted surface water. So far, the authors are unaware of long-term effects and the sustainability of this method. It is probably useful to extend the operating life of a well for a limited time.

Collars (sealing sleeves) are used to seal off spatially limited zones of defects, such as areas of pitting corrosion (Fig. 9-2), leaky casing connections (Fig. 9-3), or welding seams (Moore 1997). Several methods are available on the market. The patches used for this purpose may be thin-walled stainless steel tubes—sometimes with a vulcanized outer rubber cover—or may consist of corrugated iron sleeves. The rubber cover is helpful in closely fitting the patch and minimizing contact corrosion. In a first step, the patch is picked up by the inflation of a packer. The assembly is then lowered into the well. On reaching the location of the defect

Repair, Reconstruction, and Decommissioning of Wells 287

Figure 9.2 Pitting corrosion of a formerly plastic-coated steel tube [⌀ 400 mm (1.31 ft)] recovered during a well reconstruction. *Photograph: Christoph Treskatis.*

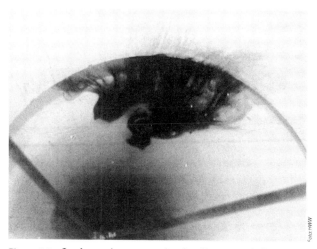

Figure 9.3 Leaky casing connection leading to an inflow of dark shallow groundwater rich in humic substances. *Photograph: Lutz Nolte, NBB, Hamburg, Germany.*

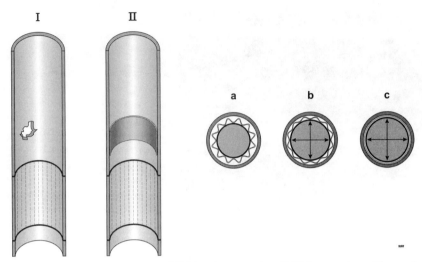

Figure 9.4 Patching of a damaged Tubing (cross sections I, II) by pressing a Sleeve of corrugated Metal onto the Tubing using an expandable Packer (plan view, a-c)

to be patched—which has to be verified by camera inspection—the packer is gradually expanded to press the sleeve against the casing. Packer pressures can reach up to 700 bar (10,150 psi). Figure 9-4 shows the process in a simplified form. Depending on the length of the patch and the packer, the procedure may be performed in several steps—usually from top to bottom—by repeatedly deflating and lifting the packer. The procedure is completed by inflating the packer once more at the bottom and the top of the patch with the rubber overlapping onto the casing. This creates a smooth transition between patch and casing. The force exerted by the packer must be strong enough to tightly attach the casing and sleeve in order to seal off the defect. On the other hand, the force must not cause any deformation of the casing. Experienced contractors of course know how much pressure they have to apply for which material.

Another technique for patching was developed in Germany. The applied sleeve consists of a rolled sheet plate of high-quality stainless steel (V4A) and is equipped with two compression seals (Fig. 9-5). The sleeve has a punched band of sprockets at each side. The diameter of the rolled plate at the time of insertion is about 40 mm smaller than the casing to be patched. Expansion of an inflatable packer causes expansion of the sleeve by the movement of toothed wheels along the sprocket band. Sealing is achieved by pressing the sleeve and its seals tightly against the casing. Undesired reversal (shrinking) of the expanded sleeve along the sprocket band after the installation is prevented by a lock-out mechanism. The system works for the partial sealing of casing of all types of wells and pipelines with diameters ranging

Figure 9.5 Patching of defects by expanding a rolled sleeve of metal onto the tubing. *Photographs courtesy of Uhrig Kanaltechnik,* Geisingen *Germany.*

from 150 to 700 mm (0.5 to 2.3 ft). Problems may occur when the casing shows irregularities, e.g., around connections, and when the casing is not fully circular (oval). This may prevent a tight fit of the sleeve and compromise its sealing function.

If more than one defect is present, it may be necessary to reline the entire tubing. Several techniques have been developed in Australia for this purpose (Figs. 9-6, 9-7). One method is to reline by installing consecutive and overlapping metal collars over the entire casing length (Fig. 9-6).

In boreholes with rough surfaces, e.g., open holes in consolidated rocks, the installation of collars is difficult. In this case a combination of relining and grouting of the annulus is used. One collar is set to the bottom by an expandable packer. The new annulus between the former casing or borehole wall above this collar is then grouted to the top (Fig. 9-7). Variations of this setup are available for different conditions, including small-diameter wells and additional relining of the screen.

In deep wells with telescoping diameters, leaky seals between casings of different diameters cannot be patched as described above. In this case a reducing tubing is inserted over the leaky seal and patched into place with an expandable packer (Fig. 9-8).

9.2.2 Complete reconstruction

Complete reconstructions are basically equivalent to the construction of a new well at the identical location as the previous one. Although this requires additional costs for removal of the old well, it may be reasonable when existing infrastructure (electricity supply, pipelines) can be used and existing protection zones and water rights can be retained, thus avoiding new applications. A full reconstruction may include:

- Complete removal of screen and casing
- Removal of annular filling material
- Reequipment with casing, screen, gravel pack, and annular seal

290 Chapter Nine

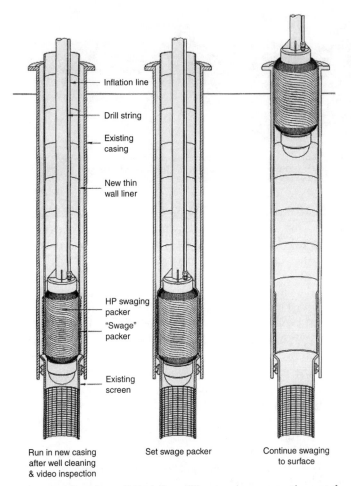

Figure 9.6 Complete relining of a well by pressing consecutive metal sleeves onto the existing casing using an expandable packer. *Drawing courtesy of Inflatable Packers Int. Pty., Perth, Australia.*

As shown in Sec. 3.9, incrustations may reach several decimeters into the gravel pack or even fill the annulus completely (Houben 2006). In the latter case it is compulsory to remove the entire annular filling and scrape the former borehole wall.

Injections to seal the annulus can be facilitated from the outside via jetting spears. From the inside of the well, this can be reached through perforations. Both methods may substantially damage the well structure, and special precautions must be taken. Perforation/injection schemes should not be applied on older wells equipped with casing made of PVC, stoneware, or laminated plywood. This method has advantages in deep wells with steel casings and especially when localized defects are targeted.

Repair, Reconstruction, and Decommissioning of Wells 291

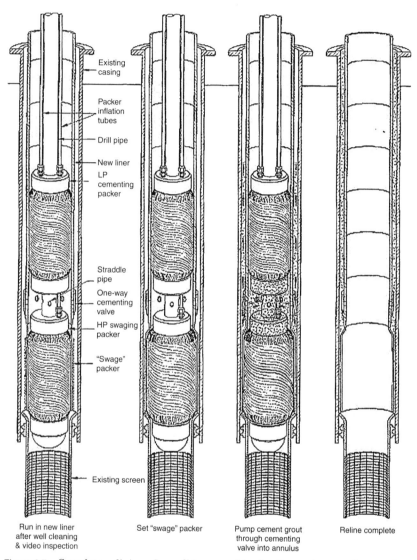

Figure 9.7 Complete relining of a well by a combined collar and cementing routine using an expandable packer and cementing valve. *Drawing courtesy of Inflatable Packers Int. Pty., Perth, Australia.*

Redrilling requires an annulus of sufficient thickness, a stable casing, and a plumb and straight borehole. Protective casing is either present (lost drill casing) or inserted during the reconstruction (Fig. 9-9). The scheme starts by boring or sucking out the annular filling materials. The interface between the annulus and the surrounding formation should also be cleared. This ensures the removal of colmation layers or even plugging

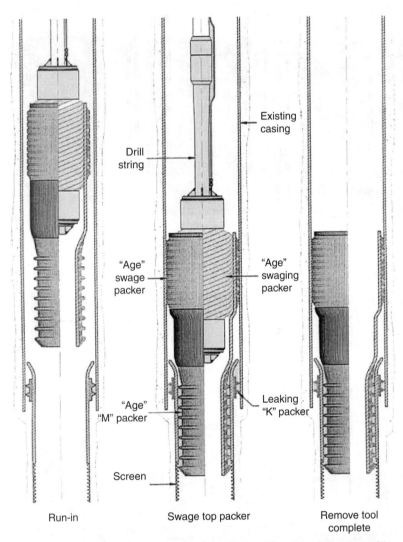

Figure 9.8 Sealing of leaky tubing connections in wells with telescoping tubing diameters by setting a reducing tube over the connection using an expandable packer. *Drawing courtesy of Inflatable Packers Int. Pty., Perth, Australia.*

dating back to the days of the original well construction (remnants of the mud cake of the drilling fluid). The excavated annulus should be stabilized with a bentonite suspension. The well casing and the screen can then be cut and recovered (Fig. 9-10). This scheme is useful for steel wells but cannot be applied with brittle material such as PVC, because the casing might collapse. In this case the whole well —including casing and screen— can be destroyed by large-diameter drilling. Recovered screens and

Figure 9.9 Casing driver (left) and "polyp gripper" (right) used in well reconstruction for excavation of the annulus. *Photographs: Christoph Treskatis.*

piezometers often give good insight into the types and thicknesses of incrustations that caused the decline in well yield (Fig. 9-11).

Casings and slotted screens made of steel can often be pulled at once—at least in wells of shallow depth. Material of lower tensile strength, e.g., PVC or wire-wound screens, in many cases must be removed section by section. Figure 9-12 shows the remnants of a PVC screen from a well of 30-m (98-ft) depth that could not be pulled due to the high jacket friction. It had to be excavated step by step with a clamshell. Following this, the well was equipped with a new casing, screen, and annular filling material.

The authors found several wells screened in confined deeper aquifers which suffered from incomplete annular seals. In one case, in a row of five wells, only one showed elevated nitrate concentrations, although all the wells seemed to be protected by a thick and impermeable confining layer. It turned out that the nitrate came from an upper aquifer and

Figure 9.10 Casing and well screen recovered during a reconstruction. *Photographs: Christoph Treskatis.*

294 Chapter Nine

Figure 9.11 Heavily ochre-incrusted piezometer [Ø 50 mm (2 in)] recovered from the former annulus of a well during reconstruction. *Photograph: Christoph Treskatis.*

Figure 9.12 Shattered PVC tubing [Ø 250 mm (10 in)] recovered during well redrilling. *Photograph: Christoph Treskatis.*

seeped through an incomplete annular seal. Although the area of the seal is quite small, one should always keep in mind that the highest gradient between two separate aquifers occurs directly at the pumped well. Therefore, even small flow paths allow substantial amounts of near-surface water to enter the well. In this case, this resulted in nitrate concentrations of up to 45 mg/liter. In another case, a well screened in the third aquifer unexpectedly showed pesticides. As the cautious reader may suspect by now, a leaky annular seal was the culprit. Both wells were reconstructed by sealing the annulus. Removal of the casing was not necessary. The measures undertaken were a full success, as the nitrate and pesticide pollutions ceased immediately.

In both these cases, reconstruction was more cost-efficient than constructing a new well. A new well, however, would probably have been more efficient in yield than prolonging use of the old one. On the other hand, even with reconstruction, the old well would still have required some costly protective measures, as German regulations require that the area of abandoned wells be restored to conditions as they were before the well was constructed.

In most cases it is impossible for contractors to provide a realistic quote, because technical difficulties during complete reconstructions are common but cannot be predicted or quantified. For this reason, quotes often contain only calculations on an hourly basis. Shallow wells and shallow annular seals are of course less difficult to handle than deep wells, yet the positioning of the dividing line between shallow and deep wells remains unclear. In extreme cases, the cost of full reconstructions may exceed the cost for a new well by several orders of magnitude ($\rightarrow$ 6.5, 6.6). This can be the case especially for wells with large diameters or severely corroded casing/screen on the brink of collapse.

If the estimated reconstruction costs reach about 50% of the cost of a new well, one should consider drilling a new well and use the remaining money to close the old well properly. As mentioned before, a new well at a different location imposes additional costs due to the necessary relocation of the water pipeline and the electricity supply.

9.3 Decommissioning

The aim of decommissioning is to put the well out of action without jeopardizing future use of the location. A variety of decommissioning techniques is currently in use (Table 9-5):

- *Backfilling of the well interior* can only be applied when the annular seal is in place and intact, or when aquitards are absent. Backfill materials include gravel, sand, clay-cement suspensions, or expandable clay pellets according to the geological strata.

TABLE 9.5 Decommissioning Techniques and Their Applicability to Wells of Various Materials

x possible o limited – not possible		backfilling according to geology	recovery		injection or repair of annular seal		
Material			redrilling cutting recovery	millcutting	injection by spears from annulus	injection by perforations from inside	
steel	uncoated	x	x	–	x	x	
	coated	x	x	–	x	x	
stainless steel		x	x	–	x	x	in open
plastic		x	o	x	o	o	borehole
stoneware		x	–	x	o	o	
laminated wood		x	o	x	o	o	

Redrawn from Rübesamen & Nolte (1999).

- *Recovery of the well casing and screen and backfilling of the borehole* according to the geological formations. This method is financially and technically reasonable only for shallow wells.
- *Redrilling of the complete well and subsequent backfilling* of the borehole is necessary when the aquitards cannot be otherwise reinstated.
- *Mill cutting of the well* is an alternative to redrilling and is applicable for shallow wells with PVC, stoneware, or laminated wood tubing. The borehole is backfilled after the procedure according to the original geological log.
- *Perforation of the casing* at the depth of the defect or absent annular seal, grouting of the annulus with a bentonite-cement suspension, and subsequent backfilling of the well interior can be applied even on very deep wells. Perforations are usually achieved by using explosive charges. Tools for this purpose are well known in the oil industry.

Current U.S. standards for locating and decommissioning abandoned wells are given in ASTM D6285-99 (2005) and D5299-99 (2005).

Depending on the potential hazard to groundwater from the abandoned well, reconstructions can be more or less complex. Shallow wells with intact annular seals can be backfilled with material according to the original geological profile (Fig. 9-13). The above-ground components should be removed and the ground surface reestablished. The removal of the above-ground installations and the well shaft usually is a straightforward, low-risk measure. It includes removal of the technical installations and fittings, separation from the water pipeline and the electricity supply, and the demolition of the well-head housing (if present). The shaft should be backfilled with material of low permeability. Backfilling in layers minimizes ground settling. Finally, aggradation of humic topsoil and grass sowing reestablish the ground surface.

Wells screened in several aquifers or with leaky annular seals require much more attention. Their reconstruction poses far greater risks in terms of technical and financial difficulties. They often include multistaged measures and can include perforation/injection and even redrilling. In many cases the authorities are the key players who decide if and how a well has to be treated.

Figures 9-14 and 9-15 show some practical examples of wells in Western Germany which were decommissioned—or reconstructed—by perforation and subsequent grouting of the annulus. In the first case the original drillhole had penetrated an aquitard separating two aquifers. Since the yield of the deeper aquifer was too low, the well was screened only in the upper aquifer. The complete borehole was

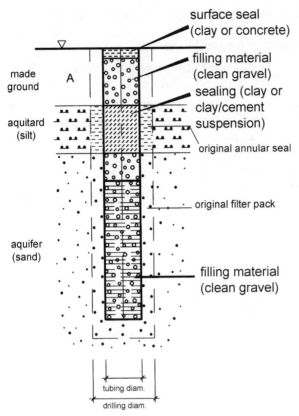

Figure 9.13 Filling of an abandoned well with intact annular seal. *Modified after Freie and Hansestadt Hamburg (1999a, 1999b).*

filled with permeable material (drilling debris). In the second case, the well was screened in the lower aquifer but the protective aquitard was not restored by sealing the annulus. In both cases the authorities demanded that the short-circuits between the aquifers be closed. In the well shown in Fig. 9-14, a small drillhole (146 mm, 0.5 ft) was drilled through the end cap. The original casing was backfilled with grout, and a packer and a detonation chord were installed. Upon explosion, the grout was injected into the annulus. The shattered original well was relined with a screen of smaller diameter (200 mm, 0.66 ft) and a new gravel pack. In the second case (Fig. 9-15) the screen was backfilled with gravel. This backfill was sealed with a clay plug. The well casing was then perforated at the depth of the aquitard from the inside, again by directed blasting. Afterwards a clay-cement suspension was injected into the annulus by another

Repair, Reconstruction, and Decommissioning of Wells 299

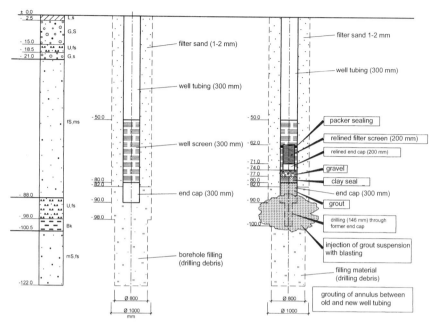

Figure 9.14 Restoration of a severed aquitard through blast perforation and grouting from a drillhole through the bottom of the old well with additional relining.

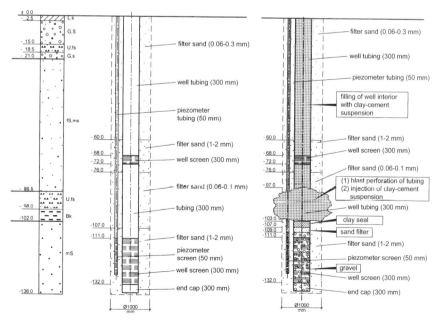

Figure 9.15 Decommissioning of a well with restoration of an aquitard severed during drilling by grout injection through blast perforations.

blasting to reinstate the permeability of the aquitard. The remaining well interior was backfilled with the same grout. Another example of perforation with subsequent grout injection and backfilling of the well interior is shown in Fig. 9-16.

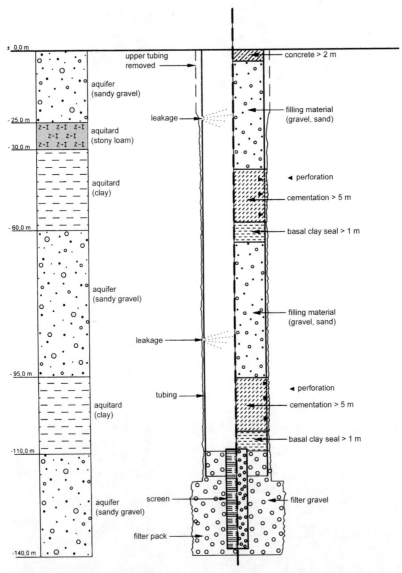

Figure 9.16 Reinstatement of an aquitard severed during drilling through blast perforation and grouting. *Modified after Freie and Hansestadt Hamburg (1999a, 1999b).*

Repair, Reconstruction, and Decommissioning of Wells 301

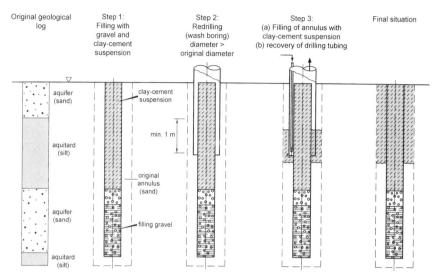

Figure 9.17 Decommissioning of a well with missing annular seal by filling, partial redrilling, and grouting of annulus. *Modified after Freie and Hansestadt Hamburg (1999a, 1999b).*

Another setup for decommissioning a shallow well with missing annular seal is shown in Fig. 9-17. The well interior is first backfilled according to the present geological formations. In a second step, the upper part of the original drillhole is redrilled to a larger diameter. Finally, both the inside and the annulus of the new drillhole are grouted from bottom to top. Casing and screen of the old well remained in place.

Another, more drastic method is shown in Fig. 9-18. In this case, the entire well, including casing, screen, and annulus, was destroyed and recovered (in pieces) by using a large-diameter auger. In some cases, protective casing is used during such redrilling. The newly created borehole is then backfilled with material of low permeability (grout, clay, cement). Nowadays even steel casings (up to 6-mm thickness) can be redrilled by this technique. The casing is cut up in sections of about 6 m (20 ft). After removal of the annular fill, the cut sections can be retrieved from the borehole (Fig. 9-10). Maximum depths realized so far were up to 300 m (985 ft). Figure 9-19 shows a perforator used for downhole cutting of the casing. A sand/water suspension (e.g., 1- to 2-mm grain size) is ejected at high velocities through nozzles and acts on the casing surface similar to the action of a blowpipe. The perforator is rotated to ensure a fully circular cut. It is essential that the perforator be aligned in the middle of the tube (e.g., by using centralizers); otherwise, incomplete cuts may occur. The annulus behind the casing to be cut must be fully backfilled at this time. Otherwise the casing might start to vibrate, which will severely diminish the cutting action.

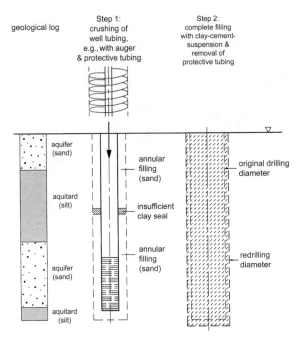

Figure 9.18 Decommissioning of a well suffering from a leaky annular seal by redrilling and grouting of the borehole. *Translated from Freie and Hansestadt Hamburg (1999a, 1999b).*

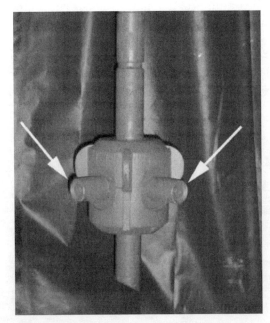

Figure 9.19 Perforator used for downhole cutting of casing: the sand/water suspension exits through nozzles indicated by arrows. *Photograph: Fangmann Salzwedel, Salzwedel, Germany.*

Chapter 10

Practical Well Rehabilitation

10.1 Preparation

10.1.1 Problem definition

We need to know our enemy to defeat him. It is therefore advisable to identify the problems well in advance of well rehabilitation or reconstruction. This will provide the opportunity to select an appropriate rehabilitation or reconstruction technique and leaves enough time for the contractor to adapt his equipment to the specific case.

Evaluation of the yield curve development, camera inspections, and—if needed—geophysical logging will help to identify the prevalent problems ($\rightarrow$ 5). If needed, consultants and rehabilitation companies can be addressed to assist in evaluating the data and in elaborating recommendations for further steps to be taken.

If the well suffers from structural damages and resulting problems, e.g., sand intake, repair and reconstruction are the options of choice ($\rightarrow$ 9). Mechanical plugging ($\rightarrow$ 4) and incomplete desanding call for mechanical rehabilitation measures ($\rightarrow$ 7). Mineral incrustations ($\rightarrow$ 3) can often be removed by mechanical rehabilitations. However, when mechanical measures alone are unlikely to succeed, samples of the incrustation should be collected to select an appropriate chemical remediation agent.

10.1.2 Tender documents

The preparation of tender documents is a tedious but nevertheless very important task. The list of conditions should be complete, and each condition should be defined with as much detail as possible. This will help to minimize potential future claims for more money. Seemingly trivial conditions such as setting up the site, disposing of waste, pumping tests, etc., must not be forgotten. In some cases, the contractor will have to involve subcontractors for special tasks, e.g., camera inspections and geophysical logging. Both sides should state well in advance how they plan to treat these issue.

Of course there will always be conditions that cannot be anticipated in detail. Disassembly of the pump and riser pipe may be a quick process, but it can also be very tiresome and time-consuming when items are corroded or heavily incrusted. Therefore contractors like to include conditions to be calculated on an hourly basis. On the other hand, an "all inclusive" tender with one general condition, "well rehabilitation," is also not useful.

Appendix C shows a tender form for a combined mechanical and chemical well rehabilitation as an example.

10.1.3 Selection of the executing company

The main argument for consideration in selecting a contractor is of course the total price. However, the costs quoted by tenderers should not be the only aspect of the selection process. More elaborate methods are probably more efficient but will cost more. On the other hand, a cost–benefit analysis might also include a comparison to the cost of a new well. If the cost for rehabilitation is close to half the cost of constructing a new well and the existing well is old and producing at significantly less than 50% of its original yield, one might consider abandoning the rehabilitation option and opting for a new well.

The selection of a contractor is usually as much a matter of trust as a matter of cost. Many companies have fancy brochures, pushy salespeople, and promise heaven on earth. Caution is advised when companies claim that their method will work on all wells without even having had a look at the problems. There is no "one size fits all" in well rehabilitation. The careful contractor will discuss the prevalent problems of the well with the customer and adjust his techniques to the well's individual conditions.

Questions the well owner might ask the contractor (and himself) include:

- What is the training level of the personnel? Are formally trained drillers present, or only people trained on the job? Is the company certified?
- Are the personnel used to and trained to handling chemicals? (only if applicable)
- Does the company have experience with the present well type?
- What could possibly go wrong during the measure?
- Can I meet the water demand if the well is out of service for a longer period of time (due to collapse, damage, microbial pollution, etc.)
- What is needed on-site to perform this type of rehabilitation? (external water or electricity supply)

- Is there enough space around the well and in the shaft to accommodate and handle the rehabilitation equipment?
- How and where can I safely discard waste and (waste) water?

Contacts to other well owners and reference lists provided by the contractors should be used. A short phone call will tell you how well a particular method has worked elsewhere and how the rehabilitation company has performed.

10.1.4 Approval by authorities

Purely mechanical rehabilitations usually do not require a permit. However, a brief notification of the planned measures is not a bad idea. In some countries—or federal states, etc.—it may be necessary to get a permit from the (local) authorities for substantial changes to the well layout (reconstruction) or the injection of chemicals (chemical rehabilitation). For some parts of the world, information on this topic is available on the Internet and application forms can be downloaded. Some authorities may require the payment of fees before granting approval. Approval may be granted with certain restrictions. One possible constraint is that the whole process and especially the safe disposal of all solid and liquid wastes be monitored and documented.

The most serious concerns of authorities usually are related to possible pollution of the well water, especially when the well is to be used for drinking water supply. The well must not be reconnected to the distribution network unless it fulfills all legal requirements. Any chemicals used during rehabilitation must have been completely removed and safely disposed off. No traces of them are allowed in the water. Bacterial pollution is also not acceptable (→ 10.4.2). Authorities therefore often require one or more analyses by a certified laboratory prior to granting approval for reconnection. Natural and rehabilitation-induced changes to water chemistry which do not affect water hygiene may be tolerated (→ 10.4.1).

The application for a chemical rehabilitation should include at least the following information:

- Name and location of well(s)
- Motivation for the use of chemicals
- Type and amount of chemicals to be used
- Parameters to be monitored during treatment and pumping
- On-site wastewater treatment and chemicals to be used for it
- Disposal concept for wastewater and sludges (including estimates of volumes)

An application for reconstruction should include following information:

- Name and location of well(s)
- Motivation for the reconstruction
- Proposed reconstruction technique
- Sketch drawings: well before and after rehabilitation

10.2 Execution

10.2.1 General recommendations for rehabilitation sites

A tidy site is not an end in itself. Pumps and riser pipes removed from the well prior to rehabilitation should be stored safely and should not be dumped onto the ground. Pipe racks and plastic tarpaulins are recommended. Submersible cameras, geophysical probes, and rehabilitation equipment must be kept clean to prevent the introduction of potentially harmful microorganisms into the well. Figure 10-1 shows an example of a rehabilitation site.

The well should remain open only when really necessary; otherwise, leaves and soil particles might be blown into it by the wind. Some animals seem to be magically attracted to holes in the ground. The authors have personally assisted in the rescue of mice and frogs from well shafts. Both plant matter and animals can cause microbial pollution ($\rightarrow$ 10.4.2).

Wells are often located in remote areas and the crew has to sleep on-site in trailers (caravans) or cabins. Both accommodation and the (mobile) toilet facilities must be kept outside the immediate protection zone of the well(s). Solid and liquid waste must not be disposed at the site. Lubricants and fuels need to be stored on-site. Again, they must be kept outside the inner protection zone and are best stored in appropriate containers and inside lockable cabins.

Vandalism is always a problem ($\rightarrow$ 4.6). If the crew has to leave the site, the well shaft must be closed—at least provisionally—and equipment as well as chemicals have to be stored safely. Lockable containers may be used for this.

Spillage of motor and hydraulic oils is always possible. Oil-absorbing material must therefore be kept on-site.

10.2.2 Occupational safety

Rehabilitations and reconstructions must be considered as construction sites from the viewpoint of occupational safety. Protective clothing, including hard hats, protective boots, gloves, and safety goggles—the latter where necessary—are not decorative but imperative.

Figure 10.1 Example of a rehabilitation site setup with a multi-chamber system suspended at the tripod. *Photograph: Aquaplus, Kronach, Germany.*

The best prevention against accidents is to employ trained personnel. Fully trained drillers are a much better choice than plumbers or even unskilled laborers trained on-the-job. Regularly repeated first-aid courses are always a good investment. Up-to-date first-aid kits are mandatory, and a mobile phone with preprogrammed emergency phone numbers is not a bad idea.

Well-maintained machinery is a good way to prevent accidents. The machinery should be checked on a regular basis and maintained by skilled personnel.

Hazardous gases must be expected at any time before, during, and after rehabilitation measures. The application of chemicals for rehabilitation can significantly enhance or even cause the release of substantial gas volumes. Gases can be suffocating or explosive. They hardly announce their presence—hence the term "silent killer" describes them very well. Table 10-1 summarizes the properties of some of the most common gases that can be expected in and around wells.

Carbon dioxide is probably the most common cause of accidents in shafts. Possible sources for carbon dioxide are

- Respiration of workers in shafts ("self-poisoning")
- Oxygen consumption by petroleum lamps or welding equipment (incomplete combustion can also lead to formation of carbon monoxide)
- Natural outgassing from soil and groundwater
- Degradation of biomass through disinfection [Eq. (10.1)]
- Chemical well rehabilitation of carbonate incrustations with acids [Eq. (10.2) ($\rightarrow$ 8.2.3)]
- Well rehabilitation with carbon dioxide ($\rightarrow$ 7.4.1)

The treatment of carbonate incrustations with acids results in the release of large amounts of carbon dioxide gas [Eq. (10.2)]. This can lead to effervescence and expulsion of acid to the surface. About 22.7 liters (6 gal) of gas are formed per 100 g (0.22 lb) of calcium carbonate dissolved! For this reason, the acid should be added slowly and under constant surveillance.

$$\text{"}CH_2O\text{"} + 2\ H_2O_2 \rightarrow CO_2 + 3\ H_2O \tag{10.1}$$

$$CaCO_3 + 2\ HCl \rightarrow Ca^{2+}_{(aq)} + CO_{2(g)}\uparrow + H_2O + 2\ Cl^-_{(aq)} \tag{10.2}$$

where "CH_2O" represents simplified organic matter.

Carbon dioxide is also likely to collect in shafts, etc., because of its higher density compared to air. Volume concentrations of 4–6% CO_2

TABLE 10.1 Properties of Gases That May Occur in Wells

Gas type	Formula	Density [kg/m^3]	Density [lb/ft^3]	Density relative to air [at 0°C (32°F)and 1.013 bar (14.69 psi)]
Air	—	1.292	0.081	1.00
Carbon dioxide	CO_2	1.977	0.123	1.53
Hydrogen sulfide	H_2S	1.539	0.096	1.19
Methane	CH_4	0.717	0.045	0.55

cause headaches, shortness of breath, and drowsiness; concentrations of 10% cause unconsciousness and apnea; higher concentrations usually lead to very fast loss of consciousness and death. Concentrations above 30% can have serious effects after an exposure of less than 1 min.

Hydrogen sulfide gas (H_2S) is poisonous but can be detected by the human nose even at very low concentrations by its distinctive smell of rotten eggs. The nose can detect concentrations as low as 0.025–1 ppm. Unfortunately, high concentrations of 200 ppm and more overpower our noses, resulting in an inability to sense the odor. Gas indicators are therefore indispensable. Hydrogen sulfide concentrations of >1,000 ppm (0.1%) become perilous within few minutes! Concentrations of >5,000 ppm kill humans within seconds. Because of its higher density compared to air, H_2S can collect in well shafts. Deadly incidents with hydrogen sulfide gas have been reported. Sources of H_2S are

- Outgassing from reducing soils and groundwaters
- Chemical well rehabilitation of sulfide incrustations with acids [Eq. (10.3)]

$$FeS + 2\ HCl \rightarrow Fe^{2+}_{(aq)} + H_2S \uparrow + 2\ Cl^- \qquad (10.3)$$

Methane (CH_4) is a less common well gas. Its occurrence is usually related to an outgassing from reducing soil and groundwater, especially where coal and peat layers are abundant. "Burning wells" have been known to people for a long time. Leaky natural gas pipelines may also release CH_4. Because of its low density, methane can escape from well shafts but may collect under closed lids, etc. Methane is hardly toxic. Concentrations >5 vol% are flammable, and concentrations >9.4 vol% are highly explosive. This phenomenon is known to coal miners around the world. Fire and explosions, of course, consume oxygen and produce CO_2.

People trying to recover injured persons from shafts are at extreme risk. What often happens is that first a worker in the shaft collapses. His colleague then climbs into the shaft to carry his buddy away but is soon weakened by the gas and the fatigue caused by lifting his co-worker. Then the next person enters the shaft to help his colleagues. The sad maximum number of dead workers in such a drama was seven.

The imperative thing to avoid such problems is ventilation of the shaft. Personnel entering shafts should always be in visual and audible contact with somebody outside the shaft. They should leave immediately if they experience nausea or headache. It is a good idea for them to wear a recovery garment (rope) so that they can be pulled out quickly if they should collapse. Hand-held gas indicators should be available and used. People who have to climb into a shaft to recover collapsed persons must wear a breathing apparatus with independent

Figure 10.2 Some safety precautions for personnel working in (well) shafts: (left) hard hat, light source, breathing apparatus, gloves; (right) roping. (*Photographs by Bernhard Arenz.*)

oxygen supply; a gas mask is not sufficient. Figure 10-2 shows some basic equipment.

Safety precautions which are very efficient against hazardous gases are

- Ventilate the shaft (air pump, umbrella, etc.).
- Always carry gas indicators.
- Always wear a safety vest with ropes to allow pulling out.
- Never work alone (one in the shaft, the other outside); always be in visual and audible contact with someone outside the shaft.
- Avoid sparks, do not use petroleum lamps, avoid welding in the shaft.
- Do not smoke in shafts.
- Leave the shaft immediately if you feel unwell!

Chemical rehabilitation poses special challenges to all people involved. Often they must handle concentrated caustic chemicals. Again, training is the key to safe work on-site. Personnel should know what they are doing, which chemical they are handling, and what they must do in case of a spill or an accident ($\rightarrow$ 8.6.2). The most important information is given in the safety data sheet ($\rightarrow$ 8.6.2). Adsorbing material and water to flush chemicals from skin or eyes must be readily available at all times.

The necessary measures for the transport and safe storage of chemicals on-site are described in Sec. 8.2.6.

10.2.3 Progress control

The contractor should monitor the amounts of particles or incrustation removed from the well during a rehabilitation. Claims such as "We have removed tons of material" are simply not sufficient. Although the pumping tests after rehabilitation will finally tell the whole story, the removed amounts are still of interest, especially in depth-specific rehabilitations. They can be calculated from repeated measurements of turbidity, dissolved ionic concentrations, and particle content of the water pumped after the rehabilitation step ($\rightarrow$ 5.6).

Measuring the removed material also allows an assessment of the success of the applied technique. Of course, we do not want to waste time and money on only slightly affected sections, whereas we want all plugging material to be removed from more heavily affected sections. The tender should therefore not include vague statements such as "Every section must be treated for 1 h." Instead, a stop criterion should be defined for the application, e.g., "The section shall be treated until the concentration of pumped sand decreases to 0.2 mg/liter." This gives enough flexibility to stop operation on less problematic sections early (Fig. 10-3a) and to invest the gained time in more heavily incrusted sections where the desired results were not obtained after a fixed time span (Fig. 10-3b).

10.2.4 Waste disposal

The proper disposal of liquid and solid waste is one of the biggest concerns in well rehabilitations, especially when chemicals are used. Flooding of basements as well as adverse effects on aquatic biotopes and soils must be avoided. Refer to Sec. 8.7 for more detailed information.

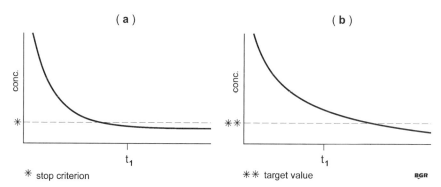

Figure 10.3 Advantages of the definition of a stop criterion (dashed line) over a fixed treatment time (t_1) for well rehabilitation: (a) treatment of section can be stopped earlier if the stop criterion is achieved; (b) applying the fixed-time concept would leave material in the section— treatment has to proceed until the stop criterion is reached.

Disposal concepts must be elaborated well in advance, and alternative emergency concepts need to be prepared. As mentioned before, permits from (local) authorities may be required for the disposal concept. The authorities may request monitoring and documentation of the chemicals and masses involved and may regulate maximum permissible concentrations for disposal, e.g., pH and turbidity after neutralization.

10.3 Sustainability of Well Rehabilitation Measures

The cost of rehabilitation can easily reach 10–20% of the price that was paid for the original well construction. Therefore the client expects—for good reason—that the yield gains obtained should last a while. Unfortunately, practical experience has shown that the improvement to well hydraulics is often short-lived and that the yield declines rather quickly. Sometimes the yield can decrease to the value before rehabilitation within weeks or a few months. To notice this phenomenon of course requires regular measurements of the well yield ($\rightarrow$ 5.2).

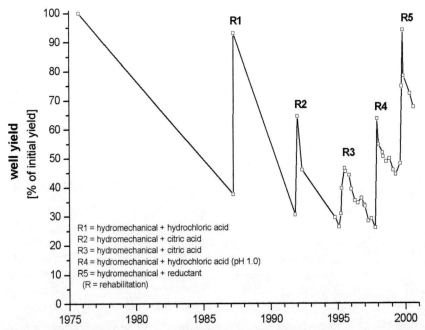

Figure 10.4 Development of the yield of a well near Krefeld (Germany) and the effects of well rehabilitations using different techniques. Note the fast decrease of well yield immediately after all rehabilitations.

Figure 10-4 shows an example of such a phenomenon. Similar observations were made by

- Gavrilko (1968, cited in Vukovic & Soro 1992), wells in Moscow (Russia)
- Moser (1978), wells in Mannheim (Germany)
- Driscoll (1989), wells in New Jersey (USA)
- Vukovic & Soro (1992), wells in Belgrade (Serbia)
- Timmer et al. (2003), wells in Bergambacht (The Netherlands)

The steep decline of the yield curves immediately after the rehabilitation of course needs to be considered in assessing the performance of rehabilitations. In his PhD thesis, Moser (1978) postulated that a rehabilitated well ages three to four times faster after a rehabilitation than before.

There are several potential causes for this quite unsatisfactory course of yield after rehabilitations. For wells in which the upper parts of the screen were blocked—and opened by the rehabilitation—higher amounts of near-surface oxic groundwater can now enter the well (Fig. 10-5). Evidence for this phenomenon is presented in Sec. 10.4.1. Oxic water can also pass through defective annular seals and leaky casings. The additional amounts of dissolved oxygen and potentially nitrate can then reaccelerate the oxidation of ferrous iron and thus ochre buildup. The kinetic rate laws presented in Sec. 3.2.2 predict that ochre buildup has a linear dependence on the dissolved oxygen concentration. Assuming we have excess ferrous iron, a threefold increase in oxygen ($\rightarrow$ 10.4.1) will lead to a threefold increase in ochre buildup.

In most rehabilitations, the ochre present in the gravel pack cannot be completely removed. If the surface area of the remaining ochre is larger than before the rehabilitation (Fig. 10.6), the increased surface acts as a catalyst for renewed oxidation of ferrous iron ($\rightarrow$ 3.2.3). The density of sorption sites and the catalytic efficiency, both of which are functions of the surface area, are then much higher than before. This will lead to increased ochre deposition rates.

An example will outline the background: we simplify the gravel pack as a cubic arrangement of uniform spheres of 2-mm size (Fig. 10.6). The plugged pore space (Fig. 10-6, left) leaves a small open channel of a diameter of 0.83 mm, which has a surface area of 2.60 mm^2 (at a length of 1 mm). After partial removal of the ochre through a rehabilitation, the gravel grains are assumed to be covered by a monomolecular ochre layer of negligible thickness. A pore channel of a length of 1 mm would then be equal to a surface area of $4 \times 0.25 \times 2$ mm $\times \pi = 6.28$ mm^2. This would equal an increase in active catalytic surface by a factor of ~2.4.

314   Chapter Ten

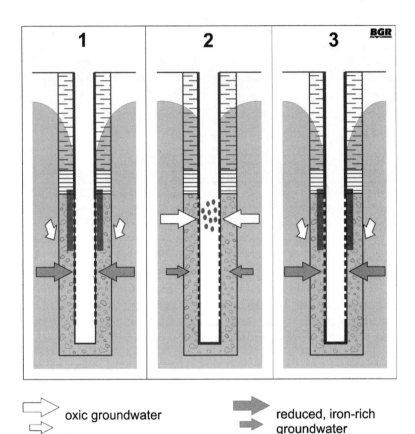

Figure 10.5 One probable cause for the low sustainability of well rehabilitations: (1) prior to rehabilitation: incrusted top of well screen, limited inflow of oxic water; (2) directly after rehabilitation: incrustations have been removed, increased inflow of oxic water through the upper screen section; (3) oxic waters caused precipitation of ochre in the upper screen sections. The size of the arrows indicates amounts of flow.

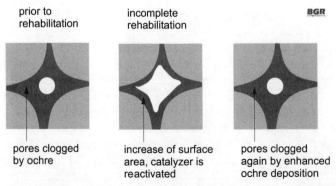

Figure 10.6 Auto-catalytic effect of iron oxide surfaces on ochre buildup in pore spaces.

The quickly precipitating ochre then refills the pore necks (Fig. 10-6, right). The increase in surface area can also be considered an increase of the available settling ground for iron bacteria.

As we can see from the considerations presented above, the success of a rehabilitation does not depend solely on the appropriate selection of the proper technique and its correct application. Even appropriate measures performed by experienced personnel can suffer from low sustainability because of the reactions described above. We should be aware that rehabilitations are only buying some time and that the life span of a well is limited.

If even repeated rehabilitations can produce only short-lived yield gains, the following well conditions should be checked:

- Screen length
- Vertical hydrochemical zonation
- Hydraulic performance of the annular seal
- Well hydraulics (entrance velocities, etc.)

Consequences arising from these studies can include:

- Decrease in pumping rate to avoid leakage from upper parts
- Shortening of the screen length to minimize mixing effects
- Refurbishment of casing and screen seals
- Decommissioning, redrilling, or relocation of the well

10.4 Rehabilitation-Induced Problems

10.4.1 Changes to water chemistry induced by rehabilitation

Rehabilitations can cause changes to water chemistry. The easiest explanation is of course the influence of remaining rehabilitation chemicals which were not completely removed after the rehabilitation. Indirect influences on the hardness of water are known, e.g., from rehabilitations with liquid carbon dioxide ($\rightarrow$ 7.4.1).

The removal of incrustations from certain screen sections through rehabilitations can change the relative amount of water entering the well from different hydrochemical zones. As the incrustation most often starts from the upper parts of the screen where oxic water enters, a removal of incrustations from this part will enhance the inflow of oxic water. Figure 10-7 shows the nitrate concentrations—unfortunately, dissolved oxygen was not measured—of well water in Germany before and after mechanical and chemical rehabilitation steps. The applied chemical did not contain nitrogen, which means that the nitrate must

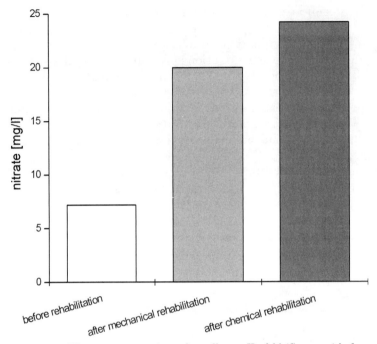

Figure 10.7 Nitrate concentrations of a well near Krefeld (Germany) before and after a combined mechanical and chemical rehabilitation. *Modified after Houben (2001).*

have originated from the groundwater. Camera inspections had shown that the upper screen sections were much more incrusted than deeper parts. The removal of these incrustations opened flow paths for more shallow, and more oxic, groundwater to enter.

10.4.2 Secondary microbial pollution

Occasionally, well water will show an elevated count of microorganisms after rehabilitations. Because water for public supply must be free of germs, this specific water cannot be distributed to customers. The well must be kept out of operation until its quality is set back to legal standards. Of course, this can be a major hassle for the well owner, and quite often a controversy develops with the contractor over whose fault it is.

Parameters used to assess the microbial state of a water sample are the numbers of bacterial colonies at 22°C (71.6°F) and at 37°C (98.6°F). They are sum parameters which do not account for types and numbers of individual bacteria, but rather the number of visible bacterial colonies grown on a substrate from 100 ml (0.0264 gal) of water incubated for about 48 h. Numbers below 100 colonies per 100 ml are usually tolerated, because groundwater is never purely sterile (→ 3.2). The presence of coliform bacteria (especially *Escherichia coli*) and of *Enteroccus,* on the other hand, cannot be tolerated. They are indicators of an influx of fecal contamination. Their colony numbers must be zero. It should be noted that the determination of microbial counts in water is a tricky business. The authors once encountered completely diverging results on samples from a previously rehabilitated well from two different laboratories: one claimed that the samples contained more than 100 colonies and the other lab found no bacteria at all. Microbiological samples are strongly influenced by the sampling procedure and the type of substrate used in the lab.

Microbial growth always has to be expected with the use of organic rehabilitation chemicals, especially when the chemical was not completely removed (Schoenen & Eisert 1987). Common rehabilitation chemicals such as citric acid and ascorbic acid are excellent substrates for many types of heterotrophic microorganisms. Remaining substrates will then lead to explosive growth of such bacterial populations which can persist for weeks and sometimes months. One particularly nasty example in Germany lasted 6 months (Bauer et al. 1993).

Figure 10-8 shows the typical course of a shorter microbial pollution due to a chemical rehabilitation involving the use of citrate. After a

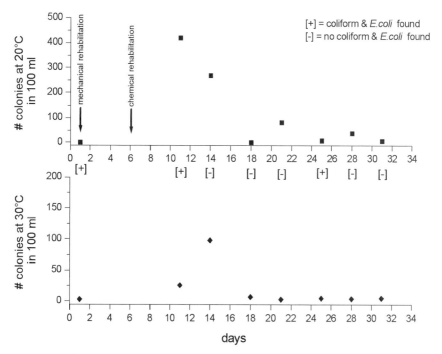

Figure 10.8 Development of the microbial pollution of a well in Germany after application of an organic chemical for rehabilitation.

short lag phase, which the bacteria needed to adapt and procreate, their number slowly decreased again following removal of their nutrient substrate by extended pumping.

Secondary microbial pollutions of well water sometimes occur after purely mechanical rehabilitations or chemical rehabilitations with inorganic chemicals. A possible cause could be the disruption of biofilms inside the well. Other microorganisms then utilize this biomass, e.g., polysaccharides, for their growth.

One gram of top soil usually contains between 1 million (10^6) and 1 billion (10^9) bacterial cells. If soil material gets into the well, there is a high probability that at least one of the bacteria will like its new surroundings and start to procreate. Again, dead or living animals or plant material (leaves, pollen) can always be carriers of microorganisms, too. Therefore the recommendations given in Sec. 10.2 should be closely followed to prevent the input of bacteria and bacterial substrate from outside the well as much as possible.

There are three principal pathways to overcome rehabilitation-induced microbial pollution:

- Disinfection
- Pumping to waste
- Sit and wait

From practical experience we know that disinfections are usually not adequate, because these only fight the symptoms (bacteria) and not the cause (nutrients). The killed-off bacteria are soon replaced by new colonies floating by from the aquifer, which then multiply quickly. Details of disinfection are discussed in Sec. 11.1.

Doing nothing and waiting until the bacteria have completely devoured their substrate—or until it has been washed away into the aquifer—is an option, but only for those with patience. Weeks or months might be necessary to complete the process.

The quickest but most expensive method is to abstract the nutrient-containing water and to discard it until the bacterial substrate has been completely removed. Of course, this consumes a lot of electricity for the pumps, and the water has to be disposed of properly.

Figure 10-9 shows the most important pathways for microbial pollution during and after well rehabilitations. If the pumped water is infiltrated too close to the well site it may be pumped back and recycled, especially when the soil and aquifer are very permeable.

Simple backfilling of the chemicals into the well without circulation will cause most of it to exit through nonincrusted parts of the screen into the aquifer where it will dissipate (Figs. 8-14, 10-9). After the

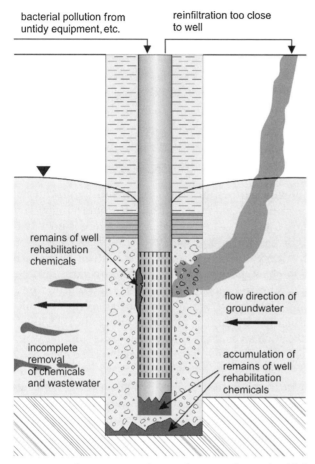

Figure 10.9 Important pathways for secondary microbial pollution induced by well rehabilitations.

pump has been turned on again, parts of these remains may be pumped back into the well and may suddenly cause water quality problems.

The well sump and gravel underneath the end cap of the sump are common causes of concern. Because all solutions used for chemical rehabilitations have a higher density than water, at least some will sink to the bottom of the well. It is therefore essential to clean the well sump after a rehabilitation. Complete removal of chemicals from the borehole underneath the sump is impossible. Such features must therefore be avoided during the construction of the well.

10.4.3 Mechanical damage and well collapse

Mechanical damage after rehabilitations can be attributed to a variety of processes:

- Recovery and insertion of pump and riser pipe
- Insertion and recovery of cameras, geophysical probes, and rehabilitation equipment
- Material, e.g., tools, falling into the well
- (Localized) high mechanical stress during mechanical well rehabilitations

The first three phenomena can usually be recognized by distinctive scratches, drag marks, and localized indentations. If camera inspection after the rehabilitation is omitted—for cost reasons—there is no way to find out if the contractor has worked properly.

The use of too much energy during mechanical rehabilitations mostly affects the weak parts such as coatings (Fig. 7-11), screen slots, and slot bridges. If high-pressure jetting nozzles get stuck, they will act on a small area of the screen. The high stresses involved can easily cut holes or destroy screen slot bridges through a blowpipe effect ($\rightarrow$ 7.3.5).

Aggressive chemicals, such as acids, can cause corrosion and may also damage coatings. Unfortunately, the effects of corrosion take some time to develop. It is therefore very difficult or even impossible to attribute some corrosion features to an acid rehabilitation performed a few years earlier.

The structural damage caused by ageing—amplified through rehabilitations—can cause the collapse of a well. In some cases this occurs during the rehabilitation, leading to burial of the equipment and rendering both well and equipment a complete loss. The authors of this book are aware of a few of unfriendly lawsuits which ensued after such incidents. In many cases the well owner complains that his well has been destroyed as a result of the incompetence of the contractor. It should not be forgotten that some wells are kept stable merely by rust, and even the slightest touch will cause their collapse. A proper assessment of the well's stability prior to rehabilitation is therefore mandatory. Camera inspections and geophysical methods are useful for this purpose ($\rightarrow$ 5).

10.5 After the Well Rehabilitation

10.5.1 When is the well due for its next rehabilitation?

The successful German soccer coach Sepp Herberger once said that "after a game is before the next game." As we know now, rehabilitations do not stop the ageing processes and are just a way of buying more time. Additionally, rehabilitations can actually increase the speed of well ageing

(→ 10.3). Therefore we have to start thinking of the next rehabilitation when we have just finished the present one. The wise well owner will continuously monitor the well's yield to get an idea of the speed of ageing. If we consider a decrease in yield of 20–25% of the original as a critical margin, we will know somewhat in advance when to release the next call for tenders just by looking at the slope of the yield curve (Fig. 10-4).

10.5.2 Insufficient yield gain after rehabilitation—what are the next steps?

Sometimes the gain in well yield obtained through rehabilitations is much lower than anticipated by the well owner and remains a far cry from the initial yield. In such cases, additional rehabilitations (steps) are often applied to overcome this unsatisfactory situation. One probable cause is the use of an inappropriate and too weak rehabilitation technique. If even several contractors with different equipment cannot get the well back to life, it is time to write off the well and start thinking about reconstruction or construction of a completely new well. Causes for such stubborn behavior of wells include:

- Hardened incrustations (→ 3.2.7)
- Incrustations/plugging out of reach of mechanical rehabilitation, e.g., at the borehole wall

Reconstruction at the original sites of such wells must include the complete removal of the annular filling and scraping of the former borehole wall. Otherwise the new well might involuntarily inherit the problems of the old one.

In a few cases, yield decreases were noticed after rehabilitation, probably due to resettling of the gravel pack. In this case it is almost impossible to improve the well yield to a satisfactory degree.

Chapter 11

Prevention

11.1 Microbicidal Techniques

11.1.1 Introduction

Specialized bacteria are closely associated with the formation of incrustations, especially those consisting of iron/manganese oxides and sulfides ($\rightarrow$ 3.1, 3.2, 3.5, 3.6). The bacteria involved are already present in small numbers in the natural groundwater environment. Because of limitations of nutrient resupply—due to the slow flow rate of groundwater—their numbers are quite small. Once some of the bacteria reach the well with the groundwater flow, however, the situation changes dramatically. The high flow rates and the resulting cumulative nutrient supply turns wells into ideal breeding ground. The number of microorganisms thus increases rapidly, with the known consequences. Keeping the growth of these populations down therefore seems to be a feasible technique to limit incrustation buildup. One has to be aware that even after a total disinfection, resettling from the aquifer by the same bacteria occurs within days or weeks as a result of the flow of pumped water.

Microorganisms are quite adaptive. Constant or repetitive stress may stimulate them to produce massive amounts of microbial slime (Mehmert 1995). This sticky slime cover is meant to protect the bacteria from chemical attack or mechanical dislodgement. Therefore disinfections may become less effective over time or require increasing amounts of chemicals. If the dead biomass is not removed after the disinfection of a well, it may be used by following bacterial generations as a substrate.

11.1.2 Chemical disinfection

Strong oxidizing agents are the weapons of choice to fight microbial biomass and slimes (disinfection $\rightarrow$ 8.2.6). Common chemicals are hydrogen peroxide (H_2O_2) and chlorine-containing agents such as hypochlorite, chlorite, chlorate, and/or perchlorate ($NaClO$, $NaClO_2$, $NaClO_3$, $NaClO_4$) salts. Free chlorine gas is by far the most effective oxidant, but it is rarely used because of the dangers involved in handling it. Oxidants kill microorganisms by simply transforming ("mineralizing") their biomass into carbon dioxide [Eq. (11.1)].

$$\text{"C(H}_2\text{O)"} + 2\ H_2O_2 \rightarrow CO_2 + 3\ H_2O \qquad (11.1)$$

Oxidants have a disadvantage in that they also cause oxidation—and subsequent flocculation—of dissolved reduced iron and manganese in the well interior [Eq. (11.2)].

$$2\ Fe^{2+} + H_2O_2 + 2\ H_2O \rightarrow 2\ FeOOH + 4\ H^+ \qquad (11.2)$$

Experiments in which chlorine-containing chemicals were injected continuously from several piezometers installed at distances of a few meters from a pumping well proved to be ineffective (Mogg 1972).

The water works of the City of Berlin, Germany, have for some decades applied a regular disinfection scheme using hydrogen peroxide (Schmolke, in Wicklein et al. 2002). A 1% H_2O_2 solution is distributed evenly over the whole screened interval and left to react for about 24 h. After this, water has to be abstracted—and discarded—for about 1 h to remove possible remnants of the hydrogen peroxide and flocculated iron/manganese turbidity. Because of the instantaneous resettlement of bacteria, intervals as short as 2 weeks between disinfections are needed to control microbial settlement. Similar experiences were found in the Netherlands using sodium hypochlorite (van Beek 1995). The procedure described does not completely cancel the need to rehabilitate the well, but it can certainly help to extend the time span of operation. The cost of such a scheme, however, can be substantial: personnel, chemicals, and equipment are needed, and the number of wells must be sufficient that one can always be out of service. A cost–benefit analysis has yet to be done.

In the presence of dissolved organic carbon, chlorine-containing oxidizing agents may cause the formation of light chlorohydrocarbons (LCHC), which are not desired from a hygienic point of view [Eq. (11.3)]. Instead, hydrogen peroxide is often preferred.

$$C_2H_6 + Cl_2 \rightarrow C_2H_5Cl + HCl \qquad (11.3)$$

Microbicidally active chlorine gas can also be produced in situ through electrolysis. Through the application of direct current (DC), parts of the chloride (Cl^-) dissolved in the water are transformed into free chlorine gas. The technique is well known from swimming pool and medical operations. Forward (1994) and McLaughlan (2002) describe an application in Southern Australia in which DC electrodes were installed in the well head. They were operated at 10 A DC and a flow rate of about 9 liters/min once a day after the pump had been turned off. After the free chlorine level reached 3–4 mg/l, the water was allowed to flow back into the well through a bypass in the reflux valve. The water treated by this process must contain sufficient amounts of chloride. The minimum electrical conductivity needed was 6.000 mS/m—which is quite high. Accordingly, the Australian example described interception wells pumping salt water (McLaughlan et al. 1990).

A similar technique—known from early on (Ehrmann 1958) but now readapted from dentistry—was applied by Riekel and Hinze (2002) on numerous wells in Botswana. In this case, water of normal conductivity was pumped from the wells and brought to 0.2% NaCl concentration. The water was then treated electrochemically, with the anode and cathode situated in separate cells. This resulted in the formation of short-lived but very reactive atomic and molecule radicals. The procedure yields very high oxidation potentials of up to +900 mV. Common constituents detected at the anode comprised ClO_2, ClO^*, $HClO$, $HClO^*$, O_3^*, ClO_2^*, $H_2O_2^*$, OH^*, HO_2, and HO_2^* (*designates a free radical). At the cathode, OH^*, H_3O_2, H_2O_2, H_2, and H^* were detected. The solution from the anode (anolythe) was then injected into the well, where it remained for about 2 h. The radicals contained in the anolythe destroyed bacterial biomass (slimes) and penetrated into the cell walls to kill off the bacteria. The anolythe volume injected usually equaled the volume of the borehole (2–5 m^3, 70–177 ft^3). After the reaction time, destroyed bacterial mass was removed from the well by pumping.

Another common technique used in swimming pools employs "drop pellet chlorinators." The technique is also used occasionally for water wells in the United States (Driscoll 1989). At preprogrammed intervals, the system releases pellets or tablets containing solid chlorine-containing chemicals, e.g., hypochlorite, from a storage container into the stagnant well. The tablet then slowly releases microbicidal free chlorine into the water.

Because chlorine is a very strong oxidizing agent, it can severely enhance the corrosion of metal. The use of calcium-based chlorine carriers, e.g., calcium hypochlorite, should be avoided in groundwater of elevated (carbonate) hardness. The reaction of calcium with dissolved carbonate can result in massive precipitation of calcium carbonate. Mansuy (1998) describes an example in which a pump was completely incrusted by this process.

11.1.3 Thermal disinfection ("Pasteurization")

People have known for a long time that unwanted microorganisms can be killed by heating or even boiling. Immersion heaters are probably not useful tools to use in wells, but the injection of hot water or even steam has been performed for a long time. One of the oldest patents on well rehabilitation in Germany is actually based on the injection of hot steam (→ 7.4.2).

Another method for "pasteurizing" the well interior was developed in Belgium (Agie de Selsaten 2000). The well tubing is inductively heated to temperatures of ca. 70°C (158°F) for about 30 min, a time period considered to be sufficient to kill off bacteria. Only wells with tubing made of electrically conductive metals can be treated this way. The distance between the probe and the tubing must be very small to ensure a sufficient transfer of energy.

11.1.4 Ionizing irradiation

Ionizing irradiation is sometimes used in the food industry and in water treatment for preservation and disinfection. Possible sources of hard ionizing radiation are radioactive substances. Gamma-ray-emitting radionuclides, e.g., cobalt-60 and cesium-137, are especially suited. Cobalt-60 has the decisive advantage of being a solid metal which can be handled quite easily. For application in wells it is coated with nickel metal and encapsulated in stainless steel probes. These are not installed in the well interior but rather in tubes inside the gravel pack of the well. Usually, two probes—spaced 180° apart—are used (Fig. 11-1). The chloride content of the groundwater should not exceed 200–300 mg/l because otherwise corrosion of the stainless steel and subsequent release of the radiation source might occur. The half-life of cobalt-60 is 5.27 years. The radiation sources must be replaced after about 10 years (two half-lives) to ensure a sufficient radiation dose. The tubes must therefore allow retrieval and reinsertion. At the same time, they must be tightly closed and secured to preclude theft or accidental removal. The energy released during the radioactive decay of cobalt-60 is too low to stimulate secondary radionuclide-generating reactions in the water, so indirect contamination of the water is not possible. The radiation dosages for the pumped water and the well at rest are about 4 or 1,200 Gy, respectively (Gy = gray unit, 1 Gy = 100 rad = 1 J/kg). For comparison: the food industry applies radiation dosages of up to 10,000 Gy.

The method described above was developed in the former East Germany (GDR) in the early 1970s and installed into 760 wells (Wissel & Gerstner 1973; Hübner 1994). Cobalt-60 was used as the radiation source. The scheme aimed at inhibition of microbial growth and at decreasing the volume of incrustations through dewatering (Wissel et al. 1985). This large-scale experiment was abandoned in 1990 after German reunification. Irradiation of food was illegal in unified Germany at that time. From a practical point of view it had been successful,

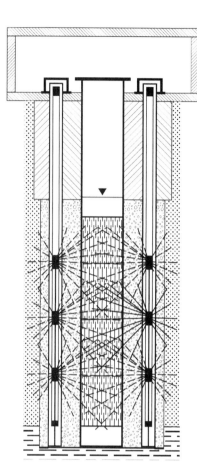

Figure 11.1 Ionizing irradiation of a well by two cobalt-60 probes with multiple sources (spaced 180° apart) installed in the gravel pack. *Redrawn from a brochure from Gamma-Service Recycling, Leipzig, Germany, with permission.*

because the buildup of ochre was much lower in most irradiated wells. Some wells showed no improvement; these must have been affected by purely chemically controlled incrustation. The recovery of the probes from the wells was complicated and costly, especially in some collapsed wells (Jakisch 1995).

Because of new EU legislation, the method is now legal again and available on the market. Nevertheless, the method still has some serious disadvantages that limit the widespread application of this otherwise elegant technique:

- Additional cost for installation, probes, and disposal
- Additional security precautions needed to prevent accidental removal or theft of radioactive material
- Extensive legal approval procedure

- Applicable only during construction of new wells
- Low social acceptance of radioactivity in conjunction with food

Only a few wells have been equipped since then, all being used for production of process water and not drinking water.

Microbicidal ionizing irradiation can also be created by ultraviolet (UV) light. The effect is of course restricted to spaces that can be penetrated by light, i.e., the well interior. Turbidity and suspended matter in the water are already a problem for above-ground UV applications and hence even more so for wells. Nevertheless, the method is available on the U.S. market.

11.2 Chemical and Physical Methods

11.2.1 Addition of reaction inhibitors

The precipitation of metals from solutions can be significantly decreased by complexing the dissolved metal ions. Molecules capable of complexing are called ligands. The have one or more electron pairs to be used in attachment to the metal ion and its fixation as a complex ($\rightarrow$ 8.2.2).

The precipitation of ochre is a serious issue in the dewatering of wetlands; see, e.g., Spencer et al. (1963). Drainage pipes are often clogged by ochre incrustations. To solve this problem, experiments were performed to force the reducing water to flow through bedding material mixed with ligand-rich organic substances before entering the drainage screen (Kuntze & Scheffer 1974; Wheatley et al. 1984). The resulting complexed iron was expected not to react with oxygen in the drainage. Plant material such as oak shavings, coniferous bark, bark of the mimosa tree, and acorn caps of Mediterranean oaks (trillo) proved to be quite effective. In addition, organic material has a reducing effect which consumes oxygen.

The installation of such material in wells, e.g., in the gravel pack, seems improbable. The potential for release of reductive substances is of course limited. This would require means to resupply the material, which is very difficult. Complexed iron requires massive additional conditioning steps in water treatment.

The precipitation of calcium carbonate scale is sensitive to the presence of impurity ions (inhibitors), such as iron and zinc, even at trace concentration levels. Experiments on heated metal surfaces by Macadam and Parsons (2003, 2004) showed that the most effective inhibitor is zinc, while copper and iron have little effect. In their experiments, a dosage of 5 mg/l zinc reduced scale formation by 35%. Such amounts effectively preclude the application of such a technique in water wells because of (drinking) water regulations.

11.2.2 Acidification

Decreasing the pH means increasing the solubility of iron. This concept has been used in the United States to battle ochre incrustations in wells used for dewatering of a lignite open-pit mine (Henke et al. 1991). Each well was equipped with a feeder tube which continuously injected a 10% hydrochloric acid solution (HCl) at rates of 5–20 ml/min, with some polyphosphate added as a corrosion inhibitor. Using the acid, the pH inside the well could be kept down to 3–4. The increased solubility of iron and decreased activity of iron bacteria in this range extended the service life of the well by up to 500%. This method is not recommended for drinking water wells because of possible legal constraints, the elevated corrosion potential, and the additional cost for water treatment (deacidification).

11.2.3 Electrochemical methods

The formation of iron and manganese ochre is based on redox processes. The redox pair includes the oxidation of ferrous to ferric iron and the reduction of molecular oxygen. [Eqs. (11.4), (11.5)].

$$Fe^{2+} \rightarrow Fe^{3+} + e^- \tag{11.4}$$

$$O_2 + 4\,H^+ + 4\,e^- \rightarrow 2\,H_2O \tag{11.5}$$

The redox potentials of the individual Eqs. (11.4) and (11.5) are +0.77 and +1.23 V, respectively. This indicates that oxygen is more noble than ferrous iron and can therefore oxidize the latter. To suppress this reaction, a less noble material could be introduced as an anode which would then consume the oxygen. Such schemes are quite common in corrosion protection ($\rightarrow$ 3.1). The less noble material ("sacrificial anode") is oxidized instead of the metal that is to be protected. Steel pipelines are commonly connected to base-metal wires made of, e.g., zinc or magnesium, which are consumed instead of iron. To protect ferrous iron, redox reactions with potentials of <0.77 V would be needed. Copper, zinc, and aluminium are possible candidates. From the experience of the authors, wells with copper tubing can still show massive ochre formation. The reason is that copper oxidizes only on its surface, and the resulting copper oxide patina protects it from further oxidation (passivation). In addition, one has to consider that ferrous iron and oxygen both occur as dissolved species in water, whereas the metal is in a solid state, which impedes its reactivity.

The deposition of ochre particles on any material is strongly dependent on the electrical charge of the surface. On conductive materials such as metals, this charge can be manipulated (polarized) by the application of an external electrical current. To prevent ochre deposition, it is therefore essential to maximize the repulsive force between surface and particle. This affects only the deposition, not the ochre formation itself.

Laboratory experiments with steel tubes showed that it is possible to repulse ochre particles by anodic as well as by cathodic polarization (Reissig et al. 1990; Fischer et al. 1990, 1993). A general disadvantage of anodic polarization is the induced material loss (corrosion) of the anode (→ 3.1). With current densities in the range of mA/m^2 this is a serious problem, which can, however, be reduced considerably by applying currents in the range of $\mu A/m^2$. The titanium electrodes used showed strong deposition of iron oxides. With cathodic polarization, incrustations were found neither on the steel surface nor on the electrodes. The voltage can be adjusted so that a nonconductive protective surface layer is produced, while avoiding the production of hydrogen gas—which could induce corrosion (→ 3.1). The protective layer prevents interaction between the metal and dissolved constituents.

Polarization is of course limited to electrically conductive materials such as metals. Attachment to filter gravel cannot be prevented by this method. Applications to wells have so far escaped the authors' attention.

An electrochemical method to prevent the precipitation of calcium and magnesium carbonates on steel pipes is based on an intervention into the carbonate system (Zeppenfeld 2003). To do so, a direct current (6 V, 30 mA) is established using the pipe as an anode and a stainless steel wire inside the pipe as a cathode. The reactions at the cathode lead to the precipitation of carbonate via the production of OH^- ions [Eqs. (11.6a), (11.6b)].

$$2\ H_2O + 2\ e^- \rightarrow H_2 + 2\ OH^- \qquad (11.6a)$$

$$Ca^{2+} + HCO_3^- + OH^- \rightarrow CaCO_3 + H_2O \qquad (11.6b)$$

Reactions at the anode create an acidic, CO_2-rich environment which promotes the dissolution of carbonates and the detachment of scale from the surface [Eqs. (11.7a), (11.7b)].

$$2\ H_2O \rightarrow O_2 + 4\ H^+ + 4\ e^- \qquad (11.7a)$$

$$HCO_3^- + H^+ \rightarrow CO_2 + H_2O \qquad (11.7b)$$

Treatments can be performed even when scale is already present, but application time must be restricted to a few minutes—otherwise, the anodic polarization of the steel tube and the buildup of CO_2 might corrode the steel. It is therefore recommended to repeat the treatment from time to time. The success of a treatment can be monitored using an amperometer, because the scale has high electrical resistance and its removal results in an increase in electrical flow. The authors of this book suspect that the efficiency of the cathode will suffer in the long term due to the electric insulation by the precipitating carbonates. It probably has to be exchanged or cleaned from time to time.

11.2.4 Magnetic methods

The use of magnets to prevent the precipitation of scale dates back to the end of the nineteeenth century. A few years ago the practice found its way into applications on pipelines and installations in households and industry. The main focus is the prevention of carbonate precipitation.

Permanent magnets and electrical induction are used to produce the magnetic fields. The benefit of magnets in scale prevention is a topic of much debate. While the U,S. Department of Energy (1998) strongly favors this technique, other groups have doubts. A scientific study with strong stationary magnetic fields from Germany by Sebold et al. (1996) states: "The results show no statistically significant influence of the magnetic field." A detailed literature survey listing advantages and disadvantages can be found in Powell (1998). The certainly necessary scientific debate on this topic is strongly disturbed by some dubious salespeople who attribute some ill-defined esoteric effects to the magnetized water.

More serious advocates ascribe the effect to an interaction of the magnetic field with the water flowing past it. Indisputably, mobile carriers of charge—such as ions dissolved in water—are affected by a force (the Lorentz force) when under the influence of a magnetic field. This force is said to weaken the hydrate shell of hydrogen carbonate ions. This supposedly allows the ions to move closer to each other and form many very small crystal seeds. Instead of a few large seeds which may accrete to form carbonate scale, these little seeds can easily be transported away by the flowing water. This means that precipitation as such is not prevented, but rather shifted to finer-grained crystals.

Other explanations consider changes in the surface charge of the carbonate particles or even influences of iron ions mobilized from the tubing material through corrosion to be the cause of the effect. Applications of magnetic fields for scale prevention in wells have not been published according to our knowledge.

11.3 Prevention through Planning, Construction, and Operation

11.3.1 Drilling techniques, materials, and operation

Successful prevention begins before constructing a well. Detailed geological and hydrochemical site investigations allow one to avoid areas with elevated incrustation potential, e.g., high-iron or corrosive groundwaters and poorly sorted aquifers.

The influence of different drilling techniques on the susceptibility to clogging has not been studied—or at least published—in much detail. Timmer et al. (2003) compared the temporal development of the yield of wells in the Netherlands drilled using cable tool and rotary (reverse circulation) drilling techniques (Fig. 11-2). They found that both well types are prone to ageing and resulting yield losses. The significantly better performance of wells drilled by cable tool can probably be explained by a smaller thickness of the filter cake that develops at the borehole wall during the drilling operation. During cable tool drilling, the borehole wall is separated from the mobilized fines of the penetrated formation by the protective casing. During rotary drilling, a suspension of formation fines and additives, especially bentonite and CMC, are infiltrated into the open borehole wall and form a low-permeability filter cake. While this layer is useful during drilling in limiting drilling fluid losses, its incomplete removal during well development will constitute a good starting point for more clogging. Scraping (back-reaming) the borehole wall of such drillholes prior to screen and gravel pack insertion is therefore recommended when standard development techniques may not be sufficient.

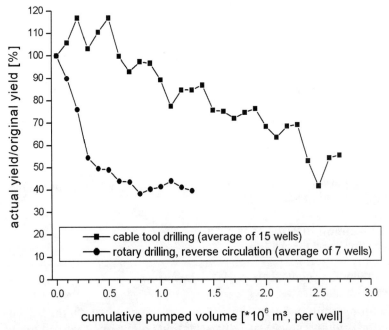

Figure 11.2 Temporal development of the yield of wells in the Netherlands drilled using cable tool and rotary (reverse circulation) drilling techniques. *Modified after Timmer et al. (2003).*

Careful dimensioning of the well components, especially of the gravel pack and the screen slots, is another way to extend the life span of a well. The gravel pack must be fine enough to retain sand from the aquifer but also coarse enough to allow desanding and to minimize head losses. The general aim is to minimize turbulent flow, which can enhance incrustation buildup through degassing and mixing ($\rightarrow$ 2.2). A calculation of relevant parameters such as the Reynolds-number should be included in the planning process (Williams 1981). Every measure to minimize the flow velocity and turbulence should be applied (Driscoll 1989). These include:

- Adjustment of gravel pack to geological formation
- Adjustment of screen slot size to gravel pack
- Consideration of vertically inhomogeneous geological stratifications
- Use of a multiple gravel pack for difficult formations
- Adjustment of drillhole and well radius ($\rightarrow$ 2.2)

The material used for the well screen has negligible influence on the formation and deposition of incrustations. Materials with a smooth surface, such as steel, stainless steel, PVC, or copper, attract the same amount of incrustation as materials with rougher surfaces. The smoothness is simply a matter of scale. What seems smooth to the human eye is a mountainous landscape with perfect hideouts for bacteria and very fine-grained incrustation particles in the microscopic view.

From many years of experience it is known that wire-wound screens (Johnson screens) are easier to rehabilitate because of their high entrance area and the geometry of the rods. On the other hand, resin-bound gravel packs have proven to be virtually impossible to rehabilitate. Their use should therefore be restricted to short-term applications such as dewatering.

As we saw in Sec. 3.2, the formation of incrustations is commonly a result of the mixing of waters from differing hydrochemical zones over a screen interval. This has incited Dutch researchers to propose an extraction scheme in which the oxic and the reducing zones are pumped separately by two partially penetrating wells, one shallow and one deep [van Beek & Brandes (1977), cited in Appelo & Postma (1996); Fig. 11-3]. Detailed investigations of the hydrochemical zonations are a necessary prerequisite. Changes in the vertical position of the zones due to pumping-induced drawdown have to be considered. The scheme did not prove to be as efficient as expected and was abandoned. Experience with horizontal and radial collector wells points in a similar direction. Although they extract water predominantly from one layer, they still experience incrustation formation, albeit slower.

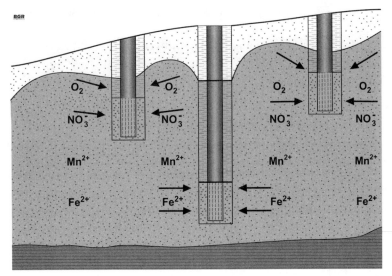

Figure 11.3 Separate extraction of oxic and reducing groundwater by multiple partially penetrating wells. *Redrawn after van Beek & Brandes (1977), cited in Appelo & Postma (1996).*

Well operation practices have some potential to minimize ageing processes. Excessive pumping should be avoided at all costs, because this will only induce turbulence, mixing, oxygen input, and CO_2 degassing (→ 3.2, 3.3). One crucial aspect to be avoided in well operation is to let the drawdown reach the screened interval. This is usually a kiss of death for the well, as it leads to extremely accelerated incrustation buildup.

If mechanical clogging by particles is the main cause of yield decline, changes in flow rate and flow direction, e.g., by repeatedly switching the pump on and off, can facilitate the breakup of particle bridges and the entrainment of plugging material (Gruesbeck & Collins 1982; Bouwer 2002). Practical experience from the Netherlands also points in this direction (Fig. 4-17).

11.3.2 Suction flow control devices (SFCDs)

As we have seen in Sec. 2.2.3, the inflow of water into a well screen is quite heterogeneous. The pump intake represents the point of lowest pressure in the system. For this reason, a relatively large amount of water enters the well through the upper parts of the screen. This is responsible for locally increased flow velocities, including all the unwanted consequences such as sand intake and incrustation formation. To avoid this uneven distribution, suction flow control devices (SFCDs) were developed in Germany in the late 1970s (Spranger 1978; Ehrhardt

& Pelzer 1992). The original idea goes back to the late 1950s (Truelsen 1958). The first generation of SFCDs consisted simply of a screen tube attached to the lower end of the pump which was inserted into the well ("well inside a well"). In other arrangements, the SFCD is not connected to the pump but is suspended independently.

The success obtained using the first generation of SFCDs was rather limited. Therefore a second generation with some modifications was developed (Albrecht & Ehrhardt 1987; Pelzer & Smith 1990). The perforation was now distributed in such a way that more openings were placed in the deeper parts of the SFCD than in the upper parts (Fig. 11-4). This was expected to force more water to enter the well at the bottom and decrease inflow in the upper screen section. Vertical flow was now restricted to the inside of the SFCD, whereas the inflow through the filter pack was nearly horizontal (Fig. 11-4). The decreased peak flow velocities induced by the SFCD imposed less drag force of the water and thus less suffossion ($\rightarrow$ 4.1). Encouraging results were obtained, especially in the control of sand intake. Some SFCDs were fitted with an additional resin-bound gravel pack.

The effect of an SFCD on the vertical distribution of well inflow was assessed by comparing the flow meter curves of a well in northern Germany with and without installed SFCDs (Fig. 11-5). Quite clearly, the peak inflow

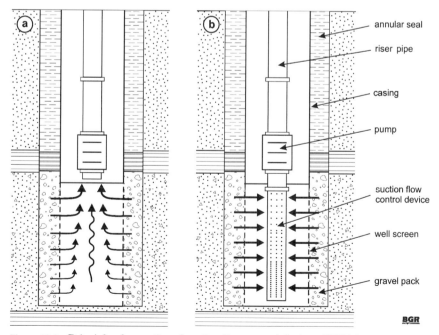

Figure 11.4 Principle of operation of suction flow-control devices (SFCDs).

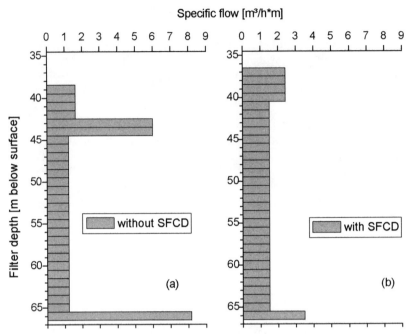

Figure 11.5 Flow-meter curves of a water well in northern Germany without and with a suction flow-control device. *From Houben (2006).*

at the top and the bottom of the screen can be decreased significantly—by a factor of approximately 2—by the installation of an SFCD.

SFCDs are an additional impediment in the well that the pumped water has to pass. This implies additional energy losses and cost. Therefore short discharge tests were performed on the same well as in Fig. 11-5. The head loss as a measure of energy loss was observed again with and without an SFCD (Fig. 11-6a). Surprisingly, the calculated well yield of a freshly rehabilitated well was affected only marginally by the SFCD. The loss caused by the SFCD was compensated by the gain in the reduction of turbulent flow. Some manufacturers claim that SFCDs can even decrease head losses. The observations described above were made for a freshly rehabilitated clean well. Figure 11-6b tells a different story: in this case the discharge test was performed after several months of operation. In this time a certain amount of ochre had precipitated in the well—and obviously onto the SFCD as well. Now the clogged SFCD presented a much more pronounced impediment to water flow, which caused higher entrance losses. The SFCD actually caused more flow impediment than the incrusted well screen. SFCDs obviously cannot directly prevent the genesis of incrustations, but they may be useful to extend the life span of a well by creating a more even distribution of the incrustations over the screen.

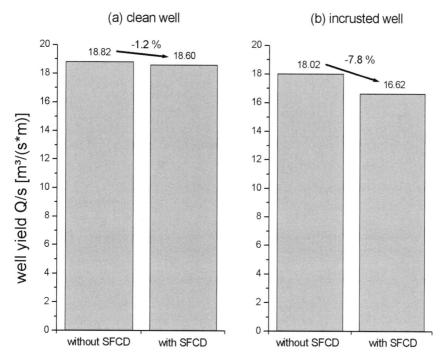

Figure 11.6 Yield of a well in northern Germany without and with a suction flow control device: (a) rehabilitated well; (b) incrusted well, after some months of operation. *From Houben (2006).*

11.3.3 Inert gases

Oxygen is one of the major components in ochre formation (→ 3.2). The ultimate source of oxygen is air, some of which is always contained inside the well and in the soil pores above the water table. Diffusion and turbulence constantly replenish dissolved oxygen in shallow groundwater that is being consumed in redox processes such as iron and manganese oxidation. Eliminating the oxygen supply by exchanging air for some inert gas would diminish the ochre potential. In practice, it is only possible to exchange the air inside the well, not in the soil.

The construction setup is quite simple (Fig. 11-7). The well head has to be rendered gas-proof, and inert gas has to be pumped into the well interior. An excess pressure of 0.3 bar (4.35 psi) is usually sufficient to prevent incursion of air. Gas bottles are a convenient tool to provide and store the gas. The well casing above the pumping water level must be completely gas-proof (no leaky seals or corrosion cavities).

The gas pressure inside the well may fluctuate due to natural and pumping-induced changes in the well-water level. Therefore a device to measure and monitor the pressure (manometer) has to be installed. It can also be used to control the amount of gas necessary to compensate

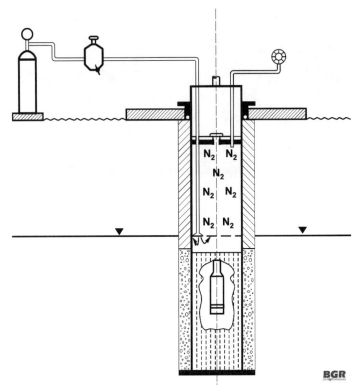

Figure 11.7 Inert gas (nitrogen, N_2) application scheme for water wells. *Modified after Mansuy (1998).*

losses. In order to prevent the gas from escaping through the screen slots, the drawdown must not reach the screen.

Nitrogen (N_2) is the inert gas of choice because it is readily available and does not alter the water chemistry the same way carbon dioxide (CO_2) does ($\rightarrow$ 7.4.1). Nitrogen is also much cheaper than noble gases, e.g., argon, which—at least in theory—could also be used. During maintenance, additional air must be pumped into the well head and shaft to prevent suffocation.

The method is used in a handful of wells in the United States (Mansuy 1998) and in Germany. Reports on practical experience and cost–benefit analyses are not available.

11.3.4 Subsurface iron removal

To decrease the cost of water treatment for reducing groundwaters, a method of subsurface iron removal has been developed in Germany (Rott et al. 1996; Appelo et al. 1999). In a first step, oxygen-enriched water is

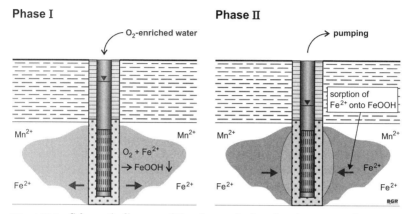

Figure 11.8 Schematic diagram of the phases of subsurface iron removal.

injected into the aquifer via the well (Fig. 11-8). Dissolved iron and manganese present in the groundwater precipitate by reaction with injected oxygen. The iron sludge which is otherwise formed during above-ground treatment thus remains underground and does not require disposal. After some time, the pump is set into operation and the water is pumped back to the surface. The fresh precipitates are able to sorb a certain amount of dissolved iron and manganese from the inflowing groundwater. The sorbed iron is oxidized in the next infiltration stage. The pumped water is therefore practically free of iron and manganese for a while. Shortly before the iron breaks through—concentrations exceed amounts that can be sorbed—the process of oxygen injection is repeated and the cycle starts again.

The ratio between pumped and injected water should be higher than 3:1 to render the scheme cost-effective. The method is currently in use in several water works in Europe and has proved to remain efficient for several years. Caution is advised for aquifers containing pyrite. This mineral consumes a lot of oxygen and may lead to unwanted alterations of water chemistry (low pH, high sulfates, mobilization of heavy metals and arsenic).

Because a large percentage of the dissolved iron and manganese is retained in the aquifer in the form of solid oxides, the potential for ochre formation in the well with such a scheme is substantially decreased and well yield decline slows down (van Beek 1989). Still, in the authors' experience, even such wells are known to require rehabilitations from time to time.

11.3.5 Gravel pack flushing devices

As we have seen in Sec. 7.5.1, large percentages of the energy put into mechanical rehabilitation are attenuated quite close to the source and do

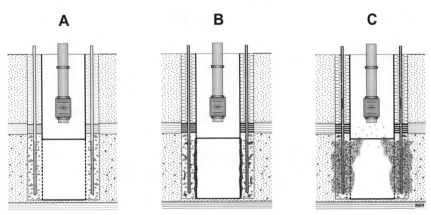

Figure 11.9 Schematic sketch of a gravel pack flushing device: (a) installed; (b) screen and gravel pack encrusted; (c) in operation, here: air/water flushing.

not reach the gravel pack. This leads to insufficient cleaning action outside the screen, where most of the incrustations and clogging particles are. Some recent patents therefore describe the installation of a few small-diameter perforated tubes into the annulus of the well (Fig. 11-9). This is of course best done during construction, prior to the insertion of the annular filling (gravel pack, annular seal) of the well. Later installation by drilling may potentially endanger the integrity of the annular seal.

The perforated tubes can be used from time to time to flush the gravel pack and remove incrustations and plugging particles. The applied pressure has to be kept low enough to avoid disruption of the gravel pack texture and dislocation of the annular seal. Flushing media can be water, air, water/air mixtures, or even carbon dioxide gas. Air is not recommended in reduced waters because of possible ochre formation. To the authors' knowledge, reports on the performance of such schemes have not been published so far.

Chapter 12

The 10 Dos and Don'ts of Well Rehabilitation

1. Take time for proper planning and dimensioning of new wells and the site investigation; this will lead to a higher yield, slower ageing effects, and easier rehabilitation.
2. Avoid false savings during construction and development of the well; higher investments now usually mean better performance.
3. After construction, fully assess your well's performance by step-drawdown tests, geophysics, flow meter, camera inspection; this provides the necessary reference data for later comparison.
4. Track your well's performance on a regular basis (step-discharge tests, camera inspections, etc.); plot these data and compare them to former data and the reference values from the time immediately after construction.
5. Know your enemy: what is the cause of any observed ageing? Incrustations (what type?), corrosion, particles?
6. If the yield has decreased by 20% of the original value, rehabilitate.
7. Do not wait too long before rehabilitating your well ("kill it before it grows"); longer waiting time means more time for incrustations to harden and reach a practically insoluble state.
8. Do not believe everything vendors recommend, select a method that is appropriate for *your* specific problem, ask around for experience with certain methods, perform dissolution tests with different chemicals on *your* type of incrustation before selecting a chemical.

9. Be aware that yield gains after rehabilitations may be short-lived, and the next rehabilitation may be needed soon.
10. Don't flog a dead horse: if a well has lost more than 50% of its yield, hardly anything will make it come back to life. Think about reconstruction!

References

Abeliovich, A. (1985). Avoiding ochre deposits in soil drainage pipes. *Agric. Water Mgmt.* 10(4). 327–334.
ADITCL (Australian Drilling Industry Training Committee Limited) (1996). *Drilling—The Manual of Methods, Applications, and Management.* Boca Raton, FL: Lewis.
Agie de Selsaten, J. (2000). Chemical free method to rehabilitate and prevent water wells from iron related bacteria contaminations and associated damaging effects—The AQUA PEM method. *Mitt. Ing. Hydrogeol.* 76: I 15–16.
Akaku, K. (1990). Geochemical study on mineral precipitation from geothermal waters at the Fushime field, Kyushu, Japan. *Geothermics* 19(5): 455–467.
Albrecht, B. & Ehrhardt, G. (1987). Saugstromsteuerungen der zweiten Generation beim Bau und Betrieb von Brunnen. *Schriftenreihe* 3. Darmstadt, Germany: Inst. Wasserversorgung TH Darmstadt.
Alford, G. & Cullimore, R. (1998). *The Application of Heat and Chemicals in the Control of Biofouling Events in Wells.* Boca Raton, FL: Lewis.
Appelo, C. A. J., Drijver, B., Hekkenberg, R., & de Jonge, M. (1999). Modeling in situ iron removal from ground water. *Ground Water* 37(6): 811–187.
Appelo, C. A. J. & Postma, D. (1996). *Geochemistry, Groundwater and Pollution.* Rotterdam: Balkema.
Appelo, C. A. J. & Postma, D. (2005). *Geochemistry, Groundwater and Pollution,* 2d ed. Leiden: Balkema.
Applin, K. R. & Zhao, N. (1989). The kinetics of Fe(II) oxidation and well screen encrustation. *Ground Water* 27(2): 168–174.
ASTM A589-96 (2001). *Standard Specification for Seamless and Welded Carbon Steel Water-Well Pipe.* West Conshohocken, PA: ASTM.
ASTM D5092-04e1. *Standard Practice for Design and Installation of Ground Water Monitoring Wells.* West Conshohocken, PA: ASTM.
ASTM D5299-99 (2005). *Standard Guide for Decommissioning of Ground Water Wells, Vadose Zone Monitoring Devices, Boreholes, and Other Devices for Environmental Activities.* West Conshohocken, PA: ASTM.
ASTM D5472-93 (2005). *Standard Test Method for Determining Specific Capacity and Estimating Transmissivity at the Control Well.* West Conshohocken, PA: ASTM.
ASTM D5521-05 (2005). *Development of Ground-Water Monitoring Wells in Granular Aquifers.* West Conshohocken, PA: ASTM.
ASTM D5737-95 (2000). *Standard Guide for Methods for Measuring Well Discharge.* West Conshohocken, PA: ASTM.
ASTM D5781-95 (2000). *Standard Guide for Use of Dual-Wall Reverse-Circulation Drilling for Geoenvironmental Exploration and the Installation of Subsurface Water-Quality Monitoring Devices.* West Conshohocken, PA: ASTM.
ASTM D5782-95 (2000). *Standard Guide for Use of Direct Air-Rotary Drilling for Geoenvironmental Exploration and the Installation of Subsurface Water-Quality Monitoring Devices.* West Conshohocken, PA: ASTM.
ASTM D5783-95 (2000). *Standard Guide for Use of Direct Rotary Drilling with Water-Based Drilling Fluid for Geoenvironmental Exploration and the Installation of Subsurface Water-Quality Monitoring Devices.* West Conshohocken, PA: ASTM.
ASTM D5784-95 (2000). *Standard Guide for Use of Hollow-Stem Augers for Geoenvironmental Exploration and the Installation of Subsurface Water-Quality Monitoring Devices.* West Conshohocken, PA: ASTM.
ASTM D5787-95 (2000). *Standard Practice for Monitoring Well Protection.* West Conshohocken, PA: ASTM.

ASTM D5872-95 (2000). *Standard Guide for Use of Casing Advancement Drilling Methods for Geoenvironmental Exploration and Installation of Subsurface Water-Quality Monitoring Devices.* West Conshohocken, PA: ASTM.

ASTM D5875-95 (2000). *Standard Guide for Use of Cable-Tool Drilling and Sampling Methods for Geoenvironmental Exploration and Installation of Subsurface Water-Quality Monitoring Devices.* West Conshohocken, PA: ASTM.

ASTM D5876-95 (2005). *Standard Guide for Use of Direct Rotary Wireline Casing Advancement Drilling Methods for Geoenvironmental Exploration and Installation of Subsurface Water-Quality Monitoring Devices.* West Conshohocken, PA: ASTM.

ASTM D5978-96 (2005). *Standard Guide for Maintenance and Rehabilitation of Ground-Water Monitoring Wells.* West Conshohocken, PA: ASTM.

ASTM D6031-96 (2004). *Standard Test Method for Logging In Situ Moisture Content and Density of Soil and Rock by the Nuclear Method in Horizontal, Slanted, and Vertical Access Tubes.* West Conshohocken, PA: ASTM.

ASTM D6034-96 (2004). *Standard Test Method (Analytical Procedure) for Determining the Efficiency of a Production Well in a Confined Aquifer from a Constant Rate Pumping Test.* West Conshohocken, PA: ASTM.

ASTM D6169-98 (2005). *Standard Guide for Selection of Soil and Rock Sampling Devices Used with Drill Rigs for Environmental Investigations.* West Conshohocken, PA: ASTM.

ASTM D6285-99 (2005). *Standard Guide for Locating Abandoned Wells.* West Conshohocken, PA: ASTM.

ASTM D6286-98 (2006). *Standard Guide for Selection of Drilling Methods for Environmental Site Characterization.* West Conshohocken, PA: ASTM.

ASTM D6724-04. (2006) *Standard Guide for Installation of Direct Push Ground Water Monitoring Wells.* West Conshohocken, PA: ASTM.

ASTM D6725-01. *Standard Practice for Direct Push Installation of Prepacked Screen Monitoring Wells in Unconsolidated Aquifers.* West Conshohocken, PA: ASTM.

ASTM F480-02. *Standard Specification for Thermoplastic Well Casing Pipe and Couplings Made in Standard Dimension Ratios (SDR), SCH 40 and SCH 80.* West Conshohocken, PA: ASTM.

ASTM G15-05. *Standard Terminology Relating to Corrosion and Corrosion Testing.* West Conshohocken, PA: ASTM.

ASTM G40-05. *Standard Terminology Relating to Wear and Erosion.* West Conshohocken, PA: ASTM.

Baboian, R. (2004). *Corrosion Tests and Standards: Application and Interpretation,* 2d ed. West Conshohocken, PA: ASTM.

Balke, K.-D., Beims, U., Heers, F. W., Hölting, B., Homrighausen, R., & Matthess, G. (2000). Grundwassererschließung. *Lehrbuch Hydrogeologie Vol. 4.* Stuttgart: Borntraeger.

Banfield, J. F. & Zhang, H. (2001). Nanoparticles in the environment. In: Banfield, J. F., & Navrotsky, A. (eds.), Nanoparticles and the Environment. *Rev. Mineral. Geochem.* 44: 1–58.

Banks, D., Cosgrove, T., Harker, D., Howsam, P., & Thatcher, J. P. (1993). Acidisation—borehole development and rehabilitation. *Quart. J. Eng. Geol.* 26 (2): 109–125.

Banks, D. & Soldal, O. (2002). Towards a policy for sustainable use of groundwater by non-governmental organizations in Afghanistan. *Hydrogeol. J.* 10: 377–392.

Banwart, S., Davies, S., & Stumm, W. (1989). The role of oxalate in accelerating the reductive dissolution of hematite (α-Fe_2O_3) by ascorbate. *Colloids Surfaces* 39: 303–309.

Barbic, F. & Savic, I. (1990). Influence of chemical regeneration on iron bacterial population and ochre deposits in tubewells. In: Howsam, P.(ed.), *Water Wells—Monitoring, Maintenance, Rehabilitation. Proc. Int. Groundwater Eng. Conf., Cranfield,* pp. 35–46, London: Spon/Chapman.

Barnes, I. & Clarke, F. E. (1969). Chemical properties of groundwater and their corrosion and encrustation effects on wells. USGS Professional Paper 498-D. Washington, DC: USGS.

Batu, V. (1998). *Aquifer Hydraulics.* New York: Wiley.

Baumann, K. & Tholen, M. (2002). Mängel an Brunnen und Grundwassermessstellen. *bbr* 53 (1): 24–34.

Baudisch, R. (1989). Verstopfung von Brunnenfiltern und Unterwasserpumpen durch Aluminiumoxide. *bbr* 40 (5): 270–274.

Bauer, R., Gatz, K. W., & Obst, U. (1993). Mikrobielle Kontaminationen durch organische Bohrspül- und Brunnenregenerationszusätze. *bbr* 44 (12): 584–589.
Baumann, K., Burde, B., & Goldbeck, J. (2003). Die "gläserne" Messstelle—Fortschritte der Bohrlochgeophysik. *bbr* 54(7): 24–32.
Baveye, P., Vandevivere, P., Hoyle, B. L., DeLeo, P. C., & Sanchez de Lozada, D. (1998). Environmental impact and mechanisms of the biological clogging of saturated soils and aquifer materials. *Crit. Rev. Environ. Sci. Technol.* 28(2): 123–191.
Bayerisches Landesamt für Wasserwirtschaft (1995). Technische und wasserrechtliche Behandlung von Brunnenregenerierungen.—Merkblatt No. 1.4/3; August 11, 1995, Munich.
Beek, C. G. E. M. van (1989). Rehabilitation of clogged discharge wells in the Netherlands. *Quart. J. Eng. Geol.* 22(1): 75–80.
Beek, C. G. E. M. van (1995). Brunnenalterung und Brunnenregenerierung in den Niederlanden. *gwf Wasser/Abwasser* 136(3): 128–137.
Beek, C. G. E. M. van, & Kooij, D. van der (1982). Sulfate-reducing bacteria in ground water from clogging and nonclogging wells in the Netherlands river region. *Ground Water* 20(3): 298–302.
Beek, C. G. E. M. van, & Kooper, W. F. (1980). The clogging of shallow discharge wells in the Netherlands river region. *Ground Water* 18(6): 578–586.
Beek, C. G. E. M. van, Vasak, L., Nieuwaal, A., Stefess, G. C., & Bakker, L. M. M. (1998). Ontwerp en onderhoud van infiltratie—en onttrekkingsmiddelen. NOBIS 96-3-06, CUR/NOBIS (Gouda).
Belzile, N. & Lebel, J. (1986). Capture of arsenic by pyrite in near-shore marine sediments. *Chem. Geol.* 54: 279–281.
Berger, H. (1997). Maßnahmen zur Brunnenregenerierung in Deutschland. In: DVGW (Hrsg.), *Stand der Brunnenbautechnik in Europa.* DVGW Schriftenreihe 89: 379–428.
Berger, H., Drews, M., & Lux, K. N. (1995). Nachweis von Regeneriereffekten bei Brunnen der Stadtwerke Wiesbaden AG. *bbr* 46 (1): 24–31.
Berner, R. A. (1970). Sedimentary pyrite formation. *Am. J. Sci.* 268: 1–23.
Berner, R. A. (1984). Sedimentary pyrite formation: an update. *Geochim. Cosmochim. Acta* 48: 605–615.
Biber, M. V., Dos Santos Alfonso, M., & Stumm, W. (1994). The coordination chemistry of weathering: IV. Inhibition of the dissolution of oxide minerals. *Geochim. Cosmochim. Acta* 58: 1999–2010.
Bichara, A. F. (1986). Clogging of recharge wells by suspended solids. *J. Irrigation Drainage Eng.* 112(3): 210–224.
Bieske, E., Rubbert, W., & Treskatis, C. (1998). *Bohrbrunnen,* 8th ed. Munich: Oldenbourg.
Blackwell, I. M., Howsam, P., & Walker, M. J. (1995). Borehole performance in alluvial aquifers: particulate damage. *Quart. J. Eng. Geol.* 28: S151–S162.
Block, M. (1999). Well closure in Arizona—opportunities to protect groundwater. *Water Well J.* 53(11): 74–76.
Borggaard, O. K. (1992). Dissolution of poorly crystalline iron oxides in soils by EDTA and oxalate. *Z. Pflanzenernähr. Bodenk.* 155: 431–436.
Bott, W. & Wilken, R.-D. (2002). Erfahrungen zur Regenerierung von Brunnen mittels Ultraschall im halbtechnischen Maßstab. *bbr* 53(11): 22–28.
Böttcher, H. (1905). Reinigen von Röhrenbrunnen mit heißem Dampf. Patent document 181 578, Imperial Germany.
Böttcher, J., Strebel, O., & Duynisveld, W. H. M. (1989). Kinetik und Modellierung gekoppelter Stoffumsetzungen im Grundwasser eines Lockergesteins-aquifers. *Geol. Jb. C* 51: 3–40.
Bouwer, H. (2002). Artificial recharge of groundwater: hydrogeology and engineering. *Hydrogeol. J.* 10(1): 121–141.
Braukmann, B., Klaus, R., Sobott, R., Weber, K. H., & Rogalsky, E. (1990). Untersuchungen zur Brunnenalterung—Analyse von Brunnenbelägen. *bbr* 41(6): 330–336.
Brown, E. D. (1942). Restoring well capacity with chlorine. *J. Am. Water Works Assoc.* 1942: 698–702.
Brown, K. L., Freeston, D. H., Dimas, Z. O., & Slatter, A. (1995). Pressure drops due to silica scaling. In: Hochstein, M. P (ed.), *Proc. 17th New Zealand Geothermal Workshop,* Auckland, New Zealand, pp. 163–167,

Brune, M., Ramke, H. G., Collins, H. J., & Hanert, H. H. (1994). Incrustation problems in landfill drainage systems. In: Christensen, T. H., Cossu, R., & Stegmann, R. (eds.), *Landfilling of Wastes*, pp. 569–605. London: Spon.

Buik, N. A. & Willemsen, A. (2002). Clogging rate of recharge wells in porous media. In: Dillon, P. J. (ed.), *Management of Aquifer Recharge for Sustainability. Proc. 4th Int. Symp. on Artificial Recharge Groundwater,* Adelaide, Australia, September 22–26, 2002, pp. 195–202. Lisse: Balkema.

Buresh, R. J. & Moraghan, J. T. (1976). Chemical reduction of nitrate by ferrous iron. *J. Environ. Qual.* 5: 320–325.

Burke, S. P. & Banwart, S. A. (2002). A geochemical model for removal of iron(II)$_{(aq)}$ from mine water discharges. *Appl. Geochem.* 17: 431–443.

Chapelle, F. H. (1993). *Ground-Water Microbiology and Geochemistry*. New York: Wiley.

Chapuis, R. P. & Aubertin, M. (2003). On the use of the Kozeny-Carman equation to predict the hydraulic conductivity of soils. *Can. Geotech. J.* 40: 616–628.

Cornell, R. M. & Schwertmann, U. (2003). *The Iron Oxides—Structure, Properties, Reactions, Occurrence and Uses*. Weinheim/New York: Wiley-VCH.

Criaud, A., Fouillac, C., & Marty, B. (1989). Low enthalpy geothermal fluids from the Paris basin. 2—oxidation-reduction state and consequences for the prediction of corrosion and sulfide scaling. *Geothermics* 18(5/6): 711–727.

Cullimore, D. R. (1989). Looking for iron bacteria in water. *Can. Water Well* 15(3): 10–12.

Cullimore, D. R. (1990). An evaluation of the risk of microbial clogging and corrosion in boreholes. In: Howsam, P. (ed.), *Water Wells—Monitoring, Maintenance, Rehabilitation. Proc. Int. Groundwater Eng. Conf., Cranfield,* pp. 25–34. London: Spon/Chapman.

Cullimore, D. R. (1993). *Practical Manual of Groundwater Microbiology*. Chelsea, Michigan, Lewis.

Cullimore, D. R. (1999). *Microbiology of Well Biofouling*. Boca Raton, FL: Lewis/CRC Press.

Cullimore, D. R. & McCann, A. E. (1978). The identification, cultivation and control of iron bacteria in groundwater. In: Skinner, F. A., & Shewan, J. M. (eds.), *Aquatic Microbiology*, pp. 220–261. New York: Academic Press.

Cullimore, D. R. & McCann, A. E. (1979). Identifying the rusty monsters. *Can. Water Well* 5(3): 10–19.

Detay, M. (1997). *Water Wells—Implementation, Maintenance and Restoration*. Chichester, UK: Wiley.

DGFZ (Dresdner Grundwasserforschungszentrum) (2003). DVGW Forschungs-vorhaben W 55/99—Brunnenregenerierung—Untersuchungen zur Bewertung von Gerätetechnik auf die Wirksamkeit in der Kiesschüttung. Report. Dresden: DGFZ.

DIN 4019 (1979). *Baugrund, Setzungsberechnungen bei lotrechter, mittiger Belastung* (Teil 1). Berlin: Beuth.

DIN 4021 (1990). *Baugrund; Aufschluss durch Schürfe und Bohrungen sowie Entnahme von Proben*. Berlin: Beuth.

DIN 4022 (1981–1987). *Baugrund und Grundwasser* (Part 1, 1987; Part 2, 1981; Part 3, 1982). Berlin: Beuth.

DIN 4922 (1978–1999). *Stahlfilterrohre für Bohrbrunnen* (Part 1, 1978; Part 2, 1981; Part 3, 1975; Part 4, 1999). Berlin: Beuth.

DIN 4924 (1998). *Sande und Kiese für den Brunnenbau*. Berlin: Beuth.

DIN 4925 (1999). *Filter- und Vollwandrohre aus weichmacherfreiem Polyvinylchlorid (PVC-U) für Bohrbrunnen* (Parts 1–3). Berlin: Beuth.

DIN 31051 (2001). *Grundlagen der Instandhaltung*. Berlin: Beuth.

DIN 50900 (2002–2006). *Korrosion der Metalle, Begriffe*. Berlin: Beuth.

DIN 50929 (1985). *Korrosion der Metalle; Korrosionswahrscheinlichkeit metallischer Werkstoffe bei äußerer Korrosionsbelastung* (Parts 1–3). Berlin: Beuth.

DIN 50930 (1993). *Korrosion der Metalle; Korrosion metallischer Werkstoffe im Inneren von Rohrleitungen, Behältern und Apparaturen bei Korrosionsbelastung durch Wässer*(Parts 1–5). Berlin: Beuth.

Domenico, P. & Schwartz, F. (1990). *Physical and Chemical Hydrogeology*. New York: Wiley.

Drieux, J. J., Girou, A., Ignatiadis, I., & Roque, C. (2002). Electro-chemical study of a scaling and corrosion sensor. Application to geothermal plants. *Geol. Jb.* SE 1: 153–162, Baria, R., Baumgärtner, J., Gerard, A., & Jung, R. (eds.). Draft Proc. Int. Conf. 4th HDR Forum, Strasbourg, France, September 28–30, 1998.
Driscoll, F. G. (1989). *Groundwater and Wells.* St. Paul, MN: Johnson.
Dupuit, J. (1863). *Etudes théorétiques et pratiques sur le mouvement des eaux dans les canaux découverts et à travers les terrains perméables,* 2d ed. Paris: Dunod.
DVGW W 111 (1997). *Technische Regeln für die Ausführung von Pumpversuchen bei der Wassererschließung.* Bonn: DVGW.
DVGW W 113 (2001). *Bestimmung des Schüttkorndurchmessers und hydrogeologischer Parameter aus der Korngrößenverteilung für den Bau von Brunnen.* Bonn: DVGW.
DVGW W 117 (1975). *Entsanden und Entschlammen von Bohrbrunnen (Vertikalbrunnen) im Lockergestein und Verfahren zur Feststellung überhöhten Eintrittswiderstands.* Bonn: DVGW.
DVGW W 119 (1982). *Über den Sandgehalt im Brunnenwasser; Bestimmung von Sandmengen im geförderten Wasser, Richtwerte für den Restsandgehalt.* Bonn: DVGW.
DVGW W 123 (2001). *Bau und Ausbau von Vertikalfilterbrunnen.* Bonn: DVGW.
DVGW W 130 (2001). *Brunnenregenerierung.* Bonn: DVGW.
DVGW W 135 (1998). *Sanierung und Rückbau von Bohrungen, Grundwassermessstellen und Brunnen.* Bonn: DVGW.
DVGW W 614 (2001). *Instandhaltung von Förderanlagen.* Bonn: DVGW.
Ehrenberg, C. G. (1836). Vorläufige Mitteilungen über das wirkliche Vorkommen fossiler Infusorien und ihre große Verbreitung. *Poggendorf's Ann. Phys. Chem.* 38: 213–227.
Ehrenreich, A. & Widdel, F. (1994). Anaerobic oxidation of ferrous iron by purple bacteria, a new type of phototrophic metabolism. *Appl. Environ. Microbiol.* 60: 4517–4526.
Ehrhardt, G., & Pelzer, R. (1992). Wirkung von Saugstromsteuerungen in Brunnen. *bbr* 43(10): 452–458.
Ehrlich, G. G., Ehlke, T. A., & Vecchioli, J. (1973). Microbiological aspects of groundwater recharge—injection of purified unchlorinated sewage effluent at Bay Park, Long Island, New York. *J. Res. U.S. Geol. Surv.* 1(3): 341–344.
Ehrlich, H. L. (2002). *Geomicrobiology,* 4th ed. New York: Marcel Dekker.
Ehrmann, H. (1958). Bakteriologische Sanierung von Trinkwasserbrunnen durch Anwendung des Katadyn-Verfahrens. *bbr* 9: 368–370.
Ellis, D. (1919). *Iron Bacteria.* London: Methuen.
Etschel, H. (2004). Mechanische Regenerierverfahren mit Wasserspülung. *bbr* 55(12): 38–41.
Etschel, C. & Schmidt, M. (2001). Das Druckwellen-Impulsverfahren für die Regenierung und Entwicklung von Brunnen. *bbr* 52(4): 30–38.
Everdingen, A. F. van (1953). The skin effect and its influence on the productive capacity of wells. *Trans. AIME* 198: 171–176.
Ferris, F. G., Tazaki, K., & Fyfe, W. S. (1989). Iron oxides in acid mine drainage environments and their association with bacteria. *Chem. Geol.* 74: 321–330.
Fischer, R., Reissig, H., Bach, J., & Gawlick, W. (1990). Verfahren zur Verhinderung von Verockerungserscheinungen an Steigleitungen, Filterbrunnen, Brunnenelementen und anderen wasserführenden Objekten. Teil 1. Gegenwärtiger Stand und theoretische Grundlagen. *Acta Hydrochim. Hydrobiol.* 18(3): 363–372.
Fischer, R., Reißig, H., & Forker, W. (1993). Elektrochemische Verfahren zur Verhinderung von Verockerungserscheinungen, Verkrustungen und Belagbildung an Brunnenelementen. *Wasserwirtschaft* 83(9): 486–489.
Fleming, I. R., Rowe, R. K., & Cullimore, D. R. (1999). Field observations of clogging in a landfill leachate collection system. *Can. Geotech. J.* 36: 685–707.
Ford, R. G., Bertsch, P. M., & Farley, K. J. (1997). Changes in transition and heavy metal partitioning during hydrous iron oxide aging. *Environ. Sci. Technol.* 31: 2028–2033.
Forward, P. (1994). Control of iron biofouling in submersible pumps in the Woolpunda salt interception scheme in South Australia. *Water Down Under 94,* Vol. 2, Part A, 169-174. Barton, Australia: Institution of Engineers.
Fountain, J. & Howsam, P. (1990). The use of high pressure water jetting as a rehabilitation technique. In: Howsam, P. (ed.), *Water wells—Monitoring, Maintenance,*

Rehabilitation. Proc. Int. Groundwater Eng. Conf., Cranfield, pp. 180–194. London: Spon/Chapman.
Freie und Hansestadt Hamburg (1999a). Merkblatt zur Qualitätssicherung: Merkblatt 8 "Sanierung und Rückbau von Grundwassermessstellen." Hamburg: Amt für Umweltschutz.
Freie und Hansestadt Hamburg (1999b). Merkblatt zur Qualitätssicherung: Merkblatt 1 "Rückbau von Förderbrunnen." Hamburg: Amt für Umweltschutz.
Fricke, S.. & Schön, J. (1999). *Praktische Bohrlochgeophysik*. Stuttgart: Enke.
Furrer, G. & Stumm, W. (1986). The coordination chemistry of weathering: I. Dissolution kinetics of δ-Al_2O_3 and BeO. *Geochim. Cosmochim. Acta* 50: 1847–1860.
Gabrielli, C., Maurin, G., Poindessous, G., & Rosset, R. (1999). Nucleation and growth of calcium carbonate by an electrochemical scaling process. *J. Crystal Growth* 200(1–2): 236–250.
Gallup, D. L. (1989). Iron silicate scale formation and inhibition at the Salton Sea geothermal field. *Geothermics* 18: 97–103.
Gallup, D. L. (1998). Geochemistry of geothermal fluids and well scales, and potential for mineral recovery. *Ore Geol. Rev.* 12(4): 225–236.
Gallup, D. L. & Reiff, W. M. (1991). Characterization of geothermal scale deposits by Fe-57 Mössbauer spectroscopy and complementary x-ray diffraction and infra-red studies. *Geothermics* 20(4): 207–224.
Geißler, W. & Wiegand, G. (1924). Verfahren zum Reinigen von inkrustierten Filtergeweben in alten Röhrenbrunnen durch stufenweises Einbringen von Säure. Patent Document 420 901, Imperial Germany.
Gottlieb, O. J. & Blattert, R. E. (1988). Concepts of well cleaning. *J. Am. Water Works Assoc.* 80(5): 34–39.
Gruesbeck, C. & Collins, R. E. (1982). Entrainment and deposition of fine particles in porous media. *Soc. Petrol. Eng. J.* 12: 847–856.
Hagen, G. (1839). Über die Bewegung des Wassers in engen cylindrischen Röhren. *Poggendorffs Ann. Phys. Chem.* 16: 423–442.
Halfmeier, F. (1960). Verfahren zur Wiederherstellung verockerter Rohrbrunnen. *bbr* 11: 348–349.
Hanert, H. (1968). Untersuchungen zur Isolierung, Stoffwechselphysiologie und Morphologie von Gallionella ferruginea Ehrenberg. *Arch. Mikrobiol.* 60: 348–376.
Hanert, H. (1973). Quantifizierung der Massenentwicklung des Eisenbakteriums Gallionella ferruginea unter natürlichen Bedingungen. *Arch. Mikrobiol.* 88: 225–243.
Hanert, H. (1974). In-situ Untersuchungen zur Analyse und Intensität der Eisen(III)-Fällung in Dränungen. *Z. Kulturtechn. Flurberein.* 15: 80–90.
Hanert, H. (1981). The genus *Gallionella*. In: Starr, M. P., Stolp, H., Trueper, H. G., Balows, A., & Schlegel, H. G. (eds.), *The Prokaryotes. A Handbook on Habitats, Isolation and Identification of Bacteria*, pp. 509–515. Berlin: Springer-Verlag.
Hansen, D. (2004). Discussion of "On the use of the Kozeny-Carman equation to predict the hydraulic conductivity of soils." *Can. Geotech. J.* 41: 990–993.
Harker, D. (1990). Effect of acidization on chalk boreholes. In: Howsam, P. (ed.), *Water Wells—Monitoring, Maintenance, Rehabilitation. Proc. Int. Groundwater Eng. Conf., Cranfield*, pp. 158–167. London: Spon/Chapman.
Hässelbarth, U. (1980). Das Verhalten von eisenspeichernden Mikroorganismen. *Schriftenreihe* 1: 29–41.
Hässelbarth, U. & Lüdemann, D. (1967). Die biologische Verockerung von Brunnen durch Massenentwicklung von Eisen- und Manganbakterien. *bbr* 18(10/11) 363–368 & 401–406.
Hässelbarth, U. & Lüdemann, D. (1972). Biological encrustation of wells due to mass development of iron and manganese bacteria. *Water Treatment Examination* 21(1): 20–29.
Helweg, O. J. (1982). Economics of improving well and pump efficiency. *Ground Water* 20(5): 556–562.
Hem, J. D. & Lind, C. J. (1983). Nonequilibrium models for predicting forms of precipitated manganese oxides. *Geochim. Cosmochim. Acta* 47: 2037–2046.
Hem, J. D. Roberson, C.,E., & Fournier, R. B. (1982). Stability of bMnOOH and manganese oxide deposition from springwater. *Water Resources Res.* 18(3): 536–570.

Henke, J. R. Norris, M., & Hall, W. D. (1991). Iron encrustation of dewatering wells—causes and remedies. *Trans. Soc. Mining Eng., AIME* 290: 1935–1941.
Herrick, D. (2000). Hard rock frack'n. *Water Well J.* 54(7): 40–42.
Herth, W. & Arndts, E. (1994). *Theorie und Praxis der Grundwasserabsenkung,* 3d ed. Berlin: Ernst & Sohn.
Hijnen, W. A. M. & Kooij, D. van der (1992). Biologische Kolmation von Schluckbrunnen unter dem Einfluß des AOC—Gehaltes des Wassers. *GWF Wasser/Abwasser* 133(3): 148–153.
Hinman, N. W. & Lindstrom, R. F. (1996). Seasonal changes in silica deposition in hot spring systems. *Chem. Geol.* 132(1–4): 237–246.
Hjulström, F. (1935). Studies of the morphological activity of rivers as illustrated by the river Fyris. *Bull. Geol. Inst. Univ. Upsala* XXV: 221–527.
Hochella, M. F. & White, A. F. (eds.) (1990). *Mineral Water Interface Geochemistry.* Rev. Mineralogy 23. Washington, DC: Mineralogy Society of America.
Houben, G. (2000). Modellansätze zu langfristigen Prognose der Entwicklung der Grundwasserbeschaffenheit—Fallbeispiel Bourtanger Moor (Emsland). Ph.D. thesis, Technical University Aachen; *Aachener Geowiss. Beitr.* 36.
Houben, G. (2001). Well ageing and its implications for well and piezometer performance. pp. 297–300. In: Gehrels, H., Peters, N., Hoehn, E., Jensen, K., Leibundgut, C., Griffioen, J., Webb, B., & Zaadnoordijk, W. J. (eds.), *Impact of Human Activity on Groundwater Dynamics.* IAHS Publ. 269. Wallingford: IAHS.
Houben, G. (2002). Der Einfluss der Brunnenverockerung auf die Qualität von Wasseranalysen und die Auswertung von hydraulischen Tests. *bbr* 53(10): 21–25.
Houben, G. (2003a). Iron oxide incrustations in wells—Part 1. Genesis, mineralogy and geochemistry. *Appl. Geochem.* 18(6): 927–939.
Houben, G. (2003b). Iron oxide incrustations in wells—Part 2. Chemical dissolution and modelling. *Appl. Geochem.* 18(6): 941–954.
Houben, G. (2004). Modeling the buildup of iron oxide encrustations in wells. *Ground Water* 42(1): 78–82.
Houben, G. (2006). The influence of well hydraulics on the spatial distribution of well incrustations. *Ground Water,* 44(5): 668–675.
Houben, G., Heuvel, M. van den, Rooijen, P. van, & Treskatis, C. (2004). Grenzüberschreitende hydrogeologische Arbeiten im deutsch-niederländischen Grenzgebiet bei Geilenkirchen-Brunssum. *Z. Angew. Geol.* 50(1): 44–50.
Houben, G., Merten, S., & Treskatis, C. (1999). Entstehung, Aufbau und Alterung von Brunneninkrustationen. *bbr* 50(10): 29–35.
Houben, G., Merten, S., & Treskatis, C. (2000a). Laborversuche zur Wirksamkeit von chemischen Mitteln zur Brunnenregenerierung. *bbr* 51(2): 41–46.
Houben, G., Schröder, C., & Treskatis, C. (2000b). Erfahrungen bei der chemischen Brunnenregenerierung mit einem neu entwickelten pH-neutralen Regeneriermittel. *bbr* 51(12): 42–47.
Houben, G. & Treskatis, C. (2003). *Regenerierung und Sanierung von Brunnen.* Munich: Oldenbourg.
Houben, G. & Treskatis, C. (2004). *Regeneracja studni.* Bydgoszcz: Proprzem-EKO.
Houben, G. Tünnermeier, T., & Himmelsbach, T. (2005). Hydrogeology of the Kabul Basin (Afghanistan), Part II: Groundwater Geochemistry and Microbiology. Hannover: BGR [available for download at www.bgr.de].
Howsam, P. (ed.) (1990a). *Water Wells—Monitoring, Maintenance, Rehabilitation. Proc. Int. Groundwater Eng. Conf., Cranfield,* London: Spon/Chapman.
Howsam, P. (1990b). Well performance deterioration—an introduction to cause processes. In: Howsam, P. *Water Wells—Monitoring, Maintenance, Rehabilitation. Proc. Int. Groundwater Eng. Conf., Cranfield,* pp. 19–24. London: Spon/Chapman.
Howsam, P. (1990c). Well performance deterioration—an introduction to cure processes. In: Howsam, P. *Water Wells—Monitoring, Maintenance, Rehabilitation. Proc. Int. Groundwater Eng. Conf., Cranfield,* pp. 151–157. London: Spon/Chapman.
Howsam, P. (1991). A well-head monitoring cell—a diagnostic tool for boreholes and tubewells with reduced yields and other operational problems. In: Stow, D. A. V., & Laming, D. J. C. (eds.), *Geosciences Development. Proc. Int. Conf. Appl. Geosci. Developing Countries,* AGID Rep. Ser. 14, pp. 131–134. London: Geological Society.

Howsam, P., Brassington, F. C., & Lucey, P. A. (1989). The examination of encrustations in an unlined borehole using the sidewall sampling technique. *Quart. J. Eng. Geol.* 22: 139–144.
Howsam, P. & Hollamby (1990). Drilling fluid invasion and permeability impairment in granular formations. *Quart. J. Eng. Geol.* 23: 161–168.
Howsam, P., Misstear, B., & Jones, C. (1995). *Monitoring, Maintenance and Rehabilitation of Water Supply Boreholes* (R 137). London: CIRIA.
Hübner, G. (1994). Irradiation prevents biological encrustation of wells. In: *Water Management Europe 1994/1995*, pp. 177–180. London.
Huerta-Diaz, M. A., Tessier, A., & Carignan, R. (1998). Geochemistry of trace metals associated with reduced sulfur in freshwater sediments. *Appl. Geochem.* 13: 213–233.
Huisman, L. & Olsthoorn, T. N. (1983). *Artificial Groundwater Recharge*. Boston: Pitman.
Hurst, W. (1953). Establishment of the skin effect and its impediment to fluid flow into a well bore. *Petrol. Eng.* 25(10): B6–B16.
Hurst, W., Clark, J. D., & Brauer, E. B. (1969). The skin effect in producing wells. *J. Petrol. Technol.* 21: 1483–1489.
Ivarson, K. C. & Sojak, M. (1978). Microorganisms and ochre deposition in field drains of Ontario. *Can. J. Soil. Sci.* 58(1): 1–17.
Jäkel, G. (1958). Untersuchung zur Minderung der Ergiebigkeit von Brunnen infolge Verockerung. Ph.D. thesis, Technical University Dresden.
Jakisch, N. (1995). Altlasten aus der technischen Anwendung von Strahlenquellen—Die ^{60}Co-Strahlenquellen in Brunnen der DDR. *Strahlenschutzpraxis* 1(3): 5–10.
Janssens, J. G., Pintelon, L., Cotton, A., & Gelders, L. (1996). Development of a framework for the assessment of operation and maintenance (O&M) performance of urban water supply and sanitation. *Water Supply* 14(1): 21–23.
Jüttner, R. & Ries, T. (2001). Umweltgerechte Brunnensanierung mit Kohlendioxid. *bbr* 52(7): 38–42.
Kalfayan, L. (2001). *Production Enhancement with Acid Treatment*. Tulsa: Penn Well.
Kester, D. R., Byrne, R. H., & Liang, Y.-J. (1975). Redox reactions and solution complexes of iron in marine systems. In: Gould, R. F. (ed.), *Marine Chemistry in the Coastal Environment*. ACS Symp. Ser. 18: 57–79.
Kiwa (2002). Onderzoek Putmanagement—Kennisplatform 16 april 2002, Stand van zaken onderzoek, augustus 2002, BTO 2002.132. Nieuwegein, The Netherlands: KIWA Water Research.
KIWA (2004a). Putregeneratie met Ultrasoon—Stand van zaken 2004 en achtergrondinformatie, BTO 2004.009. Nieuwegein, The Netherlands: KIWA Water Research.
KIWA (2004b). Putregeneratie met Roto-cavitatie—Evaluatie van het testprogramma 2002–2004, BTO 2004.040. Nieuwegein, The Netherlands: KIWA Water Research.
Klink, K. (1994). Probleme bei der Desinfektion von Brunnen. *bbr* 45(4): 13–17.
Kölle, W. (2001). *Wasseranalysen—richtig Beurteilt*. Weinheim: Wiley-VCH.
Krems, G. (1972). *Studie über die Brunnenalterung*. Bonn: Federal Ministry of the Interior.
Kruse, H. (1953). Zur Frage der Dauer des Abpumpens von Rohrbrunnen nach Beseitigung von Filterverockerungen durch Salzsäure. *GWF Gas- und Wasserfach*, Suppl. Bau und Betrieb 5: 7.
Kruseman, G. P. & De Ridder, N. A. (1994). Analysis and evaluation of pumping test data.- Int. *Institute for Land Reclamation and Improvement (ILRI) Bulletin 11*, 2d ed. Wageningen: ILRI.
Kuntze, H., & Scheffer, B. (1974). Organische Dränfilter gegen Verockerung. *Z. Kulturtechn. Flurbereinig.* 15: 70–79.
Langmuir, D. & Whittemore, D. O. (1971). Variations in the stability of precipitated ferric oxyhydroxides. *Adv. Chem. Ser.* 106: 209–234.
Lowson, R. T. (1982). Aqueous oxidation of pyrite by molecular oxygen. *Chem. Rev.* 82(5): 461–497.
LVR (Landschaftsverband Rheinland) (1998). *Brunnen der Jungsteinzeit*. Cologne: Rheinland Verlag.
Macadam, J. & Parsons, S.,A. (2003). Calcium carbonate scale control, effect of material and inhibitors. In: International Water Association, *1st IWA Conference on Scaling and Corrosion in Water and Wastewater Systems, Cranfield*, March 2003, pp. 25–27. Cranfield: Cranfield University.

Macadam, J. & Parsons, S. A. (2004). Calcium carbonate scale control, effect of material and inhibitors. *Water Sci. Technol.* 49(2): 153–159.
Manceau, A., Ildefonse, P., Hazemann, J.-L., Flank, A.-M., & Gallup, D. (1995). Crystal chemistry of hydrous iron silicate scale deposits at the Salton Sea geothermal field. *Clays Clay Minerals* 43(3): 304–317.
Maniak, U. (2001). *Wasserwirtschaft—Einführung in die Bewertung wasserwirtschaftlicher Vorhaben.* Heidelberg: Springer-Verlag.
Mansuy, N. (1992). Assessment and rehabilitation of biofouled monitoring or product well recovery. *Superfund,* pp. 527–530. Silver Spring, MD.
Mansuy, N. (1998). *Water Well Rehabilitation.* Boca Raton, FL: CRC Press.
Mansuy, N., Nuzman, C., & Cullimore, D. R. (1990). Well problem identification and its importance in well rehabilitation. In: Howsam, P. (ed.), *Water Wells—Monitoring, Maintenance, Rehabilitation. Proc. Int. Groundwater Eng. Conf., Cranfield,* pp. 87–99. London: Spon/Chapman.
Martinez, C. E. & McBride, M. B. (1998). Coprecipitates of Cd, Cu, Pb and Zn in iron oxides: solid phase transformation and metal solubility after aging and thermal treatment. *Clays Clay Minerals* 46(5): 537–545.
Martinez, C. E. & McBride, M. B. (1999). Dissolved and labile concentrations of Cd, Cu, Pb, and Zn in aged ferrihydrite-organic matter systems. *Environ. Sci. Technol.* 33: 745–750.
Martinez, C. E., Sauve, S., Jacobson, A., & McBride, M. B. (1999). Thermally induced release of adsorbed Pb upon aging ferrihydrite and soil oxides. *Environ. Sci. Technol.* 33: 2016–2020.
McDowell-Boyer, L. M., Hunt, J. R., & Sitar, N. (1986). Particle transport in porous media. *Water Resources Res.* 22(13): 1901–1921.
McLaughlan, R. G. (1996). *Water Well Deterioration—Diagnosis and Control.* Technology Transfer Publication 1/96. Sydney: UTS-NCGM.
McLaughlan, R. G. (2002). Managing Water Well Deterioration. Int. Contrib. Hydrogeol. 22. Lisse: Balkema.
McLaughlan, R. G., Knight, M. J., & Stuetz, R. M. (1993). *Fouling and Corrosion of Groundwater Wells—A Research Study.* Res. Publ. NCGM 1/93. Sydney: UTS-NCGM.
McLaughlan, R. G. & Stuetz, R. M. (1990). Tubewell fouling at a salinity interception scheme, Wakool, NSW, Australia. In: Howsam, P. (ed.), *Water Wells—Monitoring, Maintenance, Rehabilitation. Proc. Int. Groundwater Eng. Conf., Cranfield,* pp. 47–58. London: Spon/Chapman.
Mehmert, M. (1995). How bacteria complicate well rehabilitation. *Water Well J.* 49(7): 65–68.
Mehra, O. P. & Jackson, M. L. (1960). Iron oxide removal from soils and clays by a dithionite-citrate system buffered with sodium bicarbonate. *Clays Clay Minerals* 5 (7th Natl. Conf. on Clays and Clay Minerals: 317–327.
Millero, F. J. (1985). The effect of ionic interaction on the oxidation of metals in natural waters. *Geochim. Cosmochim. Acta* 49: 547–553.
Misawa, T., Hashimoto, K., & Shimodaira, S. (1974). The mechanism of formation of iron oxide and oxihydroxides in aqueous solutions at room temperature. *Corrosion Sci.* 14: 131–149.
Mitchell, G. F., Sargand, S. M., & Rueda, J. (1993). Clogging of drainage blankets in landfills. In: Brauns, J., Schuler, U., & Hcibaum, M. (eds.), *Filters in Geotechnical and Hydraulic Engineering, 1st Int. Conf. Geo-Filters,* Karlsruhe, October 20–22, 1992, pp. 97–208. Rotterdam: Balkema.
Mogg, J. L. (1972). Practical corrosion and incrustation guide lines for water wells. *Ground Water* 10(2): 6–11.
Moore, T. (1997). Swaging offers hope for damaged wells. *Water Well J.* 51(7): 52–54.
Moore, T. (1998). New well rehab technology "bursts" onto scene. *Water Well J.* 52(7): 49–51.
Moorman, J. H. N., Colin, M. G., & Stuyfzand, P. J. (2002). Iron precipitation clogging of a recovery well following nearby deep well injection. In: Dillon, P. J. (ed.), *Management of Aquifer Recharge for Sustainability. Proc. 4th Int. Symp. Artificial Recharge Groundwater,* Adelaide, September 22–26, 2002, pp. 209–214.Lisse: Balkema.
Morse, J. W. & Arakaki, T. (1993). Adsorption and coprecipitation of divalent metals with mackinawite (FeS). *Geochim. Cosmochim. Acta* 57: 3635–3640.

Morse, J. W. & Casey, W. H. (1988). Ostwald processes and mineral paragenesis in sediments. *Am. J. Sci.* 288: 537–560.

Morse, J. W. & Luther, G. W. (1999). Chemical influences on trace-metal interactions in anoxic sediments. *Geochim. Cosmochim. Acta* 63: 3373–3378.

Moser, H. (1978). Die Alterung von Vertikalfilterbrunnen, ihre technischen und wirtschaftlichen Aspekte, aufgezeigt am Beispiel der Gewinnungsanlage des Grundwasserwerkes Mannheim-Käfertal. Ph.D. thesis, Technical University Darmstadt. Darmstadt: Inst. Wasserversorgung TH Darmstadt.

Muckenthaler, P. (1989). Ergänzende Modellvorstellungen für Erosions- und Suffossionskriterien. *Wasserwirtschaft* 79(7/8): 405–409.

Munding, H. (2005). Mechanische Brunnenregenerierung nach DVGW-Merkblatt W130. *bbr* 56(3): 42–47.

Murray, J. W (1979). Iron oxides. In: Burns, R.G. (ed.), *Marine Minerals. Rev. Mineral.* 6: 47–98.

Naumann, D. (1936). Reinigung von Brunnenrohren durch Säuren und damit verbundene gesundheitliche Gefahren. *GWF Gas- und Wasserfach* 79(34): 623–624.

Niehues, B. (1999). DVGW—Umfrage "Brunnenregenerierung." *Conf. Proceedings Papers 2nd Friedrichshafener Brunnenbautage,* October 12–13, 1999. Bonn: DVGW.

Nolte, L. P., Tewes, S., & Baumann, K. (2004). Grundwassermessstellen—Pflege, Sanierung und Rückbau (Teil 1). *bbr* 55(1): 42–48.

Oberdorfer, J. A. & Peterson, F. L. (1985). Wastewater injection—geochemical and biogeochemical clogging processes. *Ground Water* 23(6): 753–761.

Obstfelder, H. J. von (1952). Das "Regenerieren" von Bohrbrunnen. *bbr* 8: 22–23.

Oles, D., & Houben, G. (1998). Greigite (Fe_3S_4) in an acid mudpool at Makiling Volcano, the Philippines. *J. Asian Earth Sci.* 16(5/6): 513–517.

Olsthoorn, T. N. (1982). The Clogging of Recharge Wells. KIWA Communications 72. Rijswijk, The Netherlands: KIWA.

Parkhurst, D. & Appelo, C. A. J. (1999). User's guide to PHREEQC (Version 2)—a computer program for speciation, reaction path, 1D-transport and inverse geochemical calculations. USGS Water Resources Invest. Rep. 99-4259. Lakewood, Colorado.

Parsons, S. A. & Doyle, J. D. (2004). Struvite scale formation and control. *Water Sci. Technol.* 49: 177–183.

Patchick, P. F. (1992). *Water Wells—Incrustation and Restoration*. Public Works 44–48. Ridgewood: Public Works Journal Corp. New Jersey

Paul, K. F. (1990). Water well regeneration—new technology. In: Howsam, P. (ed.), *Water Wells—Monitoring, Maintenance, Rehabilitation. Proc. Int. Groundwater Eng. Conf. Cranfield,* pp. 168–179. London: Spon/Chapman.

Paul, K. F. (2000). Brunnenregenerierung im Wandel der Zeit. *Proc. FIGAWA/DVGW Conf. "Brunnenbau 2000,"* Berlin, Oktober 23–24, 2000.

Pedersen, H. D., Postma, D., Jakobsen, R., & Larsen, O. (2005). Fast transformation of iron oxyhydroxides by the catalytic action of aqueous Fe(II). *Geochim. Cosmochim. Acta* 69 (16): 3967–3977.

Pelzer, R. & Smith, S. A. (1990). Eucastream suction flow control device—an element for optimization of flow conditions in wells. In: Howsam, P. (ed.), *Water Wells—Monitoring, Maintenance, Rehabilitation. Proc. Int. Groundwater Eng. Conf. Cranfield,* pp. 209–215. London: Spon/Chapman.

Perez-Paricio, A. (2001). Integrated modeling of clogging processes in artificial groundwater recharge. Ph.D. thesis, University of Barcelona, Barcelona, Spain.

Petersen, J. S., Rohwer, C., Asce, M., Albertson, M. L., & Asce, J. M. (1955). Effect of well screens on flow into wells. *Am. Soc. Civil Eng. Trans.* 120, paper 2755: 563–607.

Plujimackers, J., Kooiman, J. W., Visscher, D., & Radke, B. (2005). Kosten-Nutzen-Bewertung eines Horizontalfilterbrunnens in den Niederlanden. *bbr* 56(5): 42–47.

Plumley, T. (2000). Journey to the center of the earth—downhole video cameras provide insight to well problems. *Water Well J.* 54(9): 26–28.

Porath, G. M. & Rapsch, H.-J. (eds.) (1998). *Von Brunnen und Zucken, Pipen und Wasserkünsten. Die Entwicklung der Wasserversorgung in Niedersachsen.* Neumünster: Wachholtz.

Powell, M. R. (1998). Magnetic water and fuel treatment: myth, magic, or mainsteam science? *Sceptical 27–31 Inquirer* 22(1).

Prins, J. (2003). Suspended material in abstracted groundwater in the Netherlands, First results. KIWA Rep. KWR 03.029. Nieuwegein, The Netherlands: KIWA.
Prins, J. (2004). Onderzoek naar deeltjes en putverstopping bij put 50 van puttenveld Ritskebos. KIWA Rep. KWR 03.029. Nieuwegein: The Netherlands: KIWA.
Prins, J. (2004). Achtergrondconcentratie van deeltjes in grondwater.Resultaten van metingen in waarnemingsfilters op puttenveld C. Rodenhuis (Hydron Zuid Holland). BTO 2005.048 (s). Nieuwegein, The Netherlands: KIWA.
Pyne, R. D. G. (1995). *Groundwater Recharge and Wells: A Guide to Aquifer Storage and Recovery*. Boca Raton, FL: Lewis.
Ralph, D. E. & Stevenson, J. M. (1995). The role of bacteria in well clogging. *Water Res.* 29(1): 365–369.
Rebhun, M. & Schwarz, J. (1968). Clogging and contamination processes in recharge wells. *Water Resources Res.* 4(6): 1207–1217.
Reissig, H., Fischer, R., & Bach, J. (1990). Verfahren zur Verhinderung von Verockerungserscheinungen an Steigleitungen, Filterbrunnen, Brunnenelementen und anderen wasserführenden Objekten. Teil 2 Labor- und Freilandversuche zur Verhinderung der Verockerung mittels kathodischer und anodischer Polarisation. *Acta Hydrochim. Hydrobiol.* 18(4): 469–477.
Rider, M. H. (1996). *The Geological Interpretation of Well Logs*. Whittles Publishing Caithness (Scotland).
Riekel, T. & Hinze, G. (2002). Using chemical-free technology to clean up wells in Botswana. *Water Well J.* 56(8): 26–29.
Riempp, G. (1964). Brunnenverockerung an Vertikalbrunnen. *WWT Wasserwirtschaft-Wassertechnik* 14(4): 109–113.
Rinck-Pfeiffer, S., Ragusa, S., Sztajnbok, P., & Vandevelde, T. (2000). Interrelations between biological, chemical, and physical processes as an analog to clogging in aquifer storage and recovery (ASR) wells. *Water Res.* 34(7): 2110–2118.
Rinck-Pfeiffer, S., Dillon, P., Ragusa, S., & Hutson, J. (2002). Injection well clogging during aquifer storage and recovery (ASR) with reclaimed water. In: Dillon, P. J. (ed.), *Management of Aquifer Recharge for Sustainability. Proc. 4th Int. Symp. Artificial Recharge Groundwater*, Adelaide, September 22–26, 2002, pp. 189–194. Lisse: Balkema Australia.
Rogalsky, E. (1992). Untersuchungen zur Brunnenalterung—Analyse von Brunnenbelägen. *ESWE Schriftenreihe* 4: 7–21.
Roscoe Moss Company (1990). *Handbook of Ground Water Development*. New York: Wiley.
Rothbaum, H. P., Anderton, B. H., Harrison, R. F., Rohde, A. G., & Slatter, A. (1979). Effect of silica polymerization and pH on geothermal scaling. *Geothermics* 8(1): 1–20.
Rott, U., Meyerhoff, R., & Bauer, T. (1996). In-situ Aufbereitung von Grundwasser mit erhöhten Eisen$^-$, Mangan- und Arsengehalten. *GWF Wasser/Abwasser* 137: 358–363.
Rowe, R. K., Armstrong, M. D., & Cullimore, D. R. (2000). Mass loading and the rate of clogging due to municipal solid waste leachate. *Can. Geotech. J.* 37: 355–370.
Roy, C. (1991). La fracturation hydraulique—une methode de developpement des forages d'eau dans le socle. *Geologues* 94: 31–34.
Rübesame, K. (1996). Schonender Einsatz von Sprengstoffen zur Brunnenregenerierung. *bbr* 47(3): 18–24.
Rübesamen, U. & Nolte, L.-P. (1999). Sanierungs- und Rückbautechniken und methoden im Lockergestein. *DVGW Schriftenreihe Wasser* 93: 299–320.
Ryan, J. N. & Elimelech, M. (1996). Review colloid mobilization and transport in groundwater. *Colloids Surfaces A* 107: 1–56.
Sanyal, S. K., McNitt, J. R., Klein, C. W., & Granados, E. E. (1985). An investigation of wellbore scaling at the Miravalles geothermal field, Costa Rica. In: Proc. 10th Workshop Geothermal Reservoir Engineeing, Stanford, CA, pp. 37–44.
Saripalli, K. P., Meyer, P. D., Bacon, D. H., & Freedman, V. L. (2001). Changes in hydrologic properties of aquifer media due to chemical reactions: a review. *Crit. Rev. Environ. Sci. Technol.* 31(4): 311–349.
Saucier, R. J. (1974). Considerations in gravel pack design. *J. Petroleum Technol.* 1974: 205–212.
Saunders, A. (1996). Rejuvenating a tired water well. *Water Well J.* 50(1): 116–119.

Saunders, J. A. & Rowan, E. L. (1990). Mineralogy and geochemistry of metallic well scale, Raleigh and Boykin Church oilfields, Mississippi, U.S.A. *Trans. Inst. Mining Metall., Sec. B* 99: B54–B58.

Scheidegger, A., Bürgisser, C. S., Borkovec, M., Sticher, H., Meussen, H., & Riemsdijk, W. van (1994). Convective transport of acids and bases in porous media. *Water Resources Res.* 30(11): 2937–2944.

Schneider, E. (1983). Optische und geophysikalische Untersuchungen von Brunnen. *DVGW Schriftenreihe* 34: 331–351.

Schoenen, D. & Eisert, K., J. (1987). Eintrag von organischem Nährstoffsubstrat für Mikroorganismen in den Untergrund bei der Brunnenregeneration. *bbr* 38(3): 109–113.

Schoonen, M. A. & Barnes, H. L. (1991a). Reactions forming pyrite and marcasite from solution: I. Nucleation of FeS_2 below 100°C. *Geochim. Cosmochim. Acta* 55: 1495–1504.

Schoonen, M. A. & Barnes, H. L. (1991b). Reactions forming pyrite and marcasite from solution: II. Via FeS precursors below 100°C. *Geochim. Cosmochim. Acta* 55: 1505–1514.

Schultes, A. C. & Moses, G. (2002). Well rehabilitation is under pressure. *Water Well J.* 56(6): 18–20.

Schwertmann, U. & Cornell, R. M. (2000). *Iron Oxides in the Laboratory*, 2d ed. Weinheim/New York: Wiley-VCH.

Schwertmann, U. & Murad, E. (1983). Effect of pH on the formation of goethite and hematite from ferrihydrite. *Clays Clay Minerals* 31(4): 277–284.

Sebold, B. M. E, Franzreb, M., & Eberle, S. H. (1996). Untersuchungen zum Einfluß starker Magnetfelder auf die Kalkabscheidung aus wäßrigen Lösungen. *GWF Wasser/Abwasser* 137(14): 178–184.

Serway, R. A. & Faughn, J. S. (1992). *College Physics*, 3d ed. Fort Worth, TX: Saunders College.

Sichardt, W. (1928). *Das Fassungsvermögen von Rohrbrunnen und seine Bedeutung für die Grundwasserabsenkung, insbesondere für größere Absenkungstiefen.* Berlin: Springer-Verlag.

Sidhu, P. S., Gilkes, R. J., Cornell, R. M., Posner, A. M., & Quirk, J. P. (1981). Dissolution of iron oxides and oxyhydroxides in hydrochloric and perchloric acids. *Clays Clay Minerals* 29(6): 269–276.

Singh, B., Wilson, M. J., McHrady, W. J., Fraser, A. R., & Merrington, G. (1999). Mineralogy and chemistry of ochre sediments from an acid mine drainage near a disused mine in Cornwall, UK. *Clay Minerals* 34: 301–317.

Smith, R. C. (1963). Relation of screen design to the design of mechanically efficient wells. *J. Am. Water Works Assoc.* 55: 609–614.

Smith, S. A. (1995). *Monitoring and Remediation Wells—Problem Prevention, Maintenance, and Rehabilitation.* BocaRaton, FL: Lewis.

Snow, D. T. (1968). Fracture deformation and changes of permeability and storage upon changes of fluid pressure. *Quart. J. Colorado School of Mines* 63: 201–244.

Spencer, W. F., Patrick, R., & Ford, H. W. (1963). The occurrence and cause of of iron oxide deposits in tile drains. *Soil Sci. Soc. Am. Proc.* 27: 134–137.

Spon, R. (1999). Using well rehabilitation chemicals to improve water quality. *Water Well J.* 53(6): 26–30.

Spranger, E. (1978). Die Saugstromsteuerung. *bbr* 29: 10–16.

Steußloff, S. & Wicklein, A. (1999). Entwicklung des Sprengschockens bei Vertikal- und Horizontalfilterbrunnen. *bbr* 50(2): 18–24.

Steußloff, S., & Steinbrecher, A. (2001). Langzeiterfahrung mit Impulsverfahren. *bbr* 52(10): 20–27.

Straub, K. L., Benz, M., Schink, B., & Widdel, F. (1996). Anaerobic, nitrate-dependent microbial oxidation of ferrous iron. *Appl. Environ. Microbiol.* 62(4): 1458–1460.

Stuetz, R. M. & McLaughlan, R. G. (2003). Impact of localized dissolved iron concentrations on the biofouling of environmental wells. In: International Water Association, *1st IWA Conf. on Scaling and Corrosion in Water and Wastewater Systems,* Cranfield, 25–27 March 2003. Cranfield: Cranfield University.

Stumm, W. & Morgan, J. J. (1996). *Aquatic Chemistry,* 3d ed. New York: Wiley.

Stuyt, L. C. P. M., & Oosten, A. J. (1987). Mineral and ochre clogging of subsurface land drainage systems in the Netherlands. *Geotextiles Geomembranes* 5: 123–140.

Sung, W. & Morgan, J. L. (1980). Kinetics and product of ferrous iron oxygenation in aqueous systems. *Environ. Sci. Technol.* 14: 561–568.
Süsser, P. & Schwertmann, U. (1983). Iron oxide mineralogy of ochreous deposits in drain pipes and ditches. *Z. Kulturtechn. Flurbereinig.* 24: 386–395.
Swyter (1922). Neues Verfahren zur Reinigung von Rohrbrunnen. GWF *Gas- und Wasserfach* 65: 465–466.
Tamura, H., Goto, K., & Nagayama, M. (1976). The effect of ferric hydroxide on the oxygenation of ferrous ions in neutral solutions. *Corrosion Sci.* 16: 197–207.
Taylor, S. W., Lange, C. R., & Lesold, E. A. (1997). Biofouling of contaminated ground-water recovery wells—characterization of microorganisms. *Ground Water* 35(6): 973–980.
Thiem, A. (1870). Die Ergiebigkeit artesischer Bohrlöcher, Schachtbrunnen und Filtergalerien. *J. Gasbeleuchtung Wasserversorg.* 14: 450–467.
Thomas, D., McKibben, M. A., & Corona Ruiz, M. (1992). Sulfide scaling in Cerro Prieto geothermal wells. *Geothermal Resources Council Trans.* 16: 371–376.
Thullner, M., Mauclaire, L., Schroth, M. H., Kinzelbach, W., & Zeyer, J. (2002b). Interaction between water flow and spatial distribution of microbial growth in a two-dimensional flow field in saturated porous media. *J. Contaminant Hydrol.* 58: 169–189.
Thullner, M., Zeyer, J., & Kinzelbach, W. (2002a). Influence of microbial growth on hydraulic properties of pore networks. *Transport Porous Media* 49: 99–122.
Tillmanns, J., & Heublein, (1912). Über die kohlensauren Kalk angreifende Kohlensoure der natürlichen Wässer. *Gesundh. Ing.* 35: 669–677.
Timmer, H., Verdel, J. D., & Jongmans, A. (2000). Verstopping putten door van nature aanwezig material. H_2O 2000 (20): 24–26.
Timmer, H., Verdel, J. D., & Jongmans, A.G. (2003). Well clogging by particles in Dutch well fields. *J. Am. Water Works Assoc.* 95(8): 112–118.
Tlili, M. M., Ben Amor, M., Gabrielli, C., Joiret, S., & Maurin, G. (2003). Calcium carbonate precursors during scaling process. In: International Water Association, *1st IWA Conf. on Scaling and Corrosion in Water and Wastewater Systems, Cranfield,* March 25–27, 2003. Cranfield: Cranfield University.
Trafford, B. D., Bloomfield, C., Kelso, W. I., & Pruden, G. (1973). Ochre formation in field drains in pyritic soils. *J. Soil Sci.* 24: 453–460.
Truelsen, C. (1958). Bohrbrunnen-Dimensionierung zur Verhinderung ihrer Verockerung und Verkrustung. *GWF Wasser/Abwasser* 99(8): 185–188.
Tuhela, L., Carlson, L., & Tuovinen, O. H. (1992). Ferrihydrite in water wells and bacterial enrichment cultures. *Water Res.* 26(9): 1159–1162.
Tuhela, L., Carlson, L., & Tuovinen, O. H. (1997). Biogeochemical transformations of Fe and Mn in oxic groundwater and well-water environments. *J. Environ. Sci. Health A* 32(2): 407–426.
Tuhela, L., Smith, S. A., & Tuovinen, O. H. (1993). Microbiological analysis of iron-related biofouling in water wells and a flow-cell apparatus for field and laboratory investigations. *Ground Water* 31(6): 982–988.
Tyrell, S. F. & Howsam, P. (1990). Monitoring and prevention of iron biofouling in groundwater abstraction systems. In: Howsam, P. (ed.), *Water Wells—Monitoring, Maintenance, Rehabilitation. Proc. Int. Groundwater Eng. Conf., Cranfield,* pp. 100–106. London: Spon/Chapman.
Uhlmann, W. & Arnold, I. (2003). Iron precipitates in the Lusatian lignite district, Part 2: geochemistry and genesis of incrustation. *Surface Mining* 55(3): 276–287.
UNESCO-WWAP (2003). Water for people, water for life. The United Nations world water development report. Paris: UNESCO Publishing.
U.S. Army Corps of Engineers (1984). Engineering and design—water supply, water sources—mobilization construction, EM 1110-3-161.
U.S. Army Corps of Engineers (1992). Engineering and design—design, construction, and maintenance of relief wells, EM 1110-2-1914.
U.S. Army Corps of Engineers (1998). Engineering and design—monitoring well design, installation, and documentation at hazardous toxic, and radioactive waste sites, EM 1110-1-4000.
U.S. Army Corps of Engineers (2000). Operation and maintenance of extraction and injection wells at HTRW sites, EP1110-1-27.
U.S. Army Corps of Engineers (1998). Water supply, TI814-01.

U.S. Department of Energy (1998). Nonchemical technologies for scale and hardness control, Federal Technology Alert DOE/EE-0162.
U.S. Environmental Protection Agency (1976). Manual of water well construction practices, EPA 570/9-75-001.
Vandevivere, P. & Baveye, P. (1992). Relationship between transport of bacteria and their clogging effiency in sand columns. *Appl. Environ. Microbiol.* 58(8): 2523–2530.
Vecchioli, J. (1970). A note on bacterial growth around a recharge well at Bay Park, Long Island, New York. *Water Resources Res.* 6(5): 1415–1419.
Vukovic, M. & Soro, A. (1992). *Hydraulics of Water Wells—Theory and Application.* Littleton Colorado: Water Resources Publications.
Walter, D. A. (1997). Geochemistry and microbiology of iron-related well-screen encrustation and aquifer biofouling in Suffolk County, Long Island, New York. USGS Water Resources Invest. Rep. 97-4032.
Walter, P. (2001). Kostenbetrachtungen und Kostenanalysen beim Brunnen. *bbr* 52(5): 20–27.
Walton, W. C. (1962). Selected analytical methods for well and aquifer evaluation. *Illinois State Water Surv Bull 49,* published by Illinois State Water Survey, Champaign, IL.
Walton, W. C. (1988). *Groundwater Pumping Tests—Design and Analysis,* 3d ed. Chelsea, Michigan, Lewis.
Warner, J. W., Gates, T. K., Namvargolian, R., Miller, P., & Comes, G. (1994). Sediment and microbial fouling of experimental groundwater recharge trenches. *J. Contaminant Hydrol.* 15(4): 321–344.
Wehrli, B. (1990). Redox reactions of metal ions at mineral surfaces. In: Stumm, W. (ed.), *Aquatic Chemical Kinetics,* pp. 311–336. New York: Wiley.
Wetzel, A. (1969). *Technische Hydrobiologie—Trink-, Brauch-, Abwasser.* Leipzig: Akadem. Verlagsges.
Wheatley, R. E., Vaughan, D., & Ord, B. G. (1984). Amelioration of the ochre problem in field drainage systems using coniferous bark. In: *Proc. 7th Int. Peat Congr.,* 4: 97–105. Dublin:(Irish Natl. Peat Committee.
Wicklein, A., Steußloff, S., and co-authors (2006). Brunnen—ein komplexes System. *Kontakt & Studium 616,* 2d ed. Renningen-Malmsheim: Expert.
Wiegand, G. (1929). Inkrustierung von Brunnen und deren Beseitigung auf chemischem Wege. *GWF Gas- und Wasserfach WasserAbwasser* 72(30): 741–744.
Wilken, R. D. & Bott, W. (2002). Well regeneration by powerful ultrasound. In: Neis, U. (ed.), *Ultrasound in Environmental Engineering II.* TU Hamburg-Harburg Rep. Sanitary Eng. 35, pp. 159–172, Hamburg: TU-HH.
Wilkin, R. T. & Barnes, H. L. (1996). Pyrite formation by reactions of iron monosulfides with dissolved inorganic and organic sulfur species. *Geochim. Cosmochim. Acta* 60(21): 4167–4179.
Williams, E. B. (1981). Fundamental concept of well design. *Ground Water* 19(5): 527–542.
Williams, D. E. (1985). Modern techniques in well design. *J. Am. Water Works Assoc.* 77(9).
Williams, D. E. (2002). Rehabilitation vs. redrill—a difficult question. www.drilleronline.com/CDA/Archives/af6edd4785197010VgnVCM100000f932a8c0__.
Wissel, D., Gerstner, W. (1973). Die Gammabestrahlung der Brunnenfilter—ein wirksames Schutzverfahren gegen Brunnenverockerung. *WWT Wasserwirtschaft-Wassertechnik* 23(6): 191–200.
Wissel, D., Leonhardt, J. W., & Beise, E. (1985). The application of gamma radiation to combat ochre deposition in drilled waterwells. *Radiat. Phys. Chem.* 25: 57–61.
Yokoyama, T., Sato, Y., Nakai, M., & Tarutani, T. (1988). Chemical composition of silica scales deposited from geothermal waters in Kyushu, Japan. In: *Exploration and Development of Geothermal Resources. Int. Symp. on Geothermal Energy,* Kumamoto/Beppu, Japan, November 10–14, 1988, p. 453.
Zeppenfeld, K. (2003). "Selbstreinigung" elektrochemisch. *bbr* 54(11): 24–26.
Zeppenfeld, K. (2005). Untersuchungen über den Einfluss der Strömungsgeschwindigkeit auf die Kalkabscheidung aus calciumhaltigen Wässern. *Vom Wasser* 103(2): 3–34.
Zielinski, R. A., Bloch, S., & Walker, T. R. (1983). The mobility and distribution of heavy metals during the formation of first cycle red beds. *Econ. Geol.* 78: 1574–1589.
Zilch, K., Diederichs, C. J., & Katzenbach, R. (eds.) (2002). *Handbuch für Bauingenieure.* Berlin: Springer-Verlag.

Appendix A

Conversion of Units

TABLE A.1 Units of Time

Unit	Seconds	Minutes	Hours	Days	Years
1 second	1	0.0167	2.77×10^{-4}	1.157×10^{-5}	3.171×10^{-8}
1 minute	60	1	0.0167	6.944×10^{-4}	1.903×10^{-6}
1 hour	3,600	60	1	0.0417	1.142×10^{-4}
1 day	86,400	1,440	24	1	2.740×10^{-3}
1 year	31,536,000	525,600	8,760	365	1

TABLE A.2 Units of Length

Unit	Millimeters [mm]	Centimeters [cm]	Meters [m]	Kilometers [km]	Inches [in]	Feet [ft]	Yards [yd]	Miles [mi]
1 mm	1	0.1	0.001	1×10^{-6}	0.0397	0.00328	0.001094	6.21×10^{-7}
1 cm	10	1	0.01	0.0001	0.3937	0.0328	0.0194	6.21×10^{-6}
1 m	1,000	100	1	0.001	39.37	3.281	1.094	6.21×10^{-4}
1 km	1,000,000	100,000	1,000	1	39,370	3,281	1,093.6	0.621
Unit	mm	cm	m	km	in	ft	yd	mi
1 in	25.4	2.54	0.0254	2.54×10^{-5}	1	0.0833	0.0278	1.58×10^{-5}
1 ft	304.8	30.48	0.3048	3.05×10^{-4}	12	1	0.333	1.89×10^{-4}
1 yd	914.4	91.44	0.9144	9.14×10^{-4}	36	3	1	5.68×10^{-4}
1 mi	1,609,300	160,930	1,609.3	1.6093	63,360	5,280	1,760	1

TABLE A.3 Units of Area

Unit	Square centimeters [cm^2]	Square meters [m^2]	Square kilometers [km^2]	Hectares [ha]	Square inches [in^2]	Square feet [ft^2]	Square yards [yd^2]	Square miles [mi^2]	Acres [ac]
cm^2	1	0.0001	1×10^{-10}	1×10^{-8}	0.155	1.08×10^{-3}	1.2×10^{-4}	3.86×10^{-11}	2.47×10^{-8}
m^2	10,000	1	1×10^{-6}	1×10^{-4}	1,550	10.76	1.196	3.86×10^{-7}	2.47×10^{-4}
km^2	1×10^{10}	1,000,000	1	100	1.55×10^{9}	1.076×10^{7}	1.196×10^{6}	0.3861	247.1
ha	1×10^{8}	10,000	0.01	1	1.55×10^{7}	1.076×10^{5}	1.196×10^{4}	3.86×10^{-3}	2.471

Unit	cm^2	m^2	km^2	ha	in^2	ft^2	yd^2	mi^2	ac
in^2	6.4516	6.45×10^{-4}	6.45×10^{-10}	6.45×10^{-8}	1	6.94×10^{-3}	7.7×10^{-4}	2.49×10^{-10}	1.574×10^{-7}
ft^2	929.03	0.09290	9.29×10^{-8}	9.29×10^{-6}	144	1	0.111	3.587×10^{-8}	2.30×10^{-5}
yd^2	8,361	0.8361	8.36×10^{-7}	8.36×10^{-5}	1,296	9	1	3.23×10^{-7}	8.36×10^{-4}
mi^2	2.59×10^{10}	2.59×10^{6}	2.59	259	4.01×10^{9}	2.79×10^{7}	3.098×10^{6}	1	640
ac	4.04×10^{7}	4,047	4.047×10^{-3}	0.4047	6.27×10^{6}	43,560	4,840	1.562×10^{-3}	1

TABLE A.4 Units of Volume

Unit	Milliliters [cm³]	Liters [liter] or [l]	Cubic meters [m³]	Cubic inches [in³]	Cubic feet [ft³]	U.S. gallons [gal]	Acre-feet [ac-ft]
1 cm³	1	0.001	1×10^{-6}	0.06102	3.53×10^{-5}	2.64×10^{-4}	8.1×10^{-10}
1 liter	1,000	1	0.001	61.02	0.0353	0.264	8.1×10^{-7}
1 m³	1,000,000	1,000	1	61023	35.31	264.17	8.1×10^{-4}

Unit	cm³	Liter	m³	in³	ft³	gal	ac-ft
1 in³	16.39	0.01639	1.64×10^{-5}	1	5.79×10^{-4}	4.33×10^{-3}	1.218×10^{-8}
1 ft³	28,317	28.317	0.02832	1,728	1	7.48	2.296×10^{-5}
1 gal	3,785.4	3.785	3.78×10^{-3}	231	0.134	1	3.069×10^{-6}
1 ac-ft	1.233×10^9	1.233×10^6	1,233.5	7.527×10^7	43.560	3.26×10^5	1

TABLE A.5 Units of Flow

Unit	l/s	m³/s	m³/h	m³/d	ft³/s	ft³/d	ac-ft/d	gal/min	gal/d
1 l/s	1	0.001	3.6	86.4	0.0353	3,051.2	0.070	15.85	22,824
1 m³/s	1,000	1	3,600	86,400	35.31	3,050,784	70.05	15,850	2.282×10^7
1 m³/h	0.2778	2.778×10^{-4}	1	24	9.808×10^{-3}	847.44	0.0195	4.403	6,340.8
1 m³/d	0.0116	1.157×10^{-5}	0.0417	1	4.087×10^{-4}	35.31	8.108×10^{-4}	0.1835	264.24

Unit	l/s	m³/s	m³/h	m³/d	ft³/s	ft³/d	ac-ft/d	gal/min	gal/d
ft³/s	28.32	0.0283	101.95	2,446.8	1	86,400	1.984	448.8	646,272
ft³/d	3.28×10^{-4}	3.28×10^{-7}	1.181×10^{-3}	0.02832	1.16×10^{-5}	1	2.3×10^{-5}	5.19×10^{-3}	7.474
ac-ft/d	14.276	0.0143	51.39	1,233.4	0.5042	43,560	1	226.28	325,843.2
gal/min	0.0631	6.31×10^{-5}	0.227	5.452	2.23×10^{-3}	192.6	4.42×10^{-3}	1	1,440
gal/d	4.382	4.38×10^{-8}	1.578×10^{-4}	3.786×10^{-3}	1.55×10^{-6}	0.11337	3.07×10^{-6}	6.94×10^{-4}	1

Appendix A

TABLE A.6 Units of Velocity

Unit	cm/s	m/s	m/d	ft/s	ft/d
1 cm/s	1	0.01	864	0.0328	2,834.6
1 m/s	100	1	86,400	3.281	283,464.6
1 m/d	1.157×10^{-3}	1.157×10^{-5}	1	3.797×10^{-5}	3.281
1 ft/s	30.48	0.3048	26,334.7	1	86,400
1 ft/d	3.528×10^{-4}	3.528×10^{-6}	0.3048	1.157×10^{-5}	1

TABLE A.7 Units of Temperature

To convert temperature in degrees Fahrenheit T_F (°F) to temperature in degrees Celsius T_C (°C):

$$T_C = (T_F - 32)/1.8$$

To convert temperature in degrees Celsius T_C (°C) to temperature in degrees Fahrenheit T_F (°F):

$$T_F = (1.8 \times T_C) + 32$$

To convert Temperature in Kelvin units T_K (K) to temperature in degrees Celsius T_C (°C):

$$T_C = T_K - 273.15$$

TABLE A.8 Units of Mass and Weight

1 kg	=	2.204622621849 lb	=	35.27396194958 oz
1 lb	=	0.45359237 kg	=	16 oz
1 oz	=	0.028349523125 kg	=	0.0625 lb

(kg = kilogram, lb = pound, oz = ounce)
(1 g = 1,000 mg; 1 kg = 1,000 g; 1 ton = 1,000 kg)

TABLE A.9 Units of Force

1 N	=	0.2248 lb	=	10^5 dynes
1 lb	=	4.448 N		

(N = newton = 1 kg × m/s²; lb = pound)

TABLE A.10 Units of Work and Energy

1 J	=	0.738 ft-lb	=	0.239 cal
1 cal	=	4.186 J		
1 ft-lb	=	1.356 J		
1 kWh	=	3.60×10^6 J		

(J = joule, ft = foot, cal = calorie, W = watt, h = hour)

Conversion of Units

TABLE A.11 Units of Power

1 W	=	1 J/s	=	0.738 ft-lb/s
1 kW	=	1.341022089595 hp		
1 hp	=	0.7456998715823 kW	=	550 ft-lb/s

(hp = horsepower, s = second)

TABLE A.12 Units of Pressure

1 Pa	=	1 N/m^2	=	1.45×10^{-4} lb/in^2 (psi)
100 kPa	=	1 bar		
1 lb/in^2	=	6.895×10^3 N/m^2		
1 atm	=	1.013×10^5 N/m^2	=	14.70 lb/in^2

(Pa = pascal, in = inch, lb/in^2 = psi)

TABLE A.13 Exponential Factors

Multiplication factor	Prefix	Symbol	Example
1,000,000,000 = 10^9	giga	G	gigawatt, GW
1,000,000 = 10^6	mega	M	megawatt, MW
100,000 = 10^5			
10,000 = 10^4			
1,000 = 10^3	kilo	k	kilogram, kg
100 = 10^2	hecto	h	hectopascal, hPa
10 = 10^1	deca	d	decimeter, dm
1 = 10^0			
0.1 = 10^{-1}			
0.01 = 10^{-2}	centi	c	centimeter, cm
0.001 = 10^{-3}	milli	m	milliliter, ml
0.0001 = 10^{-4}			
0.00001 = 10^{-5}			
0.000001 = 10^{-6}	micro	μ	microgram, µg
0.000000001 = 10^{-9}	nano	n	nanometer, nm

Appendix B

How to Read a Box-Whisker Plot

Box-whisker plots are a good way to display and assess larger data sets in one diagram, e.g., the spread of concentrations of a chemical species in a catchment.

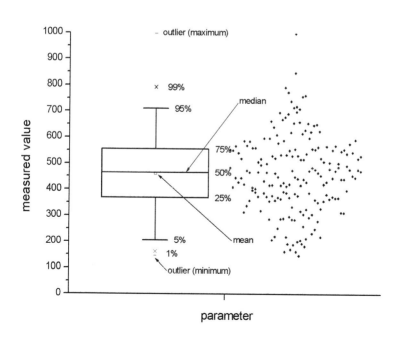

The scatter of data points on the right-hand side of the figure shows such a data set. For the sake of simplicity we will consider a smaller data set as an example.

Example

Data set ($n = 7$): 8, 3, 1, 10, 152, 5, 13

The *arithmetic mean* is simply the sum of all values divided by the number of data.

Arithmetic mean

$$(8 + 3 + 1 + 10 + 152 + 5 + 13)/7 = 192/7 = 27.43$$

From the example we can see at once that the arithmetic mean does not represent the data set very well, because it is heavily influenced by the single large value (152). Therefore we introduce the *median*. The median of a data set separates the data into two equal parts, so we have to sort the data according to value. The median is more representative of the data set when outlier values influence the data set.

Median

Sort according to value:

$$1, 3, 5, 8, 10, 13, 152$$

Find the value "in the middle":

$$\text{Median} = 8$$

Data can be separated further into *quartiles*. The first quartile is the median of the lower half of the data set. The second quartile is the median of the entire data set. The third quartile is the median of the upper half of the data. Quartiles separate the original set of data into four equal parts. Each of these contains one-fourth of the data.

The quartiles are now used to find the limits of the box of the box-whisker plot. The first quartile is the lower boundary and the third quartile is the upper boundary. The median is also shown and separates the box into two halves. The values below and above the first and third quartiles are shown as lines attached to the box (whiskers). Sometimes some data fall well outside the range of the other values. Such data are called *outliers* and can be therefore displayed outside the whiskers. The arithmetic mean can also be displayed (here: as square symbol). In the figure shown here, it falls pretty close to the median because we have a fairly uniform distribution of data.

Box-whisker plots are useful to visualize the spread of values and some statistical base information of larger data sets in one diagram. Software to generate such plots is widely available. Many of these programs can import and process data directly from common spreadsheet calculation software.

Appendix C

Example of a Tender Document for Well Rehabilitation

Specifications of the well to be rehabilitated (to be filled out by assignor of project prior to request for quotation, as a basis for subsequent calculation):

- Type of well
- Well diameter
- Borehole diameter
- Depth of well in total
- Casing material
- Screen material
- Slot
- Position of screen size
- Length of screen
- Present static water level
- Year of construction
- Operating output

Title 1: Preparative Measures

Item No. 1.1: Mobilisation, demobilisation, construction of site

General installation of the construction site with respect to all requirements to be addressed in implementing and completing the service rendered. Mobilisation and demobilisation of all equipment.

To be included: provision for water and electricity supply, as well as safeguard of the construction site with respect to all aspects to be considered. A detailed plan of the construction site is to be submitted to the assignor of the project for approval before commencement of work.

Construction documentation in the form of a rehabilitation record with daily reports and protocols.

Clearance of construction site after service rendered, including refurbishment of all areas made use of during implementation of service rendered.

Quantity: 1 Unit   Unit Rate:..........   Amount:..........

Subtotal Title 1: Preparative measures   Amount:..........

Title 2: Preliminary Measures

Item No. 2.1: Erection and dismantling of mechanical rehabilitation devices

Erection and dismantling of all mechanical rehabilitation devices to be implemented, including possible on-site transport and all secondary works.

Quantity: 1 Unit   Unit Rate:..........   Amount:..........

Item No. 2.2: Removal and reinstallation of submersible pump, riser pipe, and fittings

Removal and reinstallation after service rendered of existing submersible pump, including riser pipe and fittings as well as adequate storing during rehabilitation period.

Quantity: 1 Unit   Unit Rate:..........   Amount:..........

Item No. 2.3: Installation and dismantling of provisional discharge pipe and settling pond

Installation and dismantling of a provisional discharge pipe with the required cross section (lightweight tubes or the like) as well as a sufficiently dimensioned settling pond for leading the residue water produced during rehabilitation. A detailed plan of the course of the discharge pipe and size of the settling pond is to be submitted to the assignor of the project for approval before commencement of work. Installation of at least one water quantity measuring device on the provisional discharge pipe.

Quantity: 1 Unit   Unit Rate:..........   Amount:..........

Subtotal Title 2: Preliminary measures   Amount:..........

Title 3: Preliminary Examinations

Item No. 3.1: Camera inspection with axial and radial perspective

Camera inspection with an axially and radially focused TV-camera, measured from the upper edge of the well pipe to the depth of well in total. To be included in the service rendered is a detailed documentation of the camera inspection on videotape as well as color screen shots (format: 9.0 × 13.0 cm) or, alternatively, a CD with digital photos.

Quantity: 1 Unit   Unit Rate:..........   Amount:..........

Item No. 3.2: Assembly and disassembly of pumping-test facility

Assembly and disassembly of a complete pumping-test facility for a pumping capacity up to ..... cbm/h. The pumping-test facility is to consist of a submersible pump as well as the required riser pipes or tubes in appropriate lengths. The disposal of the test water takes place by means of the provisional discharge pipe (Item No. 2.3).

Quantity: 1 Unit   Unit Rate:..........   Amount:..........

Item No. 3.3: Execution of pumping test

Execution of pumping test for verification of the well yield.

To be included in the service rendered is a detailed documentation of the volume flow and the water level at .....-minute intervals during the entire pumping period as well as a test protocol (measuring the water level in the well and all adjacent piezometers as well as regular documentation of the volume flow), including tabular and graphical presentation of the test results.

Quantity: 1 hour/day   Unit Rate:..........   Amount:..........

Subtotal Title 3: Preliminary examinations   Amount:..........

Title 4: Mechanical Cleansing

Item No. 4.1: Cleansing of the riser pipe and the submersible pump

Cleansing of the riser pipe (inside and outside) as well as of the submersible pump.

Quantity: 1 Unit   Unit Rate:..........   Amount:..........

Item No. 4.2: Cleansing of interior wall of the well pipe (brushing)

Cleansing of the interior wall of the well pipe, by means of a specially adapted well brush, as, for example, a plastic brush with different bristle strengths with lengths corresponding to the diameter of the well pipe. Both casing and screen are to be cleaned in sections; at the same time, the removed particles are to be pumped out and removed from the well via the provisional discharge pipe.

Quantity: 1 Unit   Unit Rate:..........   Amount:..........

Item No. 4.3: Installation and dismantling of a provisional submersible pump

Installation and dismantling of an appropriate provisional pump with adequately dimensioned riser pipes in the well and connection to the provisional discharge pipe described in Item No. 2.3.

Quantity: 1 Unit   Unit Rate:..........   Amount:..........

Item No. 4.4: Intensive cleansing of the well sump

Intensive cleansing of the well sump, until all sediment removed during the mechanical cleansing stage are completely removed from the well. A final sediment net content of less than 0.1 g/liter water is to be attained.

Quantity: 1 hour/day   Unit Rate:..........   Amount:..........

Subtotal Title 4: Mechanical cleansing   Amount:..........

Title 5: Hydro-Mechanical-Chemical Rehabilitation

Item No. 5.1: Installation and dismantling of the well rehabilitation facility

Installation and dismantling of a well rehabilitation facility for mechanical well rehabilitation, consisting of a multichamber device with a circulation power of ..... cbm/h between packers (packer distance at least 1.0 m), as well as of a pump with a pumping rate of ..... cbm/h.

Quantity: 1 Unit   Unit Rate:..........   Amount:..........

Item No. 5.2: Intensive cleansing of the screen

Intensive cleansing of the well screen in sections, thereto operating of the rehabilitation facility from one working section to another. The length of a working section is equal to the sleeve distance, whereby after finishing one working section, the facility is moved by 0.5 m.

The working sections are to be cleansed until water output is clear and sand-free.

Quantity: 1 hour/day Unit Rate:..........   Amount:..........

Item No. 5.3: Supplying and mixing the rehabilitation agent

Supply of the rehabilitation agent: .... (Manufacturer: ....) including appropriate storage. Measuring instruments for determining the conductivity and the reaction products in the treated water are to be kept at hand. Clear water is to be mixed with the rehabilitation agent for each section in an appropriate receptacle, according to the specific instructions of the manufacturer.

Quantity: 1 Unit   Unit Rate:..........   Amount:..........

Item No. 5.4: Applying the rehabilitation agent

Application of the rehabilitation agent in sections, followed by enforced circulation of the rehabilitation agent in the working section.

Concentration: ..... g/liter, depending on the volume of the borehole on the length of the working section (..... kg per section)

Net length of section to be treated: ..... m

Period of exposure of the rehabilitation agent per working section: ..... minutes

Quantity: 1 Unit   Unit Rate:..........   Amount:..........

Item No. 5.5: Discharge of the treated water

After the period of exposure of the rehabilitation agent per working section has elapsed, the rehabilitation agent is to be removed immediately from the working section until the original chemical condition in the working section has been reached again. This is to be determined by means of measuring and documentation of the specific electrical conductivity and/or the pH-value (depending on the rehabilitation agent used) as well as of the concentration of the reaction products in the discharged water.

Quantity: 1 Unit   Unit Rate:..........   Amount:..........

Item No. 5.6: Treatment chemicals

Supplying and storing of all chemicals required for the neutralization on the construction site, including all necessary secondary work.

Quantity: 1 Unit   Unit Rate:..........   Amount:..........

Item No. 5.7: Treatment of waste water

Neutralization of the waste water (if acids have been employed) or discharging the precipitation of the dissolved incrustation components (if a reducing agent has been employed for the rehabilitation), including all necessary secondary work.

Quantity: 1 Unit   Unit Rate:..........   Amount:..........

Item No. 5.8: Disposal

Collection and disposal of the treated water via the provisional discharge pipe and settling pond to an adequate area for discharge or infiltration (e.g., sewer, etc.). This includes control of the sediment content in the treated water (the content is not to exceed 1 g/liter), before final discharge.

Quantity: 1 Unit   Unit Rate:..........   Amount:..........
Subtotal Title 5:
Hydro-mechanical-chemical rehabilitation   Amount:..........

Title 6: Desanding and Final Cleansing of the Well

Item No. 6.1: Installation and dismantling of the desanding facility

Installation and dismantling of a complete desanding facility for desanding performance of up to ..... cbm/h per meter desanding section along the screen pipe. The desanding facility is to be equipped with a submersible pump between packers (to be provided by the contractor). The length of a desanding section is equal to the distance between the lower and upper packers (1.0 m), whereby after finishing one desanding section, the equipment is lifted by 0.5 m.

Quantity: 1 Unit   Unit Rate:..........   Amount:..........

Item No. 6.2: Execution of the desanding

Execution of the desanding, with submersible pump between packers, for intensive desanding (packer distance 1.0 m, overlap 0.5 m) to reach the technical status "sand-free." To be included in the service rendered is a detailed documentation of the volume flow and the water level at .....-minute intervals during the entire pumping period as well as a test protocol (measuring the water level in the well pipe and all adjacent piezometric pipe as well as regular documentation of the volume flow), including tabular and graphical presentment of the test results. A protocol with an arithmetic chart showing the desanding results is to be made available.

Control and documentation of the chemical parameters according to Item No. 5.5.

Quantity: 1 hour/day   Unit Rate:..........   Amount:..........

Item No. 6.3: Cleaning the well sump

Installation and removal of an appropriate pump for removal of all sediment from the well's sump.

Water discharge to be arranged via the provisional discharge pipe and settling pond. This includes control of the sediment content in the treated water (the content is not to exceed 1 g/liter), before final discharge.

Quantity: 1 Unit   Unit Rate:..........   Amount:..........

Item No. 6.4: Final cleansing of the well (removal of fines)

Installation and removal of an appropriate pump, pumping the well free of fines. Water discharge to be arranged via provisional discharge pipe and settling pond. The residue sediment content in the cleansed water is not to exceed 0.1 g/liter.

Quantity: 1 Unit   Unit Rate:..........   Amount:..........

Subtotal Title 6:
Desanding and final cleansing of the well   Amount:..........

Title 7: Follow-Up Investigations

Item No. 7.1: Assembly and disassembly of a pump test facility

Assembly and disassembly of a complete pumping test facility for a pumping capacity up to ..... cbm/h. The pumping test facility is to consist of a submersible pump as well as the required riser pipes or tubes in appropriate lengths. The disposal of the test water takes place by means of the provisional discharge pipe (Item No. 2.3).

Quantity: 1 Unit   Unit Rate:..........   Amount:..........

Item No. 7.2: Pumping test

Execution of pumping test for quantification of well yield.

To be included in the service rendered is a detailed documentation of the volume flow and the water level at .....-minute intervals during the entire pumping period as well as a test protocol (measuring the water level in the well and all adjacent piezometers as well as regular documentation of the volume flow), including tabular and graphical presentation of the test results.

Quantity: 1 hour/day Unit Rate:..........   Amount:..........

Item No. 7.3: Camera inspection with axial and radial perspective

Camera inspection with an axially and radially focused TV-camera, measured from the upper edge of the well casing to the depth of well in total. To be included in the service rendered is a detailed documentation of the camera inspection on videotape as well as color screen shots (format: 9.0 × 13.0 cm) or, alternatively, a CD with digital photos.

Quantity: 1 Unit   Unit Rate:..........   Amount:..........

Item No. 7.4: Disinfection of the well

Supply of the disinfection agent: .... (Manufacturer: ....) including delivery and processing of the disinfectant, as well as all required secondary work. Official certification of the well for drinking-water purposes will be granted only if the well's sterility can be verified, according to the valid regulations on drinking water.

Quantity: 1 Unit   Unit Rate:..........   Amount:..........

Item No. 7.5: Documentation of the well rehabilitation

Drawing up and delivery of all documentation specified in this tender. These include the specification of the treated sections, discharge performance and sediment contents, pumping periods, and results of the control measuring during rehabilitation.

Quantity: 1 Unit   Unit Rate:..........   Amount:..........

Subtotal Title 7: Follow-up investigations   Amount in total:..........

Summary

Title 1: Preparative measures	Amount in total:..........
Title 2: Preliminary measures	Amount in total:..........
Title 3: Preliminary examinations	Amount in total:..........
Title 4: Mechanical cleansing	Amount in total:..........
Title 5: Hydro-mechanical-chemical rehabilitation	Amount in total:..........
Title 6: Desanding and final cleansing of the well	Amount in total:..........
Title 7: Follow-up investigations	Amount in total:..........

Net amount: ....
VAT/tax: ....
Gross amount: ....

Stamp   Date   Signature

Index

Abrasion, 132–134
Absorption
  of gamma rays, 170
  in impulse methods, 228, 231, 239
Accreditation, 15
Accumulation factors in one-off investments, 190
Acid mine drainage (AMD), 106
Acidification, 329
Acids in dissolution process, 247–254
  aluminum hydroxide, 98, 255
  carbonate, 255
  chemical combinations, 260
  metal sulfide, 256
Additives in drilling fluids, 260–261
Administrative laws, 10–11
ADR (Accord europeen sur le transport des marchandises dangereuses par route), 267
Aerobic heterotrophs, 102
Air entrapment, 146
Air guns, 234–235
All-inclusive tenders, 304
Aluminum hydroxides, 98–99, 255–256
Ammonia, 106
Ammonium nitrate, 232
Amortization period, 198
Amount of annuity, 191
Amphoteric substances, 252
Anaerobic corrosion, 57
Annuity factor, 191
Annuity method, 192–195
Annulus and annular seals and materials, 59
  assessing, 166
  in complete reconstruction, 290–292, 295
  core drilling into, 173
  in decommissioning, 297–298, 301–302
  settling from, 141–142

Anodic polarization, 330
Applin, K. R., 69
Approvals for rehabilitation, 305–306
Aquifers
  head loss in, 158
  protection of, 278–279
  sand intake in, 138
  yield of, 27–30
Aquitards in decommissioning, 297–300
Aragonite, 96
Area unit conversions, 359
Aristotle, 1, 101
Arithmetic mean, 366
Arsenic, 89
Arsenic hydride, 249
Ascorbic acid, 253–254
ASTM standards, 12
Attenuation in impulse methods, 228–231
Australia
  disinfection schemes in, 324–325
  incrustation samples in, 7, 98
  laws in, 11
  relining techniques in, 289
Auto-catalysis, 70–72, 314

Baboian, R., 55
Back-reaming borehole walls, 332
Backfilling in decommissioning, 295–297
Bacteria, 323
  acid dissolution for, 252
  in bioclogging, 102
  in infiltration wells, 103–104, 146
  iron-related, 76, 81
  killing. *See* Microbicidal techniques
  from rehabilitation, 316–319
  SRB. *See* Sulfate-reducing bacteria (SRB)
Balke, K.-D., 39
Banwart, S. A., 70
Barbic, F., 252

379

Barckefeldt, Mr., 1
Barite, 104
Barium, 90–91
Bases
    in aluminum hydroxide dissolution, 255
    neutralization of, 274
Batu, V., 155
Baudisch, R., 98
Belgium
    pasteurization technique in, 326
    rehabilitation results in, 240
Bentonites
    in drilling fluids, 261
    enrichment of, 166–167
    removal of, 258–259
Berger, H., 143–144, 251
Berlin
    disinfection schemes in, 324
    relining in, 286
BET technique, 174
Bichara, A. F., 148–149
Bioclogging
    causes of, 102–104
    in infiltration wells, 146–148
Biofilms, 103
Biofouling, 146
Biomass oxidation and dissolution, 256–257
Blocking process in clogging, 125
Boreholes
    geophysics, 163–171
    open borehole wells, 150–151
    in partial reconstruction, 289
    scraping, 332
    in well sizing, 38
Botswana, disinfection schemes in, 325
Bott, W., 238
Bottcher, Heinrich, 2
Box-whisker plots, 365–367
Breakdown-oriented maintenance, 42
Bridging
    in clogging, 128–129
    in mechanical filtration, 119–120
Brown, 2, 76
Brownian motion in filtration, 121–122
Brushing, 206–209
Buik, N. A., 148
Bunter Sandstone, 138
Burke, S. P., 70
Burning wells, 309

Cable tool drilling, 332
Cake filtration, 121

Calcite, 96, 101
Calcite crystals, 97
Calcium carbonate
    in carbonate incrustations, 97
    reaction inhibitors for, 328
Calcium in saturation index, 181
Calculations
    mass balancing, 73–74
    saturation index, 179–182
    unit conversions in, 357–363
Camera inspections, 153–155, 279
Capillary flow, 51–53
Capital value, 191, 193–197
Capital recovery value, 191
Carbon dioxide
    from biomass oxidation, 257
    in carbonate incrustations, 94, 255
    in neutralization process, 273
    sources and dangers of, 308–309
Carbon dioxide freezing, 222–226
Carbon in bioclogging, 102
Carbon monoxide, 308
Carbonate equilibrium, 94
Carbonate precipitation, 331
Carbonates, 94–97, 173
    dissolution of, 251, 254–255
    sludges, 275
Casings
    in complete reconstruction, 291–293
    in partial reconstruction, 284
Cathodic polarization, 330
Cavitation, enforced jet, 240–241
Celestite, 105
Cement Bond Log (CBL), 166
Cesium-137, 326
Chambered systems, 213–216
Chemical ageing processes
    aluminum hydroxide incrustations, 98–99
    bioclogging, 102–104
    carbonate incrustations, 94–97
    corrosion, 55–59
    metal sulfide incrustations, 99–101
    ochre incrustations. See Ochre incrustations
    passively incorporated components, 106–107
    spatial and temporal distribution of incrustations, 107–115
    special incrustation types, 104–106
Chemical prevention methods
    acidification, 329
    disinfection, 323–325

Chemical prevention methods (*Cont.*):
  electrochemical, 329–331
  magnetic, 331
  reaction inhibitors, 328
Chemical rehabilitation techniques
  additives in, 260–261
  applications for, 305
  combinations of chemicals, 260
  comparison of, 261–264
  in field, 264–266
  legal constraints, 245
  liquid and solid waste disposal, 271–275
  need for, 270–271
  reaction mechanisms. *See* Reaction mechanisms
  safety precautions, 266–270
Chlorinators, drop pellet, 325
Cleaning, defined, 9
Clogging
  annulus core drilling for, 173
  bioclogging, 102–104
  filter pack, 168
  from filtration, 125–132
  in infiltration wells, 146–148
Closed-circuit pumping, 213–216
CMC remnant removal, 258
Cobalt, 91–92, 326
Coliform bacteria, 316–317
Collapse after rehabilitation, 320
Collars in partial reconstruction, 286–287
Collins, R. E.
  column experiments by, 131
  on critical flow velocity, 119
  filtration study by, 123
Collisions in filtration, 121–122
Colloids, 117–118
Colmation. *See* Suffossion and colmation
Combinations of chemicals, 260
Company selection in rehabilitation, 304–305
Complete reconstruction, 289–295
Complexation, 252
Compressed fluids, 234–235
Connection assessment, 165, 167
Consolidated aquifers, 137–138
Consolidated rocks
  in open borehole wells, 150
  vertically drilled wells in, 19
Constant annuities, 191
Construction costs, 198–199
Construction Industry Research and Information Association (CIRIA), 14

Constructive wave interactions, 231
Constructive well inspection, 48
Contact corrosion, 56
Container disposal, 272
Conversion of units, 357–363
Copper, 91–92
Core drilling, 173
Cornell, R. M., 262
Corrosion
  from acid dissolution, 251
  as ageing cause, 5
  electrochemical, 55–57
  inhibitors for, 260
  microbially induced, 57–58
  nonmetallic material, 58–59
  protection schemes for, 329
  terms and definitions, 55
Costs
  construction, 198–199
  contractor, 304
Crystalline solid powders, 250
Crystallinity as dissolution factor, 264
Crystals
  goethite, 93
  piezoelectric, 235–236
Cubic dependency of permeability on porosity, 53
Cubic law, 52
Cubic packing, 122
Cullimore, D. R., 79

Damage-oriented maintenance, 42
Darcy velocity, 23–24
Darcy's law, 25
Data sheets, 45
De Ridder, N. A., 155
Decommissioning wells
  defined, 10
  overview, 277–283
  process, 295–302
Decreased yield from rehabilitation, 143, 206
Deep wells
  assessing, 165
  earthquake risk to, 144
  gravel envelopes in, 40
  history, 2
  injection schemes for, 290
  leaky seals in, 289
  ultrasound for, 236
Definitions, 9–10
Density-dependent wave reflection, 238

Deposition rate measurements for incrustation, 172
Desanding and final cleansing in tender documents, 374–375
Destructive wave interactions, 231
Desulfovibrio bacteria, 57
Dilution in dissolution process, 250
Dimensioning of wells, 37–40
Dimensionless Reynolds number, 25
DIN standards, 12
Directionally drilled wells, 22–23
Directive 91/155/EWG, 268
Discounting method, 193, 195–197
Disinfection
  chemical, 323–325
  for microbial pollution, 318
  thermal, 325–326
Disposal of waste, 271–275, 311–312
Dissolution, 246–247
  of aluminum hydroxides, 255–256
  of biomass, 256–257
  of carbonates, 254–255
  experiments for, 262–264
  of iron and manganese oxides, 247–254
  of metal sulfides, 256
Dissolved ionic concentrations, 176
Distribution coefficient in oxidation, 68
Dithionite, 263–264
Dolomite, 96–97
Do's and Don't's of rehabilitation, 341–342
Drag in suffossion, 119
Drain wells, 19
Drawdown
  in incrustations, 113
  in step-discharge tests, 156, 158–161
Dresden, rehabilitation tests in, 241
Drilling fluid remnant removal, 258–259, 261
Drilling mud, 125–128
Drilling techniques in prevention planning, 331–334
Driscoll, F. G.
  on bacteria living conditions, 79
  on corrosion, 55
  on polyphosphate dissolution, 259
  on systematic data collection, 44–45
  wire-to-water ratio by, 161
  on yield loss, 184
Drop pellet chlorinators, 325
Duderstadt, 1
Dug wells, 18–19
Dupuit, J., 27

Dutch Water Research Institute KIWA
  cost figures from, 202
  survey by, 7
DVGW
  definitions by, 9–10
  regulations by, 14
  survey by, 4–6, 50
Dynamic loads in soil settling, 140

Earnings value, 195
Earthquakes, 144
East Germany (GDR), ionizing irradiation in, 326
ECE/TRANS/175 regulations, 267
Economics
  annuity method, 192–195
  discounting method, 193, 195–197
  financial mathematical principles, 189–192
  principles and evaluation standards, 187–189
  rehabilitation and reconstruction, 197–199
  rehabilitation and reconstruction market, 199–202
Effervescence, 270, 308
Eigenfrequency, 227
Elbe River, 6
Electrical conductivity in carbon dioxide freezing, 225
Electrochemical corrosion, 55–57
Electrochemical prevention methods, 329–331
Electrolysis in explosive gases, 233
Electromagnetic measurement of material thickness (EMDS), 165
Electrostatic forces in filtration, 121–122
Elimelech, M., 119
Ellis, D., 76
Emulsifiers, 260–261
Energy unit conversions, 362
Enforced jet cavitation, 240–241
Engineer Manuals, 14
Engineer Pamphlets, 14
Enteroccus, 317
Erkelenz, 18
Erosion
  jetting for, 216
  in suffossion, 118–119
Escherichia coli, 317
Ethylenediaminetetraacetic acid (EDTA), 252
Exopolysaccharides, 103

Explosion of gas mixtures, 233–234
Explosive charges, 231–233
Exponential factor unit conversions, 363

Feldspar, 107
Fenton's reagent, 260
Ferrihydrite, 75, 82, 262–263
Field drains, 106
Filter gravel in mass balancing, 74
Filter pack
  clogging, 168
  in infiltration wells, 149
Filtration and filtration theory
  basic, 119–120
  clogging from, 125–132
  for gravel packs, 121–125
  membrane filter index, 147–148, 174
  in suffosion and colmation, 119–125
Financial mathematical principles, 189–192
Flocculation, 258, 324
Flow
  under natural conditions, 30–37
  radial symmetric, 23–27
  unit conversions for, 361–362
Flow meter investigations, 170–171
Flow velocity
  in abrasion, 133
  in capillaries, 51–53
  in carbonate incrustations, 95
  in clogging, 130
  in mechanical filtration, 120
  in suffossion, 118
  in well sizing, 38–39
Flushing devices, 339–340
Focused Electro Log (FEL) screening method, 165–167
Follow-up investigations in tender documents, 376–377
Force unit conversions, 362
Ford, R. G., 92
Forward, P., 324–325
Fountain, J., 217–218
Freezing, carbon dioxide, 222–226
Frequencies
  for compressed fluids, 235
  for explosive gases, 234
  in resonance effects, 227–228
  in ultrasound, 236
Frequency-regulated pumps, 215

Gallionella, 76–77
Gamma-Gamma Logs (GG), 170

Gamma-ray-emitting radionuclides, 326
Gas-dynamic tracer tests (GDT), 168
Gases
  explosion of, 233–234
  hazardous, 307–310
  indicators for, 309
  pressure monitoring, 337–338
Gay-Lussac law, 95
Geochemistry of iron and manganese, 59–67
Geometric considerations in gravel packs, 123
Geophysics, borehole, 163–171
Geothermal plants, 106
German Groundwater Research Centre, 241
Germany
  acid dissolution in, 247, 251
  ageing causes in, 5–7, 9
  decommissioned wells in, 297–299
  disinfection schemes in, 324
  inert gases in, 338
  ionizing irradiation in, 326
  laws in, 11, 14
  magnetic prevention methods in, 331
  mechanical rehabilitation tests in, 241
  patching in, 288
  public water supply in, 17
  rehabilitation and reconstruction market in, 199–202
  rehabilitation timing in, 184
  relining in, 286
  steam injection in, 226
  subsurface iron removal in, 338–339
  trace elements in, 88–90
  ultrasound in, 239
  well history in, 1–3
Geyser rehabilitation method, 226
Glycolic acid, 251
Goethe, Johann Wolfgang von, 75
Goethite, 75, 83–85, 262–263
Goethite crystals, 93
Grain size in well sizing, 39–40
Gravel pack
  filtration theory for, 121–125
  flushing devices, 339–340
  in well sizing, 39–40
Gravel plots, 333
Gravel washers, 213
Gravitational settling, 121–122
Greigite, 58, 100, 256
Ground movement
  earthquakes, 144
  soil settling and subsidence, 140–144

Gruesbeck, C.
    column experiments by, 131
    on critical flow velocity, 119
    filtration study by, 123
Guidelines for rehabilitation, 341–342
Gypsum, 104

Hagen, G., 51
Hagen-Poiseuile concept, 52
Hand-held gas indicators, 309
Hanert, H., 79
Hardness contrast in abrasion, 133
Hasselbarth, U., 79
Hazardous gases, 307–310
Hazardous Materials Regulations, 267
Head losses
    in aquifers, 158
    in infiltration wells, 147–148
Heating rehabilitation methods, 226–227
Heavy metal trace elements, 86–93
Helweg, O. J., 192
Hematite, 75
Henry-Dalton law, 95
Henry's law, 68
Herberger, Sepp, 320
Heterotrophic bacteria, 102
High pressure techniques, 204, 217–220
Hinze, G., 325
History of well ageing, 1–3
Hjulstrom, F., 119
Hollandite, 88
Horizontal directionally drilled (HDD) wells, 22
Horizontal distribution of iron oxide, 108
Horizontal flow under natural conditions, 32
Horizontal wells, 17, 19–22
Hot water injection, 261
Houben, G.
    deep-cold liquid experiments by, 226
    dissolution experiments by, 262–264
    flow models by, 32
    incrustation distribution samples by, 108–109
    mass balancing equations by, 73
House of cards structure, 258–259
Howsam, P., 172, 217–218
Huisman, L.
    bacteria calculations by, 103–104
    head loss calculations by, 147
    and intense pumping, 212
Hydraulic head calculations, 25
Hydraulic oil spillage, 306

Hydraulic rehabilitation methods
    closed-circuit pumping, 213–216
    intense pumping, 210–213
    jetting, 216–221
    jetting spears, 221–222
    suffossion, 209
    surge blocks, 209–210
Hydraulics
    aquifer yield and intake capacity of wells, 27–30
    influence of ageing on, 51–53
    under natural conditions, 30–37
    radial symmetric flow to wells, 23–27
Hydro-mechanical-chemical rehabilitation in tender documents, 372–374
Hydrochemical ageing indicators, 49
Hydrochloric acid
    in chemical combinations, 260
    in dissolution process, 248, 250
Hydrogen in dissolution process, 249
Hydrogen peroxide
    in chemical combinations, 260
    as disinfectant, 324
    for oxidation, 257
Hydrogen sulfide gas
    danger from, 309
    in metal sulfide dissolution, 256
Hydrogen sulfides, 58, 99–100
Hydromechanical rehabilitation techniques, 213
Hydrous ferric oxide, 75

Imhoff cones
    in closed-circuit pumping, 214
    in intense pumping, 213
Immediate one-off investments, 190
Impulse rehabilitation methods
    background, 227–231
    compressed fluids, 234–235
    enforced jet cavitation, 240–241
    explosion of gas mixtures, 233–234
    explosive charges, 231–233
    ultrasound, 235–240
In phase waves, 231
India, laws in, 11
Inert gases, 337–338
Infiltration wells
    ageing processes in, 145–150
    bacteria in, 103–104, 146
Infrared spectroscopy (IR), 174
Initial settling, 141

Injection of hot water and steam, 226–227, 261
Inline casings, 284
Inner colmation
 in clogging, 125
 for gravel packs, 123–124
Inorganic acids, neutralization of, 273
Inorganic reductants, 253
Inspection components, 40–42
Insufficient yield gain after rehabilitation, 143, 206, 321
Intake capacity of wells, 27–30
Intense pumping, 210–213
Internal laminations, 93
International standards, 12–14
Investigation period in economics evaluation, 187–188
Ion activity product (IAP), 180
Ionic chemicals, 261
Ionic concentrations in performance assessment, 176
Ionizing irradiation, 326–328
Iron
 in carbon dioxide freezing, 225
 in incrustations. See Ochre incrustations
 in redox zones, 60–63
 subsurface removal of, 338–339
Iron bacteria, acid dissolution for, 252
Iron-related bacteria (IRB), 76, 81
Irradiation, ionizing, 326–328

Jacob, 158
Jawrowsky, Igor, 223
Jetting, 216–221
Jetting spears, 221–222
Johnson screens, 333
Juttner, R., 225

Kabul and Kabul basin
 saturation index calculations for, 181–182
 well pollution in, 19
KIWA
 cost figures from, 202
 survey by, 7
Kozeny-Carman equation, 52–53
Krefeld, 107
Krusemann, G. P., 155
Kutzing, 76

Laminar aquifer loss, 156
Landfills, 106

Legal issues
 background, 10–15
 in chemical rehabilitation, 245
Length unit conversions, 358
Lepidocrocite, 75, 82, 262–263
Leptothrix, 79–80
Ligand-based dissolution, 252
Ligands, 260
Limited quantity chemical group, 267
Liquid nitrogen, 226
Liquid waste disposal, 271–275
Location-specific influences in economics evaluation, 189
Lorentz force, 331
Low-pressure jetting, 217–218
Ludemann, D., 79

Macadam, J., 328
Mackinawite, 58, 100
Magnesium carbonate, 181
Magnetic prevention methods, 331
Maintenance
 components of, 41–44
 defined, 40
Manganese
 in carbon dioxide freezing, 225
 in incrustations. See Ochre incrustations
 in redox zones, 60–63
Marcasite, 101, 256
Market for rehabilitation and reconstruction, 199–202
Martinez, C. E., 92
Mass
 in abrasion, 133
 unit conversions for, 362
Mass balancing
 of ochre buildup, 73–74
 of removed material, 176–179
Materials in prevention planning, 331–334
Maximum oscillation velocity, 242
Maximum slope in hydraulics, 29
McBride, M. B., 92
McCann, A. E., 79
McDowell-Boyer, L. M., 119
McLaughlan, R. G.
 on aluminum hydroxide incrustation, 98
 on corrosion, 55
 disinfection schemes by, 324–325
 incrustation sample study by, 7
 yield loss threshold by, 184

Mean, 366
Mechanical ageing causes
  abrasion, 132–134
  ground movement, 140–144
  infiltration wells, 145–150
  open borehole wells, 150–151
  plant roots, 139–140
  sand intake, 134–139
  suffossion and colmation. See Suffossion and colmation
  vandalism, 144–145
Mechanical cleansing in tender documents, 372
Mechanical damage and well collapse after rehabilitation, 320
Mechanical rehabilitation
  brushing, 206–209
  comparison of methods, 241–244
  hydraulic methods. See Hydraulic rehabilitation methods
  impulse methods. See Impulse rehabilitation methods
  processes, 203–206
  thermal methods, 222–227
Median, 366
Medical operations disinfection schemes, 324
Membrane filter index (MFI), 147–148, 174
Metal sulfides, 99–101, 256
Methane, 309
Microbial biomass dissolution, 256–257
Microbial pollution, secondary, 316–319
Microbially induced corrosion, 57–58
Microbicidal techniques
  chemical disinfection, 323–325
  introduction, 323
  ionizing irradiation, 326–328
  thermal disinfection, 325–326
Microbiology of ochre formation, 76–82
Mill cutting, 297
Millero, F. J., 70
Mine drainage, 106
Mineralogy of incrustations, 75–76, 253–254
Mining, subsidence from, 141
Mischungskorrosion, 94
Mobilization
  from sand intake, 138
  of subsurface particles, 117–119
Monazite, 166
Monitoring
  gas pressure, 337–338

Monitoring (*Cont.*):
  overview, 42–50
  turbidity, 165
Monosulfides, 100, 256
Monovariable efficiency calculations, 189
Morgan, J. J., 68
Morphological clogging, 127
Moser, H., 313
Moss, Roscoe, 25–26
Motor oil spillage, 306
Muckenthaler, P., 119
Multivariable cost-benefit analysis, 189
Murad, E., 83

National standards, 12–14
Natural conditions, flow under, 30–37
Natural Gamma Ray Logs (GR), 168
Natural gas pipeline leaks, 309
Near planning horizon in economics evaluation, 187–188
Neolithic wells, 1
Netherlands
  clogging in, 126–132
  disinfection schemes in, 324
  drilling techniques in, 332, 334
  enforced jet cavitation in, 240–241
  survey in, 7, 9
  ultrasound in, 240
Neutralization
  of bases, 274
  of inorganic acids, 273
  of organic acids, 274
Neutron Gamma Logs, 168
Nickel, 91–92
Nitrate-reducing microorganisms, 81
Nitrates
  in oxidation, 69
  from rehabilitation, 315–316
  in thermal methods, 226
Nitric acids, 250
Nitrogen
  in bioclogging, 102
  in compressed fluids, 234
  in ochre prevention, 338
Nonmetallic material corrosion, 58–59
Noord Bergum wellfield, 130
Numerical flow models
  for mechanical rehabilitation, 243–244
  under natural conditions, 31–32
Nuremberg, 2

Obligate anaerobic bacteria, 101
Occupational safety, 306–310

Ochre incrustations, 58, 173
  auto-catalysis of oxidation in, 70–72
  dissolution of, 247–254
  geochemistry of, 59–67
  history of, 2
  mass balancing in, 73–74
  microbiology of, 76–82
  mineralogy of, 75–76
  oxidation in, 67–70
  prevention of. *See* Prevention
  recrystallization in, 82–86
  after rehabilitation, 313
  structure and texture of, 93
  trace element interactions with, 86–93
Oil wells, 106
Olsthoorn, T. N.
  bacteria calculations by, 103–104
  head loss calculations by, 147
One-off investments, 190–191
Open borehole well ageing processes, 150–151
Operating costs, 197
Operations, 40–42
  influence of ageing on, 51–53
  in prevention planning, 331–334
Optical and geophysical monitoring methods, 48–49
Organic acid neutralization and disposal, 274
Organic chemicals in rehabilitation, 261
Organic reductants, 253
Ostwald, Wilhelm, 82
Ostwald sequence, 82–84
Out of phase waves, 231
Outer colmation, 125, 127
Outliers, 366
Oxalic acid, 263
Oxic water, 313–314, 333
Oxidants, 269, 324
Oxidation
  auto-catalysis of, 70–72
  of biomass, 256–257
  in ochre incrustations, 67–70
Oxygen. *See* Redox and redox zones
Ozone, 257

Packer flow meters, 171
Packers
  in carbon dioxide freezing, 224
  in closed-circuit pumping, 214–215
  in intense pumping, 211–212
  in partial reconstruction, 288
Parsons, S. A., 328

Partial reconstruction, 284–289
Particle bridges, 128–129
Particle counting, 174–176
Particle mobilization
  from sand intake, 138
  in suffossion and colmation, 117–119
Particulate matter, 117
Passivation, 56
Pasteurization, 325–326
Penetration depth in closed-circuit pumping, 215–216
Perez-Paricio, A., 149–150
Performance assessment
  borehole geophysics, 163–171
  camera inspections, 153–155
  incrustation deposition rate measurements, 172
  incrustation samples, 172–174
  mass balancing of removed material, 176–179
  particle counting, 174–176
  rehabilitation timing based on, 183–185
  rehabilitation vs. reconstruction based on, 185
  saturation index calculations, 179–182
  step-discharge tests, 155–162
Permeability
  of capillaries, 51–53
  in clogging, 125, 127
Permits for rehabilitation, 305
pH
  in aluminum hydroxide, 98, 255
  with carbon dioxide freezing, 225
  in carbonate dissolution, 254
  in chemical combinations, 260
  in dissolution processes, 248–250, 262, 264
  in ochre interactions with trace elements, 86–87, 89
  in oxidation, 68–72
  in reaction mechanisms, 246
  in redox zones, 63
  for reducing agents, 261
  and waste disposal, 273
Phase of waves, 231
Phosphates, 106, 259
Phosphoric acid, 249–250
Phosphorus in bioclogging, 102
PHREEQC model, 180–183
PHREEQC-2 model, 72
Piezoelectric crystals, 235–236
Piezometers, 293–294
Plant roots, 139–140

Plastic brushes, 208
Poiseuile flow, 51
Polarization, 330
Polymers in drilling fluids, 261
Polyphosphates, 259
Polyvinylchloride (PVC)
  in complete reconstruction, 293–294
  degradation of, 59
Pores and pore spaces
  in clogging, 125–128
  in gravel packs, 122–123
Potassium permanganate, 257
Power unit conversions, 363
Practical rehabilitation
  anticipating need for, 320–321
  approvals in, 305–306
  company selection, 304–305
  general recommendations, 306
  occupational safety in, 306–310
  problem definition in, 303
  progress control in, 311
  rehabilitation-induced problems, 315–320
  sustainability of, 312–315
  tender documents in, 303–304
  waste disposal in, 311–312
Preliminary examinations in tender documents, 371
Preliminary measures in tender documents, 370–371
Preparative measures in tender documents, 370
Pressure
  damage from, 204
  gas, 337–338
  in jetting, 216–221
  unit conversions for, 363
Pressure losses in aquifer yield, 30
Prevention, 41
  chemical and physical methods, 328–331
  drilling techniques, materials, and operation, 331–334
  gravel pack flushing devices, 339–340
  inert gases, 337–338
  microbicidal techniques, 323–328
  subsurface iron removal, 338–339
  suction flow control devices, 334–337
Price and volume considerations, 199–202
Primary settling, 141
Problem identification in pump replacement, 46
Programmed procedures, 189

Progress control in rehabilitation, 311
Protective clothing, 268
Proton-assisted dissolution, 247
Pulsation
  in jetting, 219
  in rehabilitation methods. *See* Impulse rehabilitation methods
Pumping and pumping rates, 159
  closed-circuit, 213–216
  intense, 210–213
  for microbial pollution, 318
  in step-discharge tests, 155–156, 159, 162
Pumps
  in economics evaluation, 188
  frequency-regulated, 215
  under natural conditions, 33–34
  in partial reconstruction, 284
  replacement checklist for, 46
  and screen diameter, 38
PVC
  in complete reconstruction, 293–294
  degradation of, 59
Pyrite, 101, 256

Qualitative well monitoring, 47–48
Quantitative well monitoring, 46–48
Quartiles, 366
Quartz, 107, 133

Radial collector wells (RCWs), 19–22
Radial symmetric flow, 23–27
Reaction half-life in oxidation, 68
Reaction inhibitors, 328
Reaction mechanisms
  aluminum hydroxide dissolution, 255–256
  biomass dissolution, 256–257
  carbonate dissolution, 254–255
  drilling fluid remnant removal in, 258–259
  introduction, 245–247
  iron and manganese oxide dissolution, 247–254
  metal sulfide dissolution, 256
Reaction stoichiometry, 179
Recommendations on the Transport of Dangerous Goods, 267
Reconstruction
  complete, 289–295
  defined, 9–10
  overview, 277–283
  partial, 284–289
Recrystallization of oxides, 82–86

Redox and redox zones
  in chemical ageing, 60–66
  in incrustations, 113
  prevention methods, 329
  in reaction mechanisms, 246
Redox potential, 63
Redrilling
  in complete reconstruction, 291
  in decommissioning, 297
Reducing agents, pH conditions for, 261
Reductants, 253
  disposal of, 274
  precautions for, 268–269
Reductive dissolution, 253–254
References, 343–356
Reflection of waves, 229–230, 238
Refraction of waves, 230
Rehabilitation
  decreased yield from, 143, 206
  defined, 9
  practical. *See* Practical rehabilitation
Rehabilitation-induced problems
  mechanical damage and well collapse, 320
  secondary microbial pollution, 316–319
  water chemistry changes, 315–316
Relining, 284, 286, 289–291
Removed material, mass balancing of, 176–179
Repairs. *See also* Reconstruction
  components of, 41
  defined, 277
Residual value of wells, 196
Resonance effects, 227–228
Reynolds number, 25–27, 333
Rhine River, 6
Rider, M. H., 163
Riekel, T., 325
Ries, T., 225
Riser pipe, 284
Risks in reconstruction, 280, 283
Roermond earthquake, 144
Romanechite, 88
Roots, 139–140
Rotary drilling, 332
Russia, explosive gases in, 233
Rust, 56–57
Ryan, J. N., 119

Safety precautions
  in carbon dioxide freezing, 226
  for chemicals, 266–270
  in rehabilitation, 306–310

Salt
  as corrosion factor, 56
  for drilling fluid remnant removal, 258
Samples, incrustation, 5–7, 172–174
Sand grains in abrasion, 133
Sand intake, 5, 134–139
Saturation index calculations, 179–182
Saucier, R. J., 123
Savic, I., 252
Scale, 94–97, 173
  dissolution of, 251, 254–255
  sludges, 275
Scanning electron microscope (SEM) investigations, 174
Scattering in impulse methods, 228
Schneider, E., 139
Schwabach wells, 88–90
Schwertmann, U., 83, 262
Scraping borehole walls, 332
Screen slot abrasion, 133
Screens
  in complete reconstruction, 292–293
  in partial reconstruction, 284, 286
  in prevention planning, 333
  in well sizing, 37–39
Sebold, B. M. E., 331
Secondary microbial pollution, 316–319
Secondary settling, 141
Segmented Gamma Logs (SGL), 168–169
Separation of incrustations, 203–204
Servicing components, 41
Settling, soil, 140–144
Shape factor in abrasion, 133
Sichardt velocity, 29
Siderite, 97, 101
Siderocapsa, 81
Siderococcus, 79
Silicates, 105
Sleeves in intense pumping, 212
Slides, 172
Slimes, 81
  in bioclogging, 102
  dissolution of, 256–257
  in infiltration wells, 146
  production of, 323
Sludge disposal, 274–275
Slug surging, 209
Smith, S. A., 7
Snow, D. T., 51
Sodium dithionite, 263–264
Sodium hypochlorite, 256–257, 324
Soil
  infiltrations of, 273

Soil (*Cont.*):
  settling and subsidence, 140–144
  in wells, 318
Solid calcium carbonate, 273
Solid waste disposal, 271–275
Solution-controlled reactions, 246
Sorption of trace elements, 86–87
Sound waves, 227–231
Spatial distribution of incrustations, 107–115
Spatial spreading in impulse methods, 228–229
Specific energy, 243
Spencer, W. F., 328
Spillage of motor and hydraulic oils, 306
Splash zones, 56
Standards, 12–14
Static loads in soil settling, 140
Statutory laws, 10
Steady-state drawdown, 156
Steam injection, 226–227, 261
Step-discharge tests, 155–162
Stepwise rehabilitations, 204
Stoneware degradation, 59
Stress
  between grains in settling, 143
  in mechanical rehabilitations, 204
Structural damage
  overview, 279–280
  from rehabilitation, 320
Struvite, 106
Submersible pumps, 188
Subsidence, soil, 140–144
Subsurface iron removal, 338–339
Subsurface particle mobilization and transport, 117–119
Suction flow control devices (SFCDs), 334–337
Suffossion and colmation
  in closed-circuit pumping, 215–216
  filtration clogging, 125–132
  hydraulic, 209
  mechanical filtration processes in, 119–125
  subsurface particle mobilization and transport, 117–119
Sulfamic acid, 250
Sulfate-reducing bacteria (SRB), 57–58
  in carbonate incrustations, 96
  in metal sulfide incrustations, 101
  in redox zones, 60
Sulfides, 99–101, 173, 256
Sulfur, 101

Sulfuric acid
  in chemical combinations, 260
  in dissolution process, 249–250
Sung, W., 68
Surface-controlled reactions, 246–247
Surface filtration, 121
Surface waters, 17
Surge blocks, 209–210
Suspended-particle content, 176
Sustainability of rehabilitation measures, 312–315
Swimming pool disinfection schemes, 324–325

Tamura, H., 70
Technical Instructions, 14
Technical University of Delft, 130
Telescopic tubing, 289, 292
Temperature
  in carbonate incrustations, 95
  in chemical reactions, 261
  pump motor, 162
  unit conversions for, 362
Temporal distribution of incrustations, 107–115
Tender documents, 303–304, 369
  desanding and final cleansing in, 374–375
  follow-up investigations in, 376–377
  hydro-mechanical-chemical rehabilitation in, 372
  mechanical cleansing in, 372
  preliminary examinations in, 371
  preliminary measures in, 370–371
  preparative measures in, 370
Termination criterion, 198
Thermal rehabilitation methods
  carbon dioxide freezing, 222–226
  disinfection, 325–326
  steam injection, 226–227
Thermodynamical equilibrium models, 180–182
Thickness
  filter-pack, 149
  incrustation, 110–112
Thiem, A., 27
Thiobacillus ferrooxidans, 252
Thixotropy, 227
Three-dimensional diffusion of pressure, 217
Time unit conversions, 357
Time value of money, 189
Time-variation graphs, 48

Timmer, H., 332
Todorocite, 75, 88
Total annuity of investments, 192
Trace elements, ochre interactions with, 86–93
Tracer fluid logging, 165, 167
Transport
  of chemicals, safety precautions for, 266–270
  of subsurface particles, 117–119
Transport container disposal, 272
Transport of Dangerous Goods (TDG), 267
Tree roots, 139–140
Trinitrotoluene, 232
Tubing in partial reconstruction, 289, 292
Tunnel structures, 88–91
Turbidity
  in carbon dioxide freezing, 225
  in closed-circuit pumping, 214
  monitoring, 165
  in performance assessment, 176
  from sand intake, 138
Turbidity meters, 213
Turbulent flow
  in prevention planning, 333–334
  in radial symmetric flow, 25
  in step-discharge tests, 158

Ultrasound, 235–240
Unconsolidated aquifers, 137
Unit conversions, 357–363
United States
  ageing causes in, 7–8
  Army Corps of Engineers, 14
  Environmental Protection Agency, 14
  inert gases in, 338
  laws in, 11
  rehabilitation and reconstruction market in, 200

Vandalism, 144–145, 306
Velocity unit conversions, 362
Ventilation, 309–310
Vertical distribution
  of iron oxide, 108–109
  under natural conditions, 33–35
  of redox zones, 60, 63–65
Vertically drilled wells, 17, 19–20
Visual inspections, 153

Vivianite, 106
Volume unit conversions, 360

Wackersdorf well, 113
Walton, W. C., 158–159
Wash and backwash action with surge blocks, 209
Waste disposal, 271–275, 311–312
Water chemistry changes from rehabilitation, 315–316
Water table
  calculating, 25
  in soil settling, 140
Wavelengths in ultrasound, 236
Wehrli, B., 70
Weight unit conversions, 362
Well ageing
  defined, 9
  types and relative importance of, 3–9
Well cleaning fests, 1–2
Well types, 17–18
  directionally drilled, 22–23
  dug, 18–19
  horizontal and radial collector, 19–22
  vertically drilled, 19–20
Western Germany, decommissioned wells in, 297–299
Wiesbaden, acid dissolution in, 251
Wildeshausen well, 108–109
Wilken, R. D., 238
Willemsen, A., 148
Williams, D. E., 26–27, 200
Wire-to-water ratio, 161
Wood-based material degradation, 58–59
Work unit conversions, 362

X-ray diffraction (XRD), 174
X-ray fluorescence (XRF), 174

Yield after rehabilitation, 143, 206, 312–315, 321
Yield coefficient in step-discharge tests, 159–162

Zeppenfeld, K., 95
Zhao, N., 69
Zielinski, R. A., 93
Zilch, K., 187, 189
Zinc, 91–92, 97